# Prestressed Members with External Fiber-Reinforced Polymer (FRP) Tendons

Woodhead Publishing Series in Civil and Structural Engineering

# Prestressed Members with External Fiber-Reinforced Polymer (FRP) Tendons

## Design, Assessment, and Modeling

**Tiejiong Lou**

School of Civil Engineering and Architecture, Wuhan University of Technology, Wuhan, P.R. China; CEMMPRE, ARISE, University of Coimbra, Coimbra, Portugal

**Yanan Wu**

School of Civil Engineering and Architecture, Wuhan University of Technology, Wuhan, P.R. China

**Sergio M.R. Lopes**

CEMMPRE, ARISE, University of Coimbra, Coimbra, Portugal

**WP**

WOODHEAD PUBLISHING

ELSEVIER    An imprint of Elsevier

Woodhead Publishing is an imprint of Elsevier
50 Hampshire Street, 5th Floor, Cambridge, MA 02139, United States
125 London Wall, London EC2Y 5AS, United Kingdom

ISBN: 978-0-443-23877-2 (print)

ISBN: 978-0-443-23878-9 (online)

For information on all Woodhead Publishing publications
visit our website at https://www.elsevier.com/books-and-journals

*Publisher:* Matthew Deans
*Acquisitions Editor:* Glyn Jones
*Editorial Project Manager:* Namrata Aggarwal
*Production Project Manager:* Prem Kumar Kaliamoorthi
*Cover Designer:* Greg Harris

Typeset by MPS Limited, Chennai, India

# Contents

# Acknowledgment

This work was supported by the Portuguese Foundation for Science and Technology under Grant Nos. 2022.04729.CEECIND/CP1714/CT0010, UIDB/00285/2020, and LA/P/0112/2020.

# Introduction

## 1.1 Background

In civil engineering, reinforced concrete (RC) structures commonly encounter challenges including insufficient stiffness, decreased durability, and a pronounced vulnerability to corrosion. In response to these concerns, the external prestressing technology emerged in the early 20th century as an effective solution aimed at enhancing RC structural performance (Naaman, 1990; Sowlat & Rabbat, 1987; Virlogeux, 1990). The introduction of external prestressing marked a pivotal advancement, specifically tailored to mitigate the propensity of RC structures to cracks under loading. By utilizing the tension in the prestressing tendons to counterbalance internal forces imposed by external loads, this technology involves multiple advantages:

1. it effectively mitigates stress levels and limits cracking, thereby enhancing the durability and stability of the structure;
2. it facilitates reductions in cross-sectional dimensions and overall structural weight, fostering the development of lightweight, cost-effective designs without compromising the bearing capacity;
3. it supports the creation of larger spans without the need for increasing structural thickness or height and hence promotes an efficient use of materials and great architectural versatility;
4. it not only streamlines construction processes and accelerates project timelines but also simplifies maintenance tasks such as inspection, repair, and tendon replacement throughout their service life;
5. the connection of external tendons to the member through anchoring and deviation points minimizes the friction loss and, therefore, promotes optimal exploitation of prestressing effects.

External prestressing is acclaimed for its effectiveness and has been widely adopted in various construction projects, including bridges, buildings, and marine structures. Its role extends beyond initial construction to the strengthening and maintenance of existing infrastructures (Harajli, 1993; Mancarti, 1984), thereby providing a comprehensive and effective strategy for improving structural reliability and longevity, as depicted in Fig. 1.1 (Recupero et al., 2014).

Since the 1960s, externally prestressed structures have undergone significant development, marking an era of considerable advancement in structural engineering. Numerous experimental studies (Lorenc & Kubica, 2006; Ng & Tan, 2006; Park et al., 2010; Uy & Craine, 2004) complemented by detailed finite element analyses (Dall'Asta & Zona, 2005; Omran et al., 2020) have been carried out to evaluate their performance. These investigations universally underscore the cost-effectiveness of external prestressing as a strengthening or construction technology.

Prestressed Members with External Fiber-Reinforced Polymer (FRP) Tendons. DOI: https://doi.org/10.1016/B978-0-443-23877-2.00001-8

**Figure 1.1** Application of external prestressing in engineering.
From Recupero, A., Spinella, N., Colajanni, P., & Scilipoti, C. D. (2014). Increasing the capacity of existing bridges by using unbonded prestressing technology: a case study. *Advances in Civil Engineering*, *2014*(1), 840902.

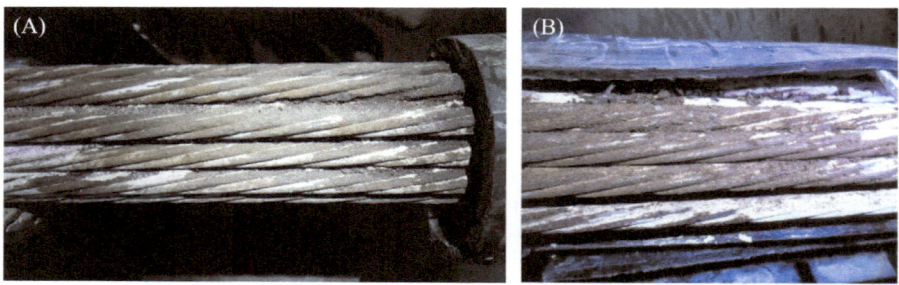

**Figure 1.2** Corrosion of external steel tendons. (A) moderate corrosion; (B) severe corrosion. From Jeon, C. H., Lee, J. B., Lon, S., & Shim, C. S. (2019). Equivalent material model of corroded prestressing steel strand. *Journal of Materials Research and Technology*, 8(2), 2450–2460.

Despite its notable advancements, the application of external prestressing to a broader extent is still impeded by a significant challenge: the susceptibility of external steel tendons to corrosion (see Fig. 1.2) (Jeon et al., 2019). Addressing this issue is paramount for this technology in structural applications. A promising strategy to overcome the corrosion challenge involves replacing steel tendons with nonmetallic alternatives such as fiber-reinforced polymer (FRP) composite materials.

## 1.2   FRP composite materials

FRP composite materials are constituted by two principal components: fibers and resin matrix. These composites are classified based on the type of fibers used, including carbon fiber-reinforced polymer (CFRP), glass fiber-reinforced polymer (GFRP), aramid fiber-reinforced polymer (AFRP), and basalt fiber-reinforced polymer (BFRP). The fibers are pivotal in enhancing the mechanical properties of these composites, such as strength, stiffness, and durability. FRPs exhibit significantly lower densities compared to conventional metallic materials, which facilitates a substantial reduction in structural weight. Moreover, the resin matrix enhances their corrosion resistance, rendering them exceptionally adaptable to corrosive environments.

Various types of FRP offer distinct mechanical properties. CFRP is recognized for its high tensile strength, elastic modulus, and excellent environmental resistance, making it suitable for high-performance applications despite its higher cost and limited ductility. AFRP exhibits favorable fatigue performance and high-temperature resistance but is limited by its lower alkalis resistance and reduced compressive and shear strengths, constraining its use in certain structures (Prasad & Talupula, 2018; Sahu et al., 2014). GFRP is advantageous in terms of strength, durability, and cost-effectiveness, especially in moist or corrosive environments. Nevertheless, its lower creep rupture limits its application. BFRP is emerging as a sustainable alternative, with good mechanical properties in varied conditions, including high temperatures and acidic or saline environments (Elgabbas et al., 2015; Hassan et al., 2016; Lopresto et al., 2011). The mechanical properties of FRPs/steel are summarized in Table 1.1.

FRPs are characterized as linear elastic materials without yielding platforms and exhibit relatively low ultimate strains compared to conventional steel reinforcement, as illustrated in Fig. 1.3. The characteristic increases the risk of brittle failure as FRPs approach their ultimate strain capacity. Additionally, the low elastic modulus of some FRPs can result in significant structural challenges, such as excessive deflection, wide crack widths, and inadequate stiffness. To overcome these issues, the application of prestressing technology to FRPs is proposed. This approach capitalizes on the inherent advantageous characteristics of FRPs, thereby enhancing the durability and performance of structures. The effectiveness of this approach has been demonstrated by El-Hacha et al. (2001), Nordin and Täljsten (2006), Wang et al. (2015), and Liu et al. (2023).

## 1.3   Prestressed concrete members with external FRP tendons

### 1.3.1   Short-term performance

As is well known, conventional RC members are prone to bending failure. The placement of FRP tendons in prestressed structures offers an effective solution to

**Table 1.1** Mechanical properties of FRPs/steel (ACI Committee 440, 2015; Banibayat & Patnaik, 2015).

| Properties | Steel rebar | Steel strand | CFRP | GFRP | AFRP | BFRP |
|---|---|---|---|---|---|---|
| Density (g·cm$^{-3}$) | 7.9 | 7.9 | 1.5–1.6 | 1.2–2.1 | 1.25–1.4 | 1.9–2.1 |
| Elastic modulus (GPa) | 200 | 180–200 | 120–580 | 35–51 | 41–125 | 41–48.5 |
| Tensile strength (MPa) | 483–690 | 1400–1890 | 600–3690 | 483–1600 | 1720–2540 | 930–1380 |
| Rupture strain (%) | 6.0–12.0 | >4.0 | 0.5–1.7 | 1.2–3.1 | 1.9–4.4 | 2.5–3.0 |
| Longitudinal thermal expansion coefficient (10$^{-6}$/°C) | 11.7 | 11.7 | −9.0–0.0 | 6.0–10.0 | −6.0--2.0 | 9.0–12.0 |
| Transverse thermal expansion coefficient (10$^{-6}$/°C) | 11.7 | 11.7 | 74–104 | 21–23 | 60–80 | 21–22 |

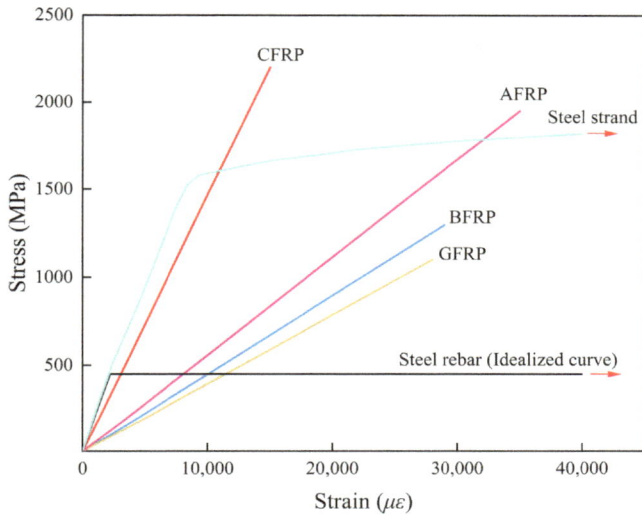

**Figure 1.3** Stress—strain curves of FRPs/steel.

enhancing bending stiffness. When the tendons are positioned externally relative to the concrete section, there occurs not only strain incompatibility between tendons and adjacent concrete but also variation in tendon eccentricity or effective depth. The phenomenon necessitates a comprehensive analysis of the entire deformation of the member.

The use of FRP tendons in unbonded or externally prestressed concrete members has been extensively addressed. An early study by Kato and Hayashida (1993) analyzed the impact of unbonded/bonded CFRP tendons on the bending properties and deformation behavior of prestressed concrete members. Their research indicated that members with unbonded CFRP tendons primarily failed due to concrete crushing, whereas those with bonded tendons were prone to tendon rupture, highlighting the potential of unbonded CFRP tendons to enhance the structural deformation capacity at the ultimate limit state. Following this, Chajes et al. (1994) identified notable enhancements in flexural capacity (increases of 36%−57%) and stiffness (increases of 45%−53%) of concrete members through the external application of various composite materials. To further affirm the effectiveness of FRP tendons, Pisani (1998) explored the practicality of replacing conventional steel strands with CFRP or AFRP tendons in prestressing applications. El-Hacha and El-Badry (2001) reported a significant 83% increase in the flexural capacity of concrete members strengthened by external CFRP cables.

The arrangement of FRP tendons, whether internal or external, bonded or unbonded, significantly impacts the bending performance of prestressed members. Saafi and Toutanji (1998) found that prestressed concrete beams with unbonded AFRP tendons exhibited 10%−20% lower flexural strength than those with bonded ones. Also, a combination of bonded and unbonded tendons resulted in greater member ductility compared to the use of bonded tendons only. To fully exploit the advantage of combining internal and external CFRP tendons, Grace and Abdel-Sayed (1998) conducted a series of bridge model tests, demonstrating a significant enhancement in the ultimate load of the bridge when using both internally bonded and externally unbonded tendons with a prestress level kept between 30% and 60% of their rupture strength. While the efficacy of external prestressing tendons in RC structures has been well established, the tendon profile (straight, draped, or parabolic) was demonstrated to affect structural responses differently. Tan and Tjandra (2007) investigated the impact of different tendon profiles on the flexural performance of continuous RC members with external prestressing. Their findings revealed that members with straight segmental tendons within the spans exhibited satisfactory behavior (ultimate loads, ductility, and service-load displacements) and that members with draped and parabolic tendons also displayed favorable failure characteristics. Importantly, the overlapping arrangement of parabolic tendons at internal supports led to an effective strengthening of both sagging and hogging moment regions in terms of flexural strength and ductile performance.

External prestressing tendons are connected to RC members through anchoring systems and deviators. The anchoring of tendons introduces the necessary prestressing to improve the structural behavior. Gao et al. (2020) developed an innovative anchoring technology that improved the reliability of strengthening of slabs with external tendons. As the member undergoes deformation under loading, the effective eccentricity of external tendons decreases, which may induce significant second-order effects. Members without deviators experienced a reduction in bearing capacity due to the continuous decrease in tendon eccentricity with increasing member deflection, as observed by Ng and Tan (2006). The provision of deviators can alter the free length of external tendons, thereby effectively mitigating the second-order effects and consequently improving the structural response. This effectiveness has been verified by Jung et al. (2013), who observed a substantial 25% increase in the ultimate load-carrying capacity with the provision of deviators. The research identified the span-to-depth ratio and deviator spacing as critical factors influencing these effects (Lou et al., 2012; Ng, 2003; Sivaleepunth et al., 2006). Tran et al. (2021) further validated the importance of these parameters through a numerical model. Their study indicated that increasing the span-to-depth ratio or deviator spacing could reduce the structural stiffness and flexural strength attributed to the exacerbating second-order effects.

The combination of experimental and theoretical research emphasizes the advantages of introducing external prestressing into concrete structures to enhance the cracking and bending resistance. The failure process of prestressed concrete members with external tendons generally progresses through three stages. Initially, members exhibit elastic behavior under applied loading. As the loading increases, cracks appear on the tensile surface and transition to the cracking elastic stage begins. Further loading causes nonprestressed steel within the members to yield, transitioning the structure into the plastic stage characterized by significant deformations. Finally, the members continue to deform until the concrete fails due to crushing.

## 1.3.2 Long-term performance

The long-term behavior of FRPs under sustained loading exhibits complexity, and is typically characterized by creep, which signifies that the strain of the material increases over time while maintaining a constant stress level, and stress relaxation, which involves the gradual decrease in stress under constant strain. Importantly, high-level sustained loading can lead to severe stress relaxation or even creep rupture of FRP tendons, critically affecting the durability and service life of structural elements. The progression of creep deformation in FRPs under sustained tension typically manifests in three distinct stages, as illustrated in Fig. 1.4. The creep curve, used to describe the creep characteristics of FRPs, exhibits good consistency (Najafabadi et al., 2018). On the other hand, the stress relaxation behavior is effectively captured by a typical tendon relaxation curve, highlighting the specific long-term response of FRPs (Zhao et al., 2020), as depicted in Fig. 1.5.

The creep behavior of FRP tendons is heavily dependent on the applied prestress levels. As prestress levels increase, the creep rate also rises, possibly leading to excessive tendon relaxation and consequent degradation in structural workability. Research by Zou (2003a) demonstrated that prestressed concrete members with CFRP tendons exhibited reduced prestress loss and deflection under sustained loads compared to conventional prestressed concrete members, attributed to the low relaxation and high elastic modulus properties of CFRP tendons. Additionally, as the concrete strength increased, members exhibited higher cracking moments, fewer cracks, and reduced creep strain. Crucially, when the stress level in CFRP tendons remained below $0.60\,f_u$ (ultimate tensile strength), the creep effects were minimal. In contrast, GFRP tendons exhibited significant deformation over time, with the creep strain at $0.40\,f_u$ being 10% higher than the initial tensile strain (Sun et al., 2017).

To mitigate the risk of creep rupture, stress limits for FRPs have been established, as detailed in Table 1.2. CFRP and AFRP are particularly suitable for prestressing

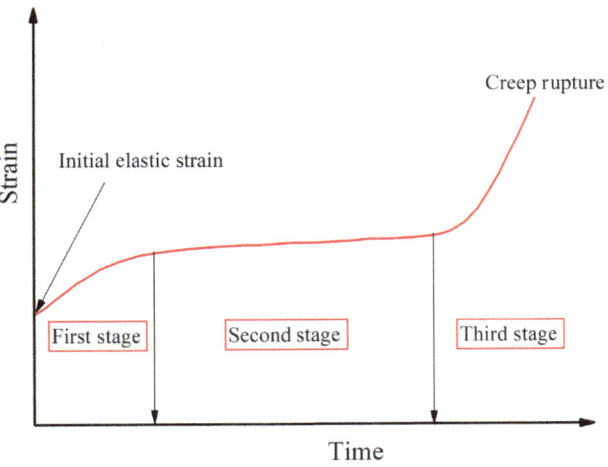

**Figure 1.4** Creep curve of FRP tendons.

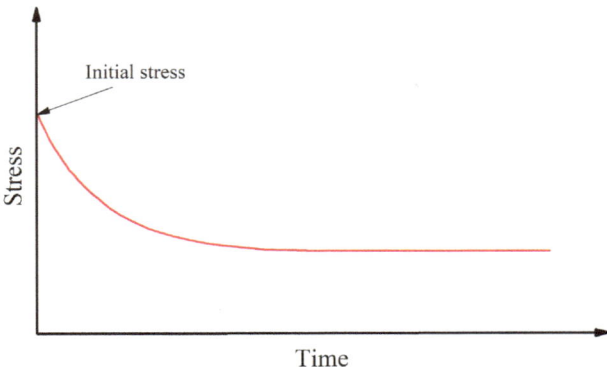

**Figure 1.5** Stress relaxation curve of FRP tendons.

**Table 1.2** Creep rupture stress limits in FRPs.

| Reference | CFRP | AFRP | GFRP | BFRP |
|---|---|---|---|---|
| Ando et al. (1997) | $0.79\,f_\mathrm{u}$ | $0.66\,f_\mathrm{u}$ | / | / |
| Yamaguchi et al. (1998) | $0.93\,f_\mathrm{u}$ | $0.47\,f_\mathrm{u}$ | $0.29\,f_\mathrm{u}$ | / |
| ACI Committee 440 (2004) | $0.70\,f_\mathrm{u}$ | $0.55\,f_\mathrm{u}$ | $0.29\,f_\mathrm{u}$ | / |
| ACI Committee 440 (2006) | $0.91\,f_\mathrm{u}$ | $0.47\,f_\mathrm{u}$ | $0.30\,f_\mathrm{u}$ | / |
| ACI Committee 440 (2017) | $0.55\,f_\mathrm{u}$ | $0.30\,f_\mathrm{u}$ | $0.20\,f_\mathrm{u}$ | / |
| Shi et al. (2015) | / | / | / | $0.54\,f_\mathrm{u}$ |
| Banibayat and Patnaik (2015) | / | / | / | $0.15\,f_\mathrm{u}$ |

applications due to their higher creep rupture stress limits, while GFRP exhibits lower limits, thereby constraining its application in prestressed structures. Studies on BFRP tendons, including those by Wang et al. (2014, 2016) and Shi et al. (2015), have highlighted their promising mechanical properties. Remarkably, after $10^3$ h of sustained loading, BFRP tendons remained at 95% of their initial tensile strength, and at a sustained stress level of 0.5 $f_\mathrm{u}$, a creep rate of 3.58% was recorded over $10^6$ h. Furthermore, pretreatment of BFRP bars significantly reduced the creep rate and increased the creep rupture stress from 0.52 $f_\mathrm{u}$ to 0.54 $f_\mathrm{u}$ after $10^6$ h, underscoring the effectiveness of this treatment in controlling creep (Shi et al., 2015; Wang et al., 2016). Additionally, the studies by Zou (2003b) and Shi et al. (2017) have demonstrated that the relaxation rate of FRP tendons followed a linear relationship with logarithmic time and that a higher initial prestress level led to a greater relaxation rate. Given the vulnerability of prestressing FRP tendons to high sustained loads and potential creep rupture, Aghani et al. (2022) recommended maintaining prestress in FRPs at a level of not higher than 0.5 $f_\mathrm{u}$ to optimize the bearing capacity and minimize the creep impact.

In engineering practice, designing prestressed structures necessitates a thorough understanding of their long-term behavior to ensure their durability and workability.

Continuous monitoring is essential to evaluate changes such as deformation, crack development, and prestress loss. Cao and Fang (2007) carried out long-term tests lasting 1001 days on the concrete box beam model with external CFRP tendons. Their findings indicated that the long-term displacement, cracking, and strains in concrete and compressive steel rebars reached a state of stabilization following approximately 2 years of sustained loading. Moreover, significant increases in crack width and deflection at long-term sustained loads were observed, ranging from 1.59 to 2.69 times and 2.32 to 2.42 times, respectively, compared to initial measurements. Xue and Liu (2021) conducted long-term experiments on prestressed concrete members with external CFRP tendons, with variables including the type of tendons, the type of rebars, and the cracking condition. They also developed a prediction method to calculate the long-term deflection. Shi et al. (2022) experimentally examined the long-term behavior of prestressed concrete members with external BFRP tendons over a 300-day period. They observed a rapid decrease in prestress within the first 20 days, followed by a stabilizing behavior. With an increase in tension control load, the prestress loss rate in the tendons substantially increased. However, under the same tensile control condition, members with external BFRP tendons exhibited a 16% lower long-term prestress loss compared to those with normal-relaxation steel strands, demonstrating the potential of BFRP tendons to mitigate the prestress loss.

## 1.4 Prestressed steel-concrete composite members

Steel-concrete composite members combine steel and concrete components to leverage the advantageous properties of both materials to enhance the bearing capacity, ductility, stability, and structural integrity. This configuration is particularly effective in reducing concrete cracking in regions experiencing sagging moments, leading to improved structural performance. However, some challenges may arise such as insufficient stiffness in the design of large-span composite bridges. Continuous composite members are susceptible to premature cracking in the hogging moment regions, which can result in functional degradation and structural damage. The application of external prestressing technology has emerged as an effective solution to these challenges, as illustrated in Fig. 1.6.

### 1.4.1 Short-term performance

An early study by Hoadley (1963) demonstrated that prestressing significantly enhanced both yield and ultimate moment capacities of composite members. Specifically, prestressed composite members exhibited increases of 25%−35% in yield moment capacities and 10%−20% in ultimate moment capacities compared to conventional composite members. The advantages of prestressing include not only enhanced ultimate strength and reduced deflection but also increased structural redundancy. These improvements are primarily attributed to the optimized utilization

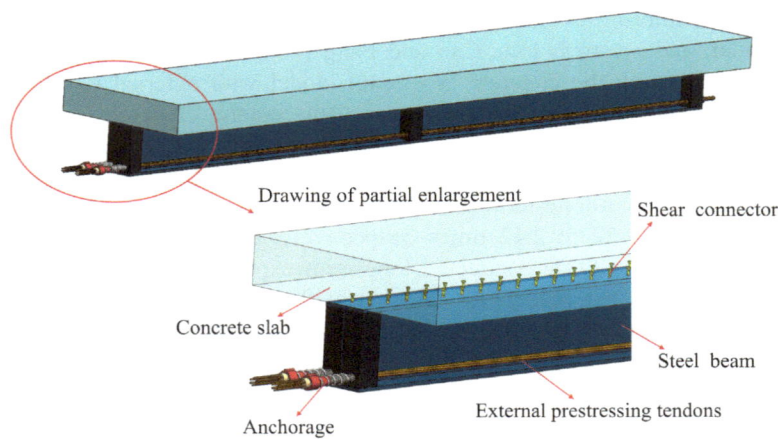

**Figure 1.6** Schematic diagram of externally prestressed steel-concrete composite beam.

of concrete slabs in prestressed constructions, which effectively redistributes applied loads and promotes a more uniform stress distribution throughout the member. To alleviate concrete cracking in the hogging moment regions, Saadatmanesh et al. (1989a, 1989b) and Ayyub et al. (1990, 1992a, 1992b) recommended that external tendons be positioned as close as possible on the lower flange of the steel beam in the sagging moment regions and on the upper flange in the hogging moment regions.

Continuous composite members exhibit complex behavior, including moment redistribution, steel beam buckling, and initial geometric imperfections. Tong and Saadatmanesh (1992) established a model by placing external tendons in the sagging moment regions and using precast prestressed concrete panels in the hogging moment regions. They noted that increasing the initial prestress from 0 to the allowable stress of tendons can reduce the moment at the internal support by 50%. Chen et al. (2009) further demonstrated that prestressing significantly enhanced the crack resistance and moment redistribution capacity, although the ultimate moment-bearing capacity in the hogging regions was heavily influenced by buckling. A recent study by Peng et al. (2024) confirmed the effectiveness of external prestressing in increasing the moment redistribution. They observed a redistribution degree of 52%−58% in the hogging moment regions. Additionally, the bearing capacity and buckling resistance of continuous composite members were significantly influenced by initial geometric imperfections, that is, the larger the initial geometric imperfections, the smaller the buckling moment resistance (Chen & Jia, 2010).

A significant mechanical characteristic in composite members is the interfacial slip, which impacts adversely the force transmission between components, reduces the stiffness, and decreases the flexural strength capacity, consequently accelerating performance degradation and causing potential failure. Ayyub et al. (1990) found that this slip could lead to increased deflection beyond predictions made based on a no-slip assumption. Zhang et al. (2019) proposed a refined model by introducing nonlinear spring elements to simulate the slip effect, while some investigators (Dall'Asta & Zona, 2005; Sousa et al., 2019; Zona, Ragni & Dall'Asta, 2008) introduced multi-DOF beam elements to account for the interfacial slip.

The flexural capacity of externally prestressed composite members is closely related to the stress state of prestressing tendons at failure. Under external loading, the stress evolution in external tendons depends on structural deformation rather than a single cross-section. Dall'asta, Ragni, and Zona (2007) and Zona, Ragni, and Dall'Asta (2009) developed an analytical approach based on shape function approximation deformation, which transformed structural analysis into a more manageable cross-sectional problem. Chen and Gu (2005) proposed a simplified approach based on compatibility and equilibrium for predicting the ultimate stress in external tendons and moment resistance of externally prestressed composite members. Zhang and Fu (2009) provided an estimation equation for the tendon stress that considered the relationship between compressive strain in the concrete slab and the bending deformation of the members. Nie et al. (2011) developed an analytical model to predict the tendon force and the structural response of continuous externally prestressed composite members under symmetric concentrated loading. The proposed formulas are capable of capturing the typical behavior characteristics during the entire loading process, including the cracking, yielding, and ultimate loads.

Finite element analysis offers a versatile tool for exploring various parameters affecting structural performance. El-Belbisi et al. (2022) and El-Belbisi et al. (2018) studied the flexural performance of prestressed composite members with external CFRP tendons, recommending that posttensioning levels be maintained within 20%−30% to mitigate early tendon rupture and enhance bearing capacity. Abu-Sena et al. (2020) assessed the ultimate load-carrying capacity of these members using ANSYS finite element models. The parametric study considered various factors such as the load type, tendon profile, span length, and initial prestress. Almeida et al. (2022) developed finite element models in ABAQUS and carried out a detailed numerical analysis of the mechanical behavior of externally prestressed composite members, taking into account variables including the tendon position, tendon shape, initial prestress, tendon length, and beam length.

## *1.4.2 Long-term performance*

Steel-concrete composite members have been widely employed in the construction of bridges and buildings. Ensuring their long-term functionality is crucial to reducing maintenance costs and enhancing project sustainability. Throughout their service life, the workability of composite members is subject to degradation due to time-dependent effects such as concrete creep and shrinkage. These effects lead to stress redistribution and, in the case of continuous members, also cause internal force redistribution (Gilbert & Bradford, 1995). Significant stress changes may occur in the interface between the steel beam and concrete slab, which can increase loads on shear connectors and induce relative slip at the interface, potentially compromising structural integrity and durability. Consideration of concrete creep and shrinkage has promoted the development of various methodologies and theoretical frameworks aimed at better understanding and predicting the time-dependent behavior of composite members. Among these, the age-adjusted effective modulus method (AEMM) and the rate of creep method (RCM) are commonly adopted by researchers. Gilbert (1989) performed a comparative analysis of the effects of concrete creep and shrinkage on the sectional behavior of composite members using AEMM and RCM. Bradford and Gilbert (1992) developed a computer program that utilized AEMM and considered interface slip effects to evaluate the impact of concrete shrinkage and creep on composite beams. Amadio and Fragiacomo (1997) introduced an optimized calculation method for concrete shrinkage and creep, considering concrete aging coefficients through AEMM.

Among various factors influencing the long-term performance of externally prestressed composite members, the prestress loss is particularly crucial. Cao et al. (2018) examined the prestress loss in prestressed steel-concrete composite box girders by long-term tests. They also proposed a predictive model for time-dependent deflections, in which the impact of prestress loss was separately involved. Xue et al. (2008) carried out an experimental and analytical study on the time-dependent behavior of prestressed composite members under 1-year sustained loading. On this basis, a proposal on calculating the long-term deflection of these members was made. Ding et al. (2010) investigated the impact of different factors on the long-term deflection of prestressed composite members. They concluded that the long-term deflection decreased with an increase in tensioning age, initial prestress, and concrete strength. They also observed that the long-term deflection was least affected by CFRP tendons compared to other tendon types, suggesting that using low-relaxation tendons could help minimize the long-term deflection of prestressed composite members.

# 1.5  Contents of the book

The application of external prestressing technology has been widespread in the construction or strengthening of various concrete or steel-concrete composite members. While traditional steel tendons are prone to corrosion, the corrosion resistance, high strength, and lightweight properties of FRP tendons make them an excellent alternative. Consequently, the integration of external prestressing technology with FRP tendons is considered an effective approach to improve structural performance. However, there are notable differences between FRP and steel tendons in mechanical properties such as elastic modulus and elongation, which can result in distinctive structural behavior of FRP prestressed members. In addition, the unbonded nature of external tendons, combined with the change in tendon eccentricity with the development of member deformation, complicates the analysis and design of prestressed members with external FRP tendons.

In this context, this book addresses some important topics related to prestressed concrete or steel-concrete composite members with external tendons. Particular emphasis is placed on the effectiveness of using FRP composites as prestressing tendons. In addition, the effects of using FRP rebars instead of steel rebars in these members are also discussed. The contents of the book include the development of numerical methods at immediate or sustained loads, evaluation of immediate or time-dependent second-order effects, prediction of stress in external tendons and flexural strength, redistribution of bending moments in continuous members, linear transformation of external cables, secondary moments resulting from prestressing in statically indeterminate members, and prediction of long-term displacement. The book covers modeling (i.e., finite element models for full-range nonlinear analysis or service-load time-dependent analysis), assessment (i.e., comprehensive assessment on short-term and long-term behavior), and design (i.e., design proposals, equations, or models for practical purposes) of prestressed members with external FRP tendons. The book is expected to be used by investigators and engineers in the fields of bridge and structural engineering. The book could also be used for a course for undergraduate and graduate students majoring in civil engineering.

# Reference

Abu-Sena, A. B., Shaaban, I. G., Soliman, M. S., & Gharib, K. A. A. M. (2020). Effect of geometrical properties on strength of externally prestressed steel-concrete composite beams. *Proceedings of the Institution of Civil Engineers - Structures and Buildings, 173*(1), 42–62.

ACI Committee 440. (2004). *Prestressing concrete structures with FRP tendons.* ACI 440.4R-04, Farmington Hills, MI.

ACI Committee 440. (2006). *Guide for the design and construction of structural concrete reinforced with FRP bars.* ACI 440.1R-06, Farmington Hills, MI.

ACI Committee 440. (2015). *Guide for the design and construction of structural concrete reinforced with FRP bars.* ACI 440.1R-15, Farmington Hills, MI.

ACI Committee 440. (2017). *Guide for the design and construction of externally bonded FRP systems for strengthening concrete structures.* ACI 440.2R-17, Farmington Hills, MI.

Aghani, K., Afshin, H., & Abedi, K. (2022). Finite element-based prediction of the long-term deflection of reinforced concrete beams strengthened with prestressed fiber-reinforced polymers. *Structures, 43*, 358–373. Available from https://doi.org/10.1016/j.istruc.2022.06.059.

Almeida, M. M. D. R., De Souza, A. S. C., De Albuquerque, A. T., & Rossi, A. (2022). Parametric analysis of steel-concrete composite beams prestressed with external tendons. *Journal of Constructional Steel Research, 189*, 107087. Available from https://doi.org/10.1016/j.jcsr.2021.107087.

Amadio, C., & Fragiacomo, M. (1997). Simplified approach to evaluate creep and shrinkage effects in steel-concrete composite beams. *ASCE Journal of Structural Engineering, 123*(9), 1153–1162. Available from https://doi.org/10.1061/(ASCE)0733-9445(1997)123:9(1153).

Ando, N., Matsukawa, H., Hattori, A., & Mashima, M. (1997). Experimental studies on the long-term tensile properties of FRP tendons. In *Proceedings of the third international symposium on non-metallic (FRP) reinforcement for concrete structures (FRPRCS-3)* (pp. 203–210).

Ayyub, B. M., Sohn, Y. G., & Saadatmanesh, H. (1990). Prestressed composite girders under positive moment. *Journal of Structural Engineering, 116*(11), 2931–2951. Available from https://doi.org/10.1061/(asce)0733-9445(1990)116:11(2931).

Ayyub, B. M., Sohn, Y. G., & Saadatmanesh, H. (1992a). Prestressed composite girders. II: Analytical study for negative moment. *Journal of Structural Engineering*, *118*(10), 2763−2782. Available from https://doi.org/10.1061/(asce)0733-9445(1992)118:10(2763).

Ayyub, B. M., Sohn, Y. G., & Saadatmanesh, H. (1992b). Prestressed composite girders. I: Experimental study for negative moment. *ASCE Journal of Structural Engineering*, *118* (10), 2743−2762.

Banibayat, P., & Patnaik, A. (2015). Creep rupture performance of basalt fiber-reinforced polymer bars. *ASCE Journal of Aerospace Engineering*, *28*(3), 04014074. Available from https://doi.org/10.1061/(ASCE)AS.1943-5525.0000391.

Bradford, M. A., & Gilbert, R. I. (1992). Composite beams with partial interaction under sustained loads. *Journal of Structural Engineering*, *118*(7), 1871−1883. Available from https://doi.org/10.1061/(asce)0733-9445(1992)118:7(1871).

Cao, G., & Fang, Z. (2007). Experimental study on the long-term behavior of concrete box girders prestressed with external CFRP tendons. *China Civil Engineering Journal*, *40*(2), 18−24.

Cao, G., Han, C., Dai, Y., & Zhang, W. (2018). Long-term experimental study on prestressed steel-concrete composite continuous box beams. *ASCE Journal of Bridge Engineering*, *23*(9), 04018067. Available from https://doi.org/10.1061/(ASCE)BE.1943-5592.0001269.

Chajes, M. J., Thomson, T. A., Januszka, T. F., & Finch, W. W. (1994). Flexural strengthening of concrete beams using externally bonded composite materials. *Construction and Building Materials*, *8*(3), 191−201. Available from https://doi.org/10.1016/S0950-0618 (09)90034-4.

Chen, S., & Gu, P. (2005). Load carrying capacity of composite beams prestressed with external tendons under positive moment. *Journal of Constructional Steel Research*, *61* (4), 515−530. Available from https://doi.org/10.1016/j.jcsr.2004.09.004.

Chen, S., & Jia, Y. (2010). Numerical investigation of inelastic buckling of steel-concrete composite beams prestressed with external tendons. *Thin-Walled Structures*, *48*(3), 233−242. Available from https://doi.org/10.1016/j.tws.2009.10.009.

Chen, S., Wang, X., & Jia, Y. (2009). A comparative study of continuous steel-concrete composite beams prestressed with external tendons: Experimental investigation. *Journal of Constructional Steel Research*, *65*(7), 1480−1489. Available from https://doi.org/ 10.1016/j.jcsr.2009.03.005.

Dall'asta, A., Ragni, L., & Zona, A. (2007). Simplified method for failure analysis of concrete beams prestressed with external tendons. *Journal of Structural Engineering*, *133*(1), 121−131. Available from https://doi.org/10.1061/(ASCE)0733-9445(2007)133:1(121).

Dall'Asta, A., & Zona, A. (2005). Finite element model for externally prestressed composite beams with deformable connection. *Journal of Structural Engineering*, *131*(5), 706−714. Available from https://doi.org/10.1061/(ASCE)0733-9445(2005)131:5(706).

Ding, M., Jiang, X. G., & Ju, J. S. (2010). Analysis of long-term deflection of externally presterssed steel-concrete composite beams. *Engineering Mechanics*, *27*(9), 94−101.

El-Belbisi, A. H., El-Shihy, A. A., & Shaaban, H. F. (2018). Strengthening of pre-stressed steel-concrete composite beams using carbon fiber tendons-A parametric study. *Nano Research & Applications*, *4*(7), 2−7.

El-Belbisi, A. H., El-Sisi, A. A., Hassan, H. A., Salim, H. A., & Shabaan, H. F. (2022). Parametric study on steel-concrete composite beams strengthened with post-tensioned CFRP tendons. *Sustainability*, *14*(23), 15792.

Elgabbas, F., Ahmed, E. A., & Benmokrane, B. (2015). Physical and mechanical characteristics of new basalt-FRP bars for reinforcing concrete structures. *Construction and*

*Building Materials*, *95*, 623−635. Available from https://doi.org/10.1016/j.conbuildmat. 2015.07.036.

El-Hacha, R., & El-Badry, M. (2001). Strengthening concrete beams with externally prestressed carbon fiber composite cables. In *Proceedings of the 5th international conference on fiber reinforced plastics for reinforced concrete structures (FRPRCS-5)* (pp. 699−708).

El-Hacha, R., Wight, R. G., & Green, M. F. (2001). Prestressed fibre-reinforced polymer laminates for strengthening structures. *Progress in Structural Engineering and Materials*, *3*(2), 111−121. Available from https://doi.org/10.1002/pse.76.

Gao, D., Fang, D., You, P., Chen, G., & Tang, J. (2020). Flexural behavior of reinforced concrete one-way slabs strengthened via external post-tensioned FRP tendons. *Engineering Structures*, *216*, 110718. Available from https://doi.org/10.1016/j.engstruct.2020.110718.

Gilbert, R. I. (1989). Time-dependent analysis of composite steel-concrete sections. *Journal of Structural Engineering (United States)*, *115*(11), 2687−2705. Available from https://doi.org/10.1061/(ASCE)0733-9445(1989)115:11(2687).

Gilbert, R. I., & Bradford, M. A. (1995). Time-dependent behavior of continuous composite beams at service loads. *Journal of Structural Engineering*, *121*(2), 319−327. Available from https://doi.org/10.1061/(ASCE)0733-9445(1995)121:2(319).

Grace, N. F., & Abdel-Sayed, G. (1998). Behavior of externally draped CFRP tendons in prestressed concrete bridges. *PCI Journal*, *43*(5), 88−101. Available from https://doi.org/10.15554/pcij.09011998.88.101.

Harajli, M. H. (1993). Strengthening of concrete beams by external prestressing. *PCI Journal*, *38*(6), 76−88. Available from https://doi.org/10.15554/pcij.11011993.76.88.

Hassan, M., Benmokrane, B., ElSafty, A., & Fam, A. (2016). Bond durability of basalt-fiber-reinforced-polymer (BFRP) bars embedded in concrete in aggressive environments. *Composites Part B: Engineering*, *106*, 262−272. Available from https://doi.org/10.1016/j.compositesb.2016.09.039.

Hoadley, P. G. (1963). Behavior of prestressed composite steel beams. *Journal of the Structural Division*, *89*(3), 21−34. Available from https://doi.org/10.1061/jsdeag.0000930.

Jeon, C. H., Lee, J. B., Lon, S., & Shim, C. S. (2019). Equivalent material model of corroded prestressing steel strand. *Journal of Materials Research and Technology*, *8*(2), 2450−2460.

Jung, W. T., Park, J. S., & Park, Y. H. (2013). An experimental study on the flexural behavior of post-tensioned concrete beams with CFRP tendons. *Applied Mechanics and Materials*, *351*, 717−721. Available from https://doi.org/10.4028/www.scientific.net/AMM.351-352.717.

Kato, T., & Hayashida, N. (1993). Flexural characteristics of prestressed concrete beams with CFRP tendons. *ACI Special Publication*, *138*, 419−439. Available from https://doi.org/10.14359/10036.

Liu, X., Yu, W., Huang, Y., Yang, G., You, W., Gao, L., & Song, J. (2023). Long-term behaviour of recycled aggregate concrete beams prestressed with carbon fibre-reinforced polymer (CFRP) tendons. *Case Studies in Construction Materials*, *18*, e01785.

Lopresto, V., Leone, C., & De Iorio, I. (2011). Mechanical characterisation of basalt fibre reinforced plastic. *Composites Part B: Engineering*, *42*(4), 717−723. Available from https://doi.org/10.1016/j.compositesb.2011.01.030.

Lorenc, W., & Kubica, E. (2006). Behavior of composite beams prestressed with external tendons: Experimental study. *Journal of Constructional Steel Research*, *62*(12), 1353−1366.

Lou, T., Lopes, S. M. R., & Lopes, A. V. (2012). Numerical analysis of behaviour of concrete beams with external FRP tendons. *Construction and Building Materials*, *35*, 970−978. Available from https://doi.org/10.1016/j.conbuildmat.2012.04.055.

Mancarti, G. D. (1984). Strengthening california steel bridges by prestressing. *Transportation Research Record*, *950*, 183−187.

Naaman, A. E. (1990). New methodology for the analysis of beams prestressed with external or unbonded tendons. *ACI Special Publication, 120*, 339–354. Available from https://doi.org/10.14359/2765.

Najafabadi, E. P., Bazli, M., Ashrafi, H., & Oskouei, A. V. (2018). Effect of applied stress and bar characteristics on the short-term creep behavior of FRP bars. *Construction and Building Materials, 171*, 960–968. Available from https://doi.org/10.1016/j.conbuildmat.2018.03.204.

Ng, C. K. (2003). Tendon stress and flexural strength of externally prestressed beams. *ACI Structural Journal, 100*(5), 644–653.

Ng, C. K., & Tan, K. H. (2006). Flexural behaviour of externally prestressed beams. Part II: Experimental investigation. *Engineering Structures, 28*(4), 622–633. Available from https://doi.org/10.1016/j.engstruct.2005.09.016.

Nie, J., Tao, M., Cai, C. S., & Li, S. (2011). Analytical and numerical modeling of prestressed continuous steel-concrete composite beams. *Journal of Structural Engineering, 137*(12), 1405–1418. Available from https://doi.org/10.1061/(ASCE)ST.1943-541X.0000409.

Nordin, H., & Täljsten, B. (2006). Concrete beams strengthened with prestressed near surface mounted CFRP. *Journal of Composites for Construction, 10*(1), 60–68. Available from https://doi.org/10.1061/(ASCE)1090-0268(2006)10:1(60).

Omran, G. M., Beygi, M. H. A., & Dehestani, M. (2020). Numerical analysis of externally prestressed concrete beams and parametric study of factors affecting their flexural performance. *International Journal of Advances in Engineering Sciences and Applied Mathematics, 12*(3-4), 142–157. Available from https://doi.org/10.1007/s12572-020-00284-4.

Park, S., Kim, T., Kim, K., & Hong, S. N. (2010). Flexural behavior of steel I-beam prestressed with externally unbonded tendons. *Journal of Constructional Steel Research, 66*(1), 125–132. Available from https://doi.org/10.1016/j.jcsr.2009.07.013.

Peng, F., Xue, W., & Bai, L. (2024). Flexural behavior of externally prestressed continuous steel-concrete composite beams. *Journal of Constructional Steel Research, 212*, 108282.

Pisani, M. A. (1998). A numerical survey on the behaviour of beams pre-stressed with FRP cables. *Construction and Building Materials, 12*(4), 221–232. Available from https://doi.org/10.1016/S0950-0618(97)00081-0.

Prasad, V. V., & Talupula, S. (2018). A review on reinforcement of Basalt and Aramid (Kevlar 129) fibers. *Materials Today: Proceedings, 5*(2), 5993–5998. Available from https://doi.org/10.1016/j.matpr.2017.12.202.

Recupero, A., Spinella, N., Colajanni, P., & Scilipoti, C. D. (2014). Increasing the capacity of existing bridges by using unbonded prestressing technology: a case study. *Advances in Civil Engineering, 2014*(1), 840902.

Saadatmanesh, H., Albrecht, P., & Ayyub, B. M. (1989a). Experimental study of prestressed composite beams. *ASCE Journal of Structural Engineering, 115*(9), 2348–2363.

Saadatmanesh, H., Albrecht, P., & Ayyub, B. M. (1989b). Analytical study of prestressed composite beams. *Journal of Structural Engineering, 115*(9), 2364–2381. Available from https://doi.org/10.1061/(asce)0733-9445(1989)115:9(2364).

Saafi, M., & Toutanji, H. (1998). Flexural capacity of prestressed concrete beams reinforced with aramid fiber reinforced polymer (AFRP) rectangular tendons. *Construction and Building Materials, 12*(5), 245–249.

Sahu, N. P., Khande, D. K., Patel, G. C., Bohidar, S. K., & Sen, P. K. (2014). Study on aramid fibre and comparison with other composite materials. *International Journal for Innovative Research in Science & Technology, 1*(7), 303–306.

Shi, J., Wang, X., Huang, H., & Wu, Z. (2017). Relaxation behavior of prestressing basalt fiber-reinforced polymer tendons considering anchorage slippage. *Journal of Composite Materials, 51*(9), 1275–1284. Available from https://doi.org/10.1177/0021998316673893.

Shi, J., Wang, X., Wu, Z., & Zhu, Z. (2015). Creep behavior enhancement of a basalt fiber-reinforced polymer tendon. *Construction and Building Materials*, *94*, 750−757. Available from https://doi.org/10.1016/j.conbuildmat.2015.07.118.

Shi, J., Wang, X., Wu, Z., Wei, X., & Ma, X. (2022). Long-term mechanical behaviors of uncracked concrete beams prestressed with external basalt fiber-reinforced polymer tendons. *Engineering Structures*, *262*, 114309. Available from https://doi.org/10.1016/j.engstruct.2022.114309.

Sivaleepunth, C., Niwa, J., Diep, B. K., Tamura, S., & Hamada, Y. (2006). Prediction of tendon stress and flexural strength of externally prestressed concrete beams. *Doboku Gakkai Ronbunshuu E*, *62*(1), 260−273. Available from https://doi.org/10.2208/jsceje.62.260.

Sousa, J. B. M., Parente, E., Lima, É. M. F., & Oliveira, M. V. X. (2019). Beam-tendon finite elements for post-tensioned steel-concrete composite beams with partial interaction. *Journal of Constructional Steel Research*, *159*, 147−160. Available from https://doi.org/10.1016/j.jcsr.2019.04.009.

Sowlat, K., & Rabbat, B. G. (1987). Testing of segmental concrete girders with external tendons. *PCI Journal*, *32*(2), 86−107. Available from https://doi.org/10.15554/pcij.03011987.86.107.

Sun, C., Zheng, Y., & Jiang, J. B. (2017). Long-term creep behavior experiment of GFRP bars under different service loads and environmental conditions. In *Sixth Asia-Pacific conference on FRP in structures (APFIS2017)*.

Tan, K. H., & Tjandra, R. A. (2007). Strengthening of RC continuous beams by external prestressing. *Journal of Structural Engineering*, *133*(2), 195−204. Available from https://doi.org/10.1061/(ASCE)0733-9445(2007)133:2(195).

Tong, W., & Saadatmanesh, H. (1992). Parametric study of continuous prestressed composite girders. *Journal of Structural Engineering*, *118*(1), 186−206. Available from https://doi.org/10.1061/(asce)0733-9445(1992)118:1(186).

Tran, D. T., Pham, T. M., Hao, H., & Chen, W. (2021). Numerical study on bending response of precast segmental concrete beams externally prestressed with FRP tendons. *Engineering Structures*, *241*, 112423. Available from https://doi.org/10.1016/j.engstruct.2021.112423.

Uy, B., & Craine, S. (2004). Static flexural behaviour of externally post-tensioned steel-concrete composite beams. *Advances in Structural Engineering*, *7*(1), 1−20. Available from https://doi.org/10.1260/136943304322985729.

Virlogeux, M. P. (1990). External prestressing: From construction history to modern technique and technology. *ACI Special Publication*, *120*, 1−60. Available from https://doi.org/10.14359/3223.

Wang, X., Shi, J., Liu, J., Yang, L., & Wu, Z. (2014). Creep behavior of basalt fiber reinforced polymer tendons for prestressing application. *Materials and Design*, *59*, 558−564. Available from https://doi.org/10.1016/j.matdes.2014.03.009.

Wang, X., Shi, J., Wu, G., Yang, L., & Wu, Z. (2015). Effectiveness of basalt FRP tendons for strengthening of RC beams through the external prestressing technique. *Engineering Structures*, *101*, 34−44. Available from https://doi.org/10.1016/j.engstruct.2015.06.052.

Wang, X., Shi, J., Wu, Z., & Zhu, Z. (2016). Creep strain control by pretension for basalt fiber-reinforced polymer tendon in civil applications. *Materials and Design*, *89*, 1270−1277. Available from https://doi.org/10.1016/j.matdes.2015.10.090.

Xue, W., Ding, M., He, C., & Li, J. (2008). Long-term behavior of prestressed composite beams at service loads for one year. *Journal of Structural Engineering*, *134*(6), 930−937. Available from https://doi.org/10.1061/(ASCE)0733-9445(2008)134:6(930).

Xue, W., & Liu, T. (2021). Time-dependent behavior of concrete beams externally prestressed with carbon fiber-reinforced polymer tendons for 1000 days. *ACI Structural Journal, 118*(3), 15−26.

Yamaguchi, T., Nishimura, T., & Uomoto, T. (1998). Creep rupture of FRP rods made of aramid, carbon and glass fibers. *Structural Engineering & Construction: Tradition, Present and Future, 2,* 1331−1336.

Zhang, N., & Fu, C. C. (2009). Experimental and theoretical studies on composite steel-concrete box beams with external tendons. *Engineering Structures, 31*(2), 275−283. Available from https://doi.org/10.1016/j.engstruct.2008.08.004.

Zhang, Q., Jia, D., Bao, Y., Dong, S., Cheng, Z., & Bu, Y. (2019). Flexural behavior of steel-concrete composite beams considering interlayer slip. *ASCE Journal of Structural Engineering, 145*(9), 04019084. Available from https://doi.org/10.1061/(ASCE)ST.1943-541X.0002374.

Zhao, J., Mei, K., & Wu, J. (2020). Long-term mechanical properties of FRP tendon−anchor systems—A review. *Construction and Building Materials, 230,* 117017. Available from https://doi.org/10.1016/j.conbuildmat.2019.117017.

Zona, A., Ragni, L., & Dall'Asta, A. (2008). Finite element formulation for geometric and material nonlinear analysis of beams prestressed with external slipping tendons. *Finite Elements in Analysis and Design, 44*(15), 910−919. Available from https://doi.org/10.1016/j.finel.2008.06.005.

Zona, A., Ragni, L., & Dall'Asta, A. (2009). Simplified method for the analysis of externally prestressed steel-concrete composite beams. *Journal of Constructional Steel Research, 65*(2), 308−313. Available from https://doi.org/10.1016/j.jcsr.2008.07.015.

Zou, P. X. W. (2003a). Long-term deflection and cracking behavior of concrete beams prestressed with carbon fiber-reinforced polymer tendons. *Journal of Composites for Construction, 7*(3), 187−193. Available from https://doi.org/10.1061/(ASCE)1090-0268(2003)7:3(187).

Zou, P. X. W. (2003b). Long-term properties and transfer length of fiber-reinforced polymers. *Journal of Composites for Construction, 7*(1), 10−19. Available from https://doi.org/10.1061/(ASCE)1090-0268(2003)7:1(10).

# Finite element modeling at immediate loads

**2**

## 2.1 Introduction

Analysis of externally prestressed members is more complicated than that of bonded prestressed members. In an externally posttensioned member, the tendons are located outside of the cross-section and the prestressing force is transferred to the member through anchorages and deviators. As external tendons are unbonded, there is no strain compatibility between these tendons and their adjacent components. Therefore the increase in stress in external tendons is member-dependent rather than section-dependent as in the case of bonded tendons. Moreover, the eccentricities of external tendons are subject to change with the development of member deformation, leading to second-order effects. Over past decades, numerous empirical/semiempirical equations for estimating the stress in external tendons have been proposed, but the determination of the accurate stress in external tendons should be based on the deformation of the entire tendons between anchorages, and an iterative procedure needs to be applied. Alkhairi and Naaman (1993) proposed an analytical model in which the change in strain in external or internal unbonded tendons was calculated by integrating the curvature in concrete at the level of the tendons between end anchorages. Similar analytical methods were also proposed by Du et al. (2008) and Ozkul et al. (2008) for nonlinear analysis of unbonded prestressed concrete members. Ayyub et al. (1990) used a strain compatibility method based on force equilibrium and deformation compatibility to determine the complete response of externally prestressed steel−concrete composite members. By using an empirical equation for computing the tendon stress, Zhang and Fu (2009) proposed an analytical method to determine the ultimate moment resistance of externally prestressed composite box girders. Based on the observation that the strain and curvature distributions at ultimate would not vary substantially under given conditions, Zona et al. (2009) developed a simplified approach for evaluating the moment capacity of externally prestressed composite beams. The aforementioned analytical or simplified approaches are able to predict the stress in external (or internal unbonded) tendons and the response of simply supported members over the entire loading history up to failure but may have limitations when applied to more complicated scenarios (e.g., statically indeterminate members).

The development of a general, reliable, and powerful computer model, which is capable of simulating the comprehensive behavior of various externally posttensioned members, is essential. The finite element method affords an effective numerical technique to approximate closely the real behavior of a member. In the finite element idealization of prestressed beams, the one-dimensional model, in which the

Prestressed Members with External Fiber-Reinforced Polymer (FRP) Tendons. DOI: https://doi.org/10.1016/B978-0-443-23877-2.00002-X

member is idealized as an assemblage of beam elements interconnected at nodes, has been usually preferred by researchers for its simplicity and computer efficiency (Campbell & Kodur, 1990; Kang & Scordelis, 1980; Marković et al., 2013). This chapter describes a finite element model based on the Euler−Bernoulli beam theory for the full-range nonlinear analysis of externally prestressed concrete or steel−concrete composite beams under immediate loads (Lou & Xiang, 2006; Lou, Lopes, & Lopes, 2016). Time-dependent effects such as concrete creep, concrete shrinkage, and tendon relaxation are not covered in this chapter but will be discussed in Chapter 12.

## 2.2 Stress−strain relationships of materials

### 2.2.1 Concrete in compression

The stress−strain law (see Fig. 2.1) suggested by Hognestad (1951) is widely used to simulate the behavior of concrete in compression. The stress−strain equation is expressed by

$$\sigma_c = f_{ck} \left[ \frac{2\varepsilon_c}{\varepsilon_0} - \left( \frac{\varepsilon_c}{\varepsilon_0} \right)^2 \right] \text{ for } 0 \leq \varepsilon_c \leq \varepsilon_0 \tag{2.1a}$$

$$\sigma_c = f_{ck} \left[ 1 - 0.15 \left( \frac{\varepsilon_c - \varepsilon_0}{\varepsilon_u - \varepsilon_0} \right) \right] \text{ for } \varepsilon_0 < \varepsilon_c \leq \varepsilon_u \tag{2.1b}$$

where $\sigma_c$ and $\varepsilon_c$ are the concrete stress and strain, respectively; $f_{ck}$ is the characteristic cylinder compressive strength of concrete; $\varepsilon_0$ is the concrete strain at peak stress, assumed equal to 0.002; $\varepsilon_u$ is the concrete ultimate compressive strain.

It is noted that Eq. (2.1) is only applicable to normal-strength concrete (NSC). Nowadays high-strength concrete (HSC) is extensively used in engineering

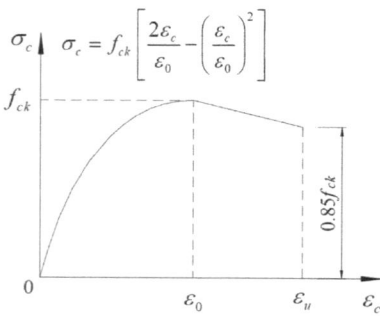

**Figure 2.1** Stress−strain diagram for concrete in compression suggested by Hognestad (1951).

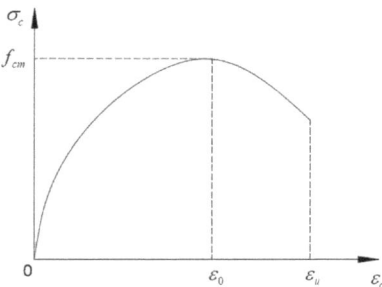

**Figure 2.2** Stress—strain diagram for concrete in compression recommended by EC2 (CEN, 2004).

structures. The compressive model recommended by EC2 (CEN, 2004) has been proven to be applicable to both NSC and HSC. The stress—strain relationship is shown in Fig. 2.2 and expressed by

$$\frac{\sigma_c}{f_{cm}} = \frac{k\eta - \eta^2}{1 + (k-2)\eta} \tag{2.2}$$

where $\eta = \varepsilon_c/\varepsilon_0$; $k = 1.05 E_c \varepsilon_0/f_{cm}$; $\varepsilon_0(\text{‰}) = 0.7 f_{cm}^{0.31} < 2.8$; $E_c$ is the modulus of elasticity of concrete (in GPa), and $E_c = 22(f_{cm}/10)^{0.3}$; $f_{cm}$ is the mean compressive strength of concrete, and $f_{cm} = f_{ck} + 8$, in MPa. Eq. (2.2) is subject to the condition that the concrete strain is not greater than the ultimate compressive strain $\varepsilon_u$, which is determined by

$$\varepsilon_u(\text{‰}) = 3.5 \text{ for } f_{ck} < 50 \text{ MPa} \tag{2.3a}$$

$$\varepsilon_u(\text{‰}) = 2.8 + 27 \left[\frac{98 - f_{cm}}{100}\right]^4 \text{ for } f_{ck} \geq 50 \text{ MPa} \tag{2.3b}$$

### 2.2.2 Concrete in tension

The concrete in tension is assumed to be linearly elastic up to cracking, followed by linear tension-stiffening, as illustrated in Fig. 2.3. The stress—strain equation is expressed as follows:

$$\sigma_c = E_c \varepsilon_c \text{ for } 0 \leq \varepsilon_c \leq \varepsilon_{cr} \tag{2.4a}$$

$$\sigma_c = f_t \left[1 - \frac{\varepsilon_c - \varepsilon_{cr}}{\varepsilon_{t0} - \varepsilon_{cr}}\right] \text{ for } \varepsilon_{cr} < \varepsilon_c \leq \varepsilon_{t0} \tag{2.4b}$$

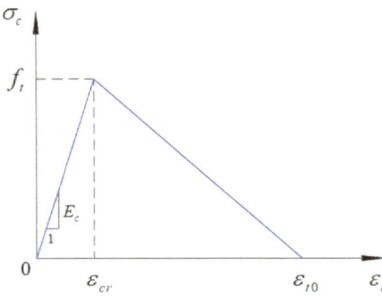

**Figure 2.3** Stress−strain diagram for concrete in tension.

where $f_t$ is the concrete tensile strength; $\varepsilon_{cr}$ is the concrete cracking strain; $\varepsilon_{t0}$ is the concrete tensile strain at the end of strain softening. $\sigma_c$ is zero when $\varepsilon_c > \varepsilon_{t0}$.

### 2.2.3 Prestressing steel

The stress−strain law recommended by Menegotto and Pinto (1973) is adopted here to simulate the behavior of prestressing steel tendons, and it is expressed as follows:

$$\sigma_p = E_p \varepsilon_p \left[ Q + \frac{1-Q}{\left\{ 1 + [\varepsilon_p E_p / (K f_{py})]^R \right\}^{1/R}} \right] \leq f_{pu} \tag{2.5}$$

where $\sigma_p$ and $\varepsilon_p$ are the stress and strain of prestressing tendons, respectively; $f_{py}$ and $f_{pu}$ are the tendon yield stress and tensile strength, respectively; $E_p$ is the tendon modulus of elasticity; and $K$, $Q$, and $R$ are coefficients, depending on the type of steel tendons. For 1860 MPa steel strands, the values of $K$, $Q$, and $R$ are 1.0618, 0.01174, and 7.344, respectively; while for 1620 MPa steel wires, those values are 1.0325, 0.00625, and 6.06, respectively.

### 2.2.4 Nonprestressed steel

An elastic and perfectly plastic law may be adopted to simulate the behavior of nonprestressed steel (steel rebars and steel beam).

### 2.2.5 FRP reinforcement

The fiber-reinforced polymer (FRP) reinforcements (prestressing tendons and reinforcing bars) are linearly elastic up to rupture.

## 2.3 Finite element formulation

Fig. 2.4 shows a plane beam element with two end nodes $i$ and $j$, by which the local coordinate system $(x, y)$ is defined. The node points are at the centroid of the concrete or composite section. Each node has three degrees of freedom, namely, axial displacement $u$, transverse displacement $v$, and rotation $\theta$. Assuming that a plane section remains plane after bending and that the shear deformation is negligible, the axial strain $\varepsilon$ at any fiber of a concrete or composite section is defined by

$$\varepsilon = \frac{\partial u}{\partial x} + \frac{1}{2}\left(\frac{\partial v}{\partial x}\right)^2 - y\frac{\partial^2 v}{\partial x^2} \tag{2.6}$$

where the second term of the right side represents the large-displacement effect.

Assume $u$ is a linear function and $v$ is a cubic polynomial. According to the updated Lagrangian description, the equilibrium equations for an element are given as follows:

$$d\boldsymbol{P}^{\mathrm{e}} = \boldsymbol{K}_{\mathrm{T}}^{\mathrm{e}}d\boldsymbol{u}^{\mathrm{e}} = (\boldsymbol{K}_1^{\mathrm{e}} + \boldsymbol{K}_2^{\mathrm{e}})d\boldsymbol{u}^{\mathrm{e}} \tag{2.7}$$

$$\boldsymbol{P}^{\mathrm{e}} = \left\{ N_i \quad V_i \quad M_i \quad N_j \quad V_j \quad M_j \right\}^T \tag{2.8}$$

$$\boldsymbol{u}^{\mathrm{e}} = \left\{ u_i \quad v_i \quad \theta_i \quad u_j \quad v_j \quad \theta_j \right\}^T \tag{2.9}$$

$$\boldsymbol{K}_1^{\mathrm{e}} = \begin{bmatrix} d_{11}/l & 0 & -d_{12}/l & -d_{11}/l & 0 & d_{12}/l \\ & 12d_{22}/l^3 & 6d_{22}/l^2 & 0 & -12d_{22}/l^3 & 6d_{22}/l^2 \\ & & 4d_{22}/l & d_{12}/l & -6d_{22}/l^2 & 2d_{22}/l \\ & & & d_{11}/l & 0 & -d_{12}/l \\ & \text{Symmetry} & & & 12d_{22}/l^3 & -6d_{22}/l^2 \\ & & & & & 4d_{22}/l \end{bmatrix}$$

$$\tag{2.10a}$$

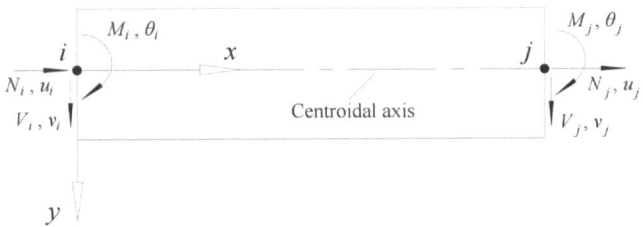

**Figure 2.4** Sketch of beam element.

$$\boldsymbol{K}_2^{\mathrm{e}} = N \begin{bmatrix} 0 & 0 & 0 & 0 & 0 & 0 \\ & 6/5l & 1/10 & 0 & -6/5l & 1/10 \\ & & 2l/15 & 0 & -1/10 & -l/30 \\ & & & 0 & 0 & 0 \\ & \text{Symmetry} & & & 6/5l & -1/10 \\ & & & & & 2l/15 \end{bmatrix} \tag{2.10b}$$

where $\boldsymbol{P}^{\mathrm{e}}$ is the element equivalent nodal loads; $\boldsymbol{u}^{\mathrm{e}}$ is the element nodal displacements; $\boldsymbol{K}_{\mathrm{T}}^{\mathrm{e}}$ is the element tangent stiffness matrix which consists of the material stiffness matrix $\boldsymbol{K}_1^{\mathrm{e}}$ and the geometric stiffness matrix $\boldsymbol{K}_2^{\mathrm{e}}$; $l$ is the length of the beam element; and

$$d_{11} = \sum_k E_k A_k \tag{2.11a}$$

$$d_{12} = \sum_k E_k y_k A_k \tag{2.11b}$$

$$d_{22} = \sum_k E_k y_k^2 A_k \tag{2.11c}$$

$$N = \sum_k \sigma_k A_k \tag{2.12}$$

where the summation symbol signifies that the cross-section is divided into a number of layers (see Fig. 2.5) to employ a layered approach and the subscript $k$ represents each layer; $E$ is the tangent modulus for materials; $A$ corresponds to area and $\sigma$ corresponds to stress.

A load-control or displacement-control incremental method is used to solve the structure equilibrium equations, which are assembled in the global coordinate system from the contributions of all elements. For each increment step, the Newton–Raphson iterative algorithm is employed to eliminate the out-of-balance loads:

$$\boldsymbol{R} = \boldsymbol{P} - \boldsymbol{Q} = \boldsymbol{P}_{\mathrm{load}} + \boldsymbol{P}_{\mathrm{pres}} - \boldsymbol{Q} \tag{2.13}$$

where $\boldsymbol{Q}$ is the structure internal resisting loads; and $\boldsymbol{P}$ is the structure equivalent nodal loads which consist of two components, namely, $\boldsymbol{P}_{\mathrm{load}}$ (equivalent loads due to applied loads) and $\boldsymbol{P}_{\mathrm{pres}}$ (equivalent loads due to external prestressing). The obtainment of $\boldsymbol{P}_{\mathrm{pres}}$ is discussed in the following section.

## 2.4 Equivalent loads due to external tendons

In this model, the current force in external tendons is transformed into equivalent loads. Different from the internal unbonded tendons where the tendon eccentricities

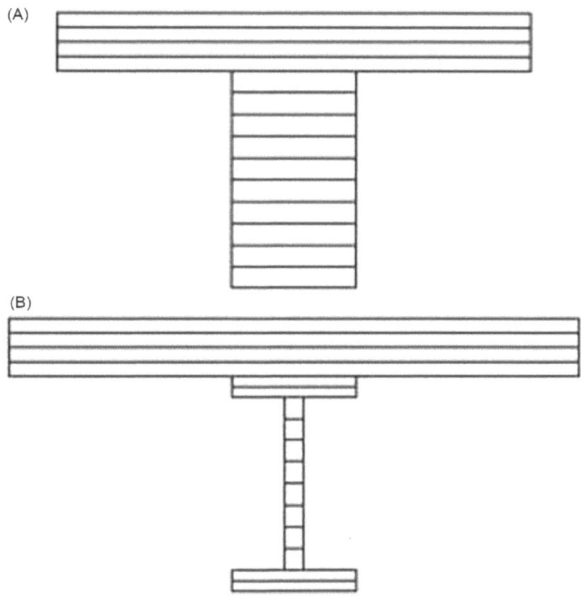

**Figure 2.5** Layered cross-section: (A) concrete section; (B) steel−concrete composite section.

remain constant during loading, the eccentricities of external tendons would change with varying member deflection. Therefore to obtain the equivalent loads of external prestressing, both the force and eccentricities of the tendons need to be determined.

For an externally prestressed member, the prestressing tendons are connected to the member by deviators and anchorages. The member is idealized as an assemblage of a series of beam elements. The deviator and anchorage points are associated with the corresponding element nodes. Consider two adjacent deviators (and/or anchorages) A and B, and their corresponding beam element nodes a and b, as shown in Fig. 2.6.

Before deformation, the deviators are related to the element nodes by

$$X_{A0} = X_{a0} - e_a \sin\beta_0 ; Y_{A0} = Y_{a0} + e_a \cos\beta_0 \tag{2.14a}$$

$$X_{B0} = X_{b0} - e_b \sin\beta_0 ; Y_{B0} = Y_{b0} + e_b \cos\beta_0 \tag{2.14b}$$

in which $(X_{A0}, Y_{A0})$ and $(X_{B0}, Y_{B0})$ = global coordinates of the deviators A and B before deformation, respectively; $(X_{a0}, Y_{a0})$ and $(X_{b0}, Y_{b0})$ = global coordinates of the beam element nodes a and b before deformation, respectively; $e_a$ and $e_b$ = tendon eccentricities at nodes a and b, respectively; $\beta_0$ = initial angle of the X-axis and a−b line.

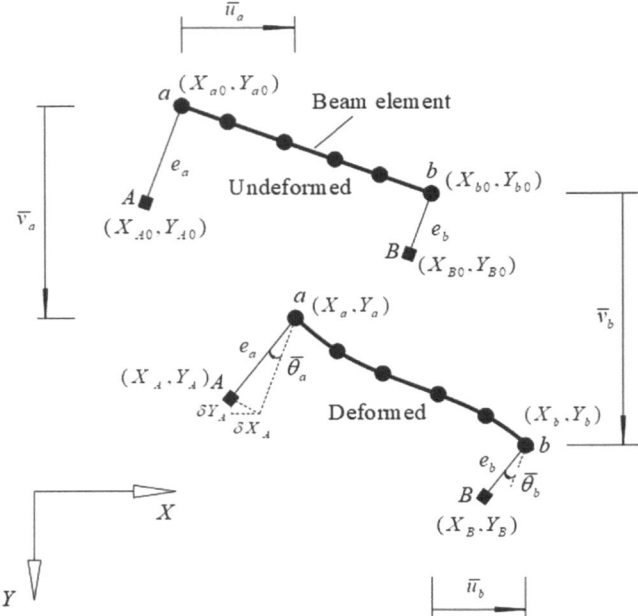

**Figure 2.6** Geometric relationship between two adjacent deviators and corresponding nodes.

At a deformed state, as can be seen in Fig. 2.6, the $X$- and $Y$-displacements at the element node $a$ (i.e., $\bar{u}_a$ and $\bar{v}_a$) or $b$ (i.e., $\bar{u}_b$ and $\bar{v}_b$ ) will pass to the corresponding deviator $A$ or $B$, while the rotation $\bar{\theta}_a$ or $\bar{\theta}_b$ will lead to additional $X$- and $Y$-displacements at the deviator. As a result, the global coordinates of the deviators $A$ and $B$ are expressed as follows:

$$X_A = X_{A0} + \bar{u}_a + \delta X_A; Y_A = Y_{A0} + \bar{v}_a + \delta Y_A \tag{2.15a}$$

$$X_B = X_{B0} + \bar{u}_b + \delta X_B; Y_B = Y_{B0} + \bar{v}_b + \delta Y_B \tag{2.15b}$$

where

$$\delta X_A = - e_a \bar{\theta}_a \cos\beta_0; \delta Y_A = - e_a \bar{\theta}_a \sin\beta_0 \tag{2.16a}$$

$$\delta X_B = - e_b \bar{\theta}_b \cos\beta_0; \delta Y_B = - e_b \bar{\theta}_b \sin\beta_0 \tag{2.16b}$$

The length of the external tendon between the adjacent deviators $A$ and $B$ can be calculated from

$$L_p = L_{AB} = \sqrt{(X_B - X_A)^2 + (Y_B - Y_A)^2} \tag{2.17}$$

Assuming that frictions between external tendons and deviators are negligible, the strain in external tendons over the full length is uniformly distributed. The tendon strain increment $\Delta\varepsilon_p$ is calculated by

$$\Delta\varepsilon_p = \left(\sum L_p - \sum L_{p0}\right) / \sum L_{p0} \tag{2.18}$$

in which the summation symbol signifies that this strain increment results from the elongation of the full length. Add $\Delta\varepsilon_p$ to the initial tendon strain $\varepsilon_{p0}$ to get the current tendon strain $\varepsilon_p$:

$$\varepsilon_p = \varepsilon_{p0} + \Delta\varepsilon_p \tag{2.19}$$

Substituting the strain $\varepsilon_p$ into the constitutive relationship for prestressing tendons yields the stress $\sigma_p$. The tendon force $N_p$ can therefore be obtained from

$$N_p = \sigma_p A_p \tag{2.20}$$

in which $N_p$ = tensile force in prestressing tendons and $A_p$ = tendon area.

The external tendon is discretized into segments. Each segment spans a beam element as demonstrated in Fig. 2.7. The tendon eccentricities $e_i$ and $e_j$ would vary with increasing deformation and should be updated at each step. To obtain the tendon eccentricities, three lines, designated as Lines 1, 2, and 3 as shown in Fig. 2.7, are considered. At each deformed state, Line 1 can be determined by the global coordinates of the two deviator points $A$ and $B$, while Line 2 or 3 can also be determined by the global coordinates of the node $i$ or $j$ and the line slope. The new coordinates of the tendon segment joint $pi$ can be accurately determined by Lines 1 and 2 while $pj$ by Lines 1 and 3, as shown in Fig. 2.7. The eccentricities $e_i$ and $e_j$ can

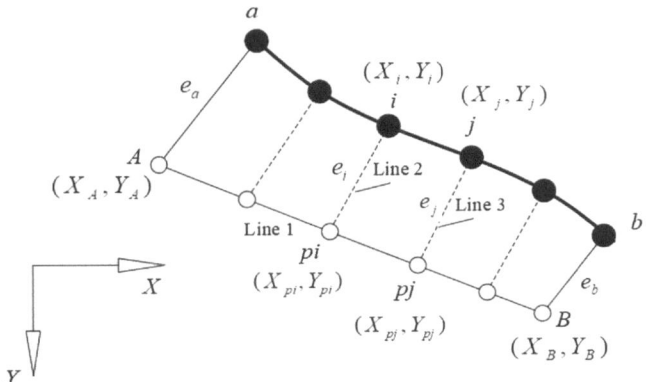

**Figure 2.7** Determination of tendon eccentricities according to tendon segment joints and beam element nodes.

therefore be updated according to the coordinates of segment joints and the corresponding beam element nodes. Thereby, the second-order effects are considered.

When $N_p$, $e_i$, and $e_j$ are determined, the equivalent nodal forces by external prestressing are obtained by

$$F_{pres}^e = \{ N_{ip} \quad V_{ip} \quad M_{ip} \quad N_{jp} \quad V_{jp} \quad M_{jp} \}^T \tag{2.21}$$

where $N_{ip} = -N_p\cos\gamma$; $V_{ip} = N_p\sin\gamma$; $M_{ip} = e_iN_p\cos\gamma$; $N_{jp} = N_p\cos\gamma$; $V_{jp} = -N_p\sin\gamma$; $M_{jp} = -e_iN_p\cos\gamma$; $\gamma$ = angle of the tendon segment and $x$-axis. The element equivalent nodal loads $P_{pres}^e$ as a result of external prestressing are just the opposite, that is, $P_{pres}^e = -F_{pres}^e$.

The structure equivalent nodal loads due to external prestressing are assembled from the contribution of all the tendon segments by

$$P_{pres} = \sum T^T P_{pres}^e \tag{2.22}$$

where $T$ is the displacement transformation matrix, which would be varying continuously during the solution process in accordance with the continuous change of the direction of the local coordinate system.

According to Eq. (2.19), the value of the tendon strain at the original undeformed state is required for the calculation of the current strain in external tendons. Also, the original stress in external tendons would be applied for the analysis of prestress transfer. Due to elastic compression, the stress in tendons after the prestress transfer is generally smaller than the original value. The tendon stress after the prestress transfer is referred to as the effective prestress ($\sigma_{pe}$), the value of which is usually known. The original tendon stress (strain) can be determined according to the effective prestress by the trial-and-error method as follows: (1) estimate approximately a value of the original tendon stress, and then transform it into equivalent nodal loads; (2) analyze the beam and, when the analysis is completed, check the difference between the tendon stress calculated and the effective prestress; (3) adjust the original tendon stress according to this stress difference, and the process is repeated until the difference vanishes.

## 2.5    Numerical examples

### 2.5.1    Simply supported externally prestressed concrete specimens

A total of 11 simply supported beam specimens are selected, including five beams with external steel tendons (Ng & Tan, 2006) and six beams with external carbon fiber-reinforced polymer (CFRP) tendons (Bennitz et al., 2012). All specimens were of simply supported beams with a T-shaped section and were subjected to third-point loading up to failure, as shown in Fig. 2.8. The bottom nonprestressed

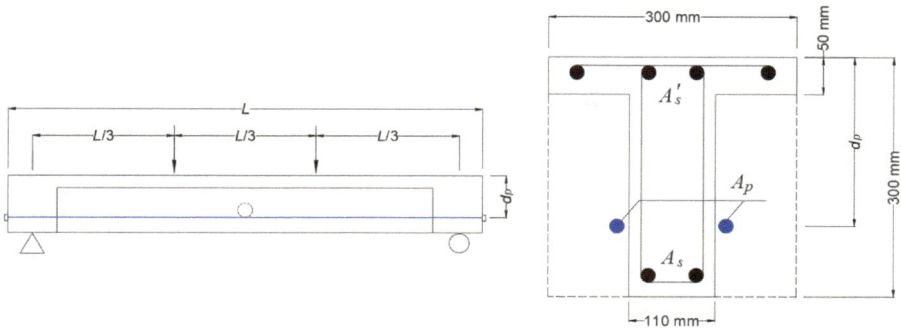

**Figure 2.8** Simply supported externally prestressed concrete specimens.

**Table 2.1** Design parameters of simply supported externally prestressed concrete specimens.

| Specimen | Tendon | $A_p$ (mm²) | $f_{pu}$ (MPa) | $E_p$ (GPa) | $\sigma_{pe}$ (MPa) | $A_s$ (mm²) | $f_y$ (MPa) | $E_s$ (GPa) | $f_{ck}$ (MPa) |
|---|---|---|---|---|---|---|---|---|---|
| ST-1 | Steel | 201 | 1900 | 193 | 764 | 402 | 530 | 200 | 34.5 |
| ST-2 | Steel | 201 | 1900 | 193 | 771 | 402 | 530 | 200 | 29.9 |
| ST-3 | Steel | 201 | 1900 | 193 | 750 | 402 | 530 | 200 | 33.2 |
| ST-4 | Steel | 201 | 1900 | 193 | 757 | 402 | 530 | 200 | 28.3 |
| ST-5 | Steel | 201 | 1900 | 193 | 760 | 402 | 530 | 200 | 25.1 |
| B2 | CFRP | 100.5 | 2790 | 158 | 895 | 402 | 560 | 172 | 35.2 |
| B3 | CFRP | 100.5 | 2790 | 158 | 1382 | 402 | 560 | 172 | 33.4 |
| B4 | CFRP | 100.5 | 2790 | 158 | 396 | 402 | 560 | 172 | 35.9 |
| B5 | CFRP | 100.5 | 2790 | 158 | 889 | 402 | 560 | 172 | 28.4 |
| B6 | CFRP | 100.5 | 2790 | 158 | 917 | 402 | 560 | 172 | 40.6 |
| B7 | CFRP | 100.5 | 2790 | 158 | 1407 | 402 | 560 | 172 | 35.9 |

reinforcement consisted of two deformed steel rebars each having 16 mm in diameter, while the top nonprestressed reinforcement consisted of four deformed steel rebars each having 8 mm in diameter. A summary of the main design parameters is given in Table 2.1, where $A_s$, $f_y$, and $E_s$ are the area, yield strength, and elastic modulus of tensile steel rebars, respectively.

The specimens with external steel tendons were designated as ST-1, ST-2, ST-3, ST-4, and ST-5. The specimens were provided with one deviator at midspan and the tendons were of a straight horizontal profile with an initial depth ($d_p$) of 200 mm. The main test variable was the span-to-depth ratio, that is, $L/d_p$ = 7.5, 9.0, 15.0, 22.5, and 30.0 for Specimens ST-1, ST-2, ST-3, ST-4, and ST-5, respectively.

The specimens with external CFRP tendons were designated as B2, B3, B4, B5, B6, and B7. The main test variables were the initial tendon depth, effective prestress, and deviator arrangement. The initial tendon depth was 200 mm for Specimens B2, B3, B6, and B7 and 250 mm for Specimens B4 and B5. Specimens B2, B3, B4, and B5 were

provided with one deviator at the midspan, while Specimens B6 and B7 had no deviators. The effective prestress varied between 396 and 1407 MPa.

According to the proposed analysis, crushing failure for all specimens occurs at
the midspan, after significant yielding of tensile rebars (formation of the plastic
hinge with sufficient rotation). This failure mode is consistent with the experimental
observation. The load—displacement curves and the development of the tendon
stress or force generated by the finite element model are compared with the experimental data in Figs. 2.9 and 2.10. It is observed that although the proposed analysis
overestimates the tendon stress and force at ultimate in some specimens (ST-1, ST-
2, ST-3, B2, B3, B4, B5, and B7), the numerical results are generally in satisfactory
agreement with the experimental ones over the entire ranges of loading up to the
ultimate limit state.

## 2.5.2  Continuous externally prestressed concrete specimens

Five continuous beam specimens tested by Tan and Tjandra (2007) are selected,
including two specimens (C2 and C2L) that were prestressed with external steel
tendons and three specimens (MCBC-1, MCBC-2, and MCBC-3) that were prestressed with external CFRP tendons. The sketch of the specimens is shown in
Fig. 2.11, and the main material properties are given in Table 2.2.

The test variable of the specimens with external steel tendons was the loading
pattern. Specimen C2 was subjected to symmetrical loads, that is, two concentrated
loads at third points were equally applied on each span of the beam. Specimen C2L
was under unsymmetrical loads, that is, two concentrated loads at third points
applied on the left-hand span were three times those applied on the right-hand span.
In these two specimens, external tendons were of draped profile through the provision of six deviators along the beam. The steel tendons had an area of 110 mm$^2$, an
ultimate strength of 1900 MPa, and an elastic modulus of 195 GPa.

On the other hand, for MCBC-1, MCBC-2, and MCBC-3, the main test variable
was the layout of external CFRP tendons (see Fig. 2.11). These tendons were placed

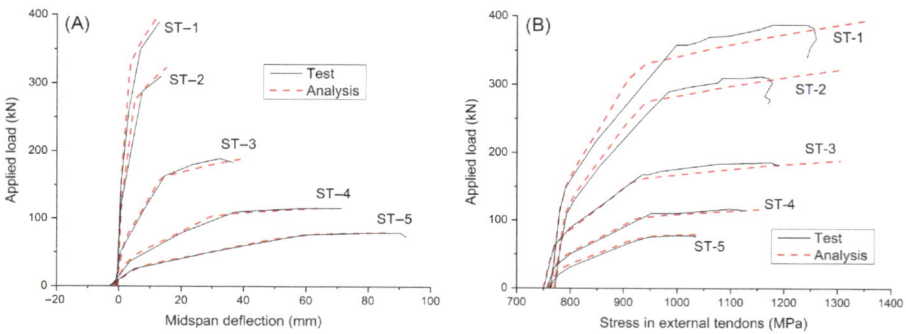

**Figure 2.9** Comparison with experimental results for simply supported prestressed concrete
specimens with external steel tendons: (A) load versus deflection; (B) load versus tendon stress.

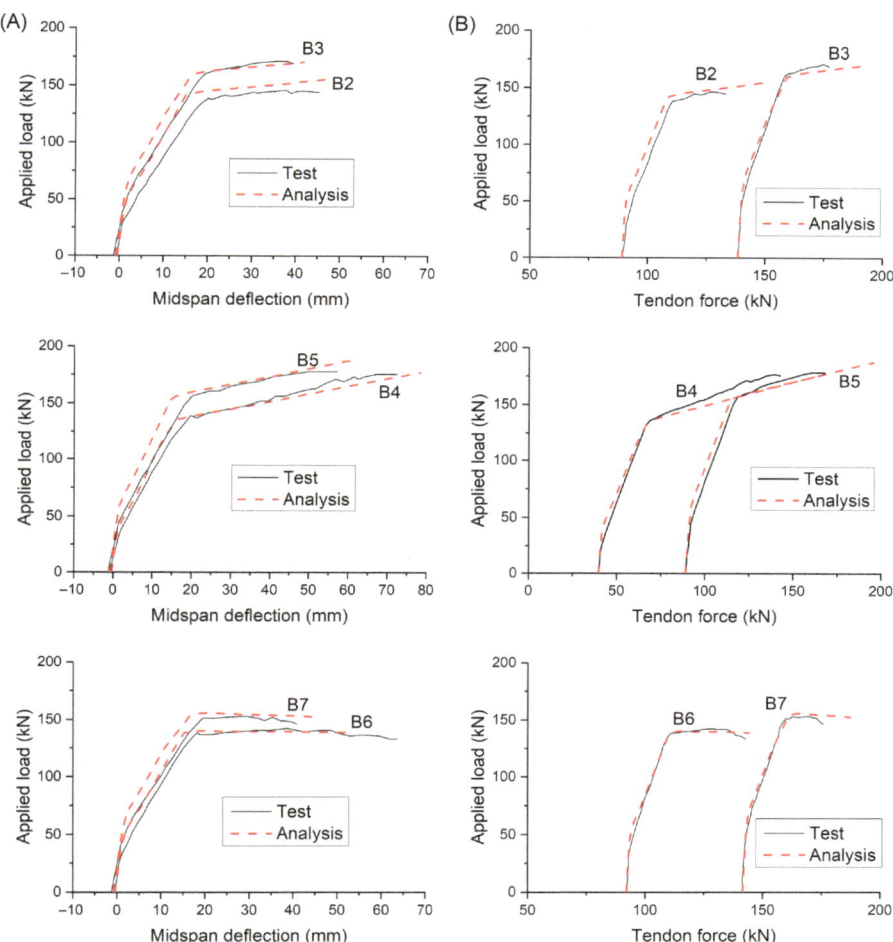

**Figure 2.10** Comparison with experimental results for simply supported prestressed concrete specimens with external CFRP tendons: (A) load versus deflection; (B) load versus tendon force.

over the sagging region only for MCBC-1, over the hogging region only for MCBC-2, and over both the sagging and hogging regions for MCBC-3. Third-point loading was symmetrically applied to the beams. For CFRP tendons, their tensile strength was 1868 MPa and elastic modulus was 139 GPa. The tendon area was 111.4 mm².

For all specimens, the initial prestress was approximately 50% of the tensile strength. The bottom longitudinal nonprestressed reinforcement consisted of two steel rebars with 16 mm in diameter each. The top one consisted of four and six steel rebars with a diameter of 10 mm each over the sagging and hogging regions, respectively. The steel yield strength was 540 MPa. The concrete strength $f_{ck}$ was 30 MPa.

Fig. 2.12 A and B compares the proposed analysis with test data for C2 and C2L regarding the load—deflection response and the development of tendon stress,

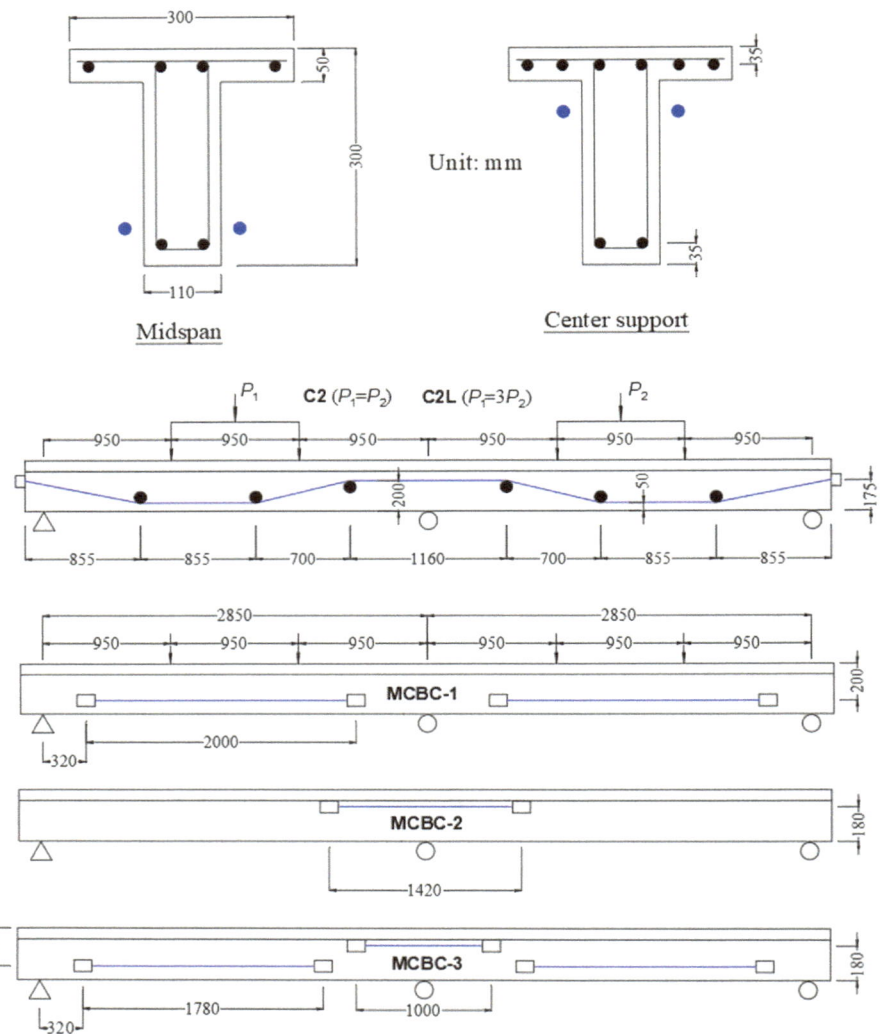

**Figure 2.11** Continuous externally prestressed concrete specimens.

**Table 2.2** Material properties for continuous externally prestressed concrete specimens.

| Specimen | Tendon | $A_p$ (mm²) | $f_{pu}$ (MPa) | $E_p$ (GPa) | $\sigma_{pe}$ (MPa) | $f_y$ (MPa) | $E_s$ (GPa) | $f_{ck}$ (MPa) |
|---|---|---|---|---|---|---|---|---|
| C2 | Steel | 110 | 1900 | 195 | 950 | 540 | 200 | 30 |
| C2L | Steel | 110 | 1900 | 195 | 950 | 540 | 200 | 30 |
| MCBC-1 | CFRP | 111.4 | 1868 | 139 | 950 | 540 | 200 | 30 |
| MCBC-2 | CFRP | 111.4 | 1868 | 139 | 950 | 540 | 200 | 30 |
| MCBC-3 | CFRP | 111.4 | 1868 | 139 | 950 | 540 | 200 | 30 |

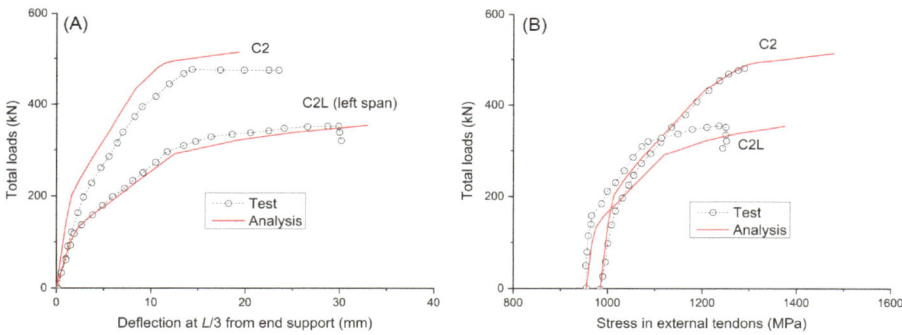

**Figure 2.12** Comparison with test data for continuous prestressed concrete specimens with external steel tendons: (A) load versus deflection; (B) load versus tendon stress.

respectively. The deflection was measured at $L/3$ ($L$ = span length) from the end support. In general, the proposed analysis shows good predictions on the global response characteristics of both specimens, but some discrepancies between test and analysis are noticed. At symmetrical loading (Beam C2), the proposed analysis exhibits a stiffer behavior, possibly attributed to that the analysis neglects the relative slip between bonded rebars and surrounding concrete. The predicted tendon stresses in both specimens appear to be greater than the test results. This may be attributed to the negligible friction loss assumed in the analysis.

According to the numerical analysis, the first cracking for Beam MCBC-2 appears at the critical sagging region, followed by the second cracking at the inner support. Reversed order of the appearance of cracking is observed for MCBC-1 and MCBC-3. Failure of the specimens is caused by concrete crushing, after the formation of plastic hinges over both the critical hogging and sagging regions. The above statements about cracking and failure of the beams are consistent with the experimental observations. The predicted load versus deflection and tendon stress characteristics are compared to experimental data in Fig. 2.13. The numerical analysis considerably overestimates the stress in external CFRP tendons for Beam MCBC-1. This may be attributed to the premature termination of the test of the specimen. Despite some discrepancies, numerical predictions and experimental data show satisfactory agreement.

### 2.5.3 Externally prestressed steel—concrete composite specimens

Two externally prestressed steel—concrete composite beam specimens are selected for the model validation. The specimens include a simply supported composite beam (BS2) tested by Chen and Gu (2005) and a two-span continuous composite beam (PCCB1) tested by Chen et al. (2009). The specimen BS2 was prestressed with straight external steel tendons with no deviators along the span (see Fig. 2.14). The tendons were placed at 30 mm above the bottom of the specimen. The thicknesses of the top flange, bottom flange, and web of the steel beam were 10, 10, and

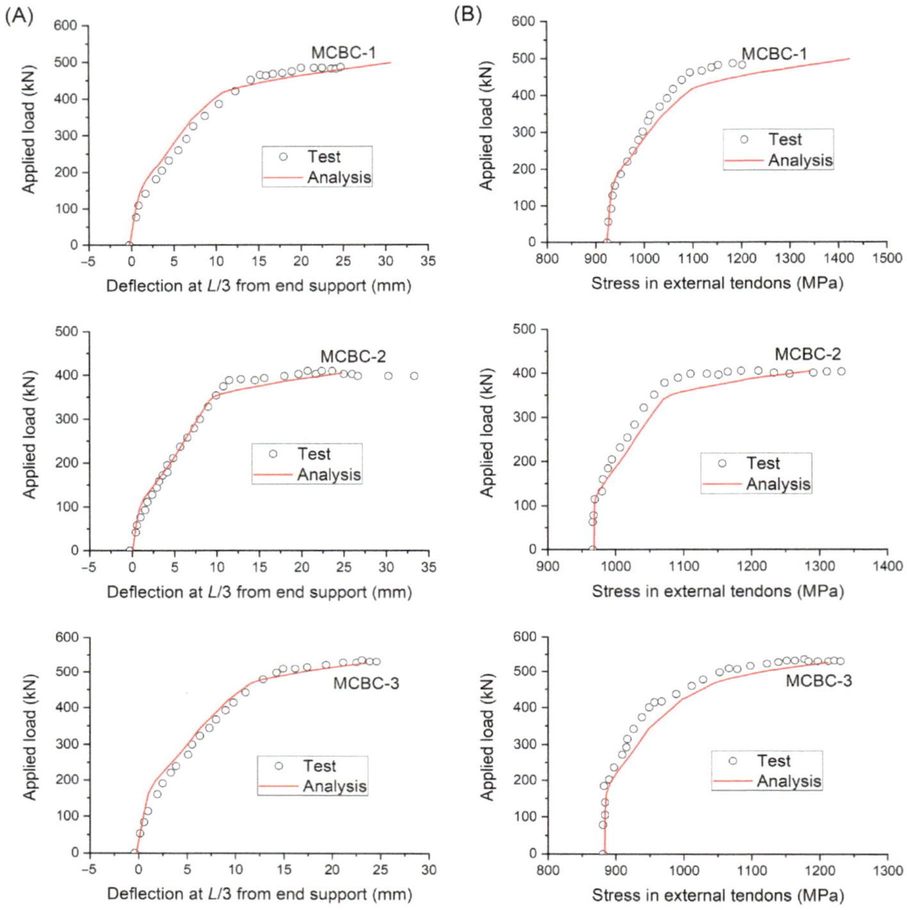

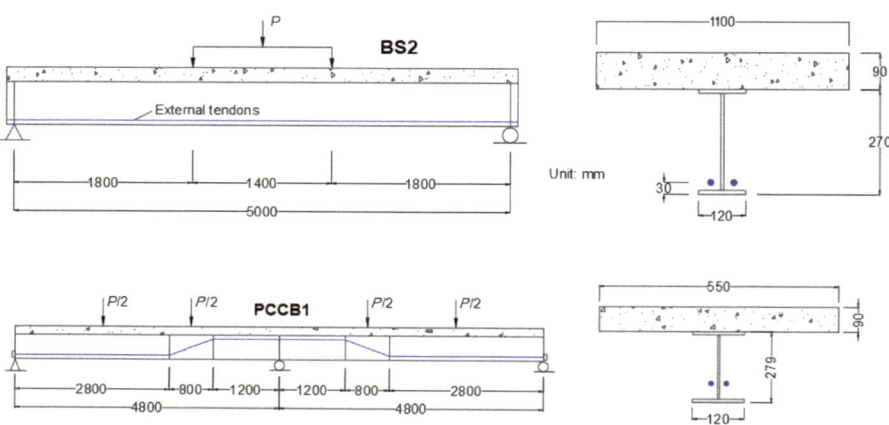

**Figure 2.13** Comparison with test data for continuous prestressed concrete specimens with external CFRP tendons: (A) load versus deflection; (B) load versus tendon stress.

**Figure 2.14** Externally prestressed steel–concrete composite beam specimens.

6 mm, respectively. In Specimen PCCB1, the steel tendons were straight horizontal over the positive (30 mm above the bottom flange) and negative moment regions (30 mm below the top flange) and draped at the deviator points of the span (see Fig. 2.14). The main material properties of the specimens are presented in Table 2.3.

Fig. 2.15 shows a comparison between the analysis and test results regarding the moment-deflection response and moment versus total tendon force response for the simply supported composite specimen (BS2), while the predicted load−deflection response and stress increase in external tendons for the continuous composite specimen (PCCB1) were compared with the experimental results in Fig. 2.16. It is seen in the figures that despite some discrepancies, the finite element analysis captures well the trend of the evolution of the deflection and tendon force/stress. Note that for the continuous composite specimen (PCCB1), there was an obvious discrepancy between the experimentally obtained tendon stresses measured from the west and middle segments, as illustrated in Fig. 2.16B. The tendon stress evolution predicted by the finite element analysis is in good agreement with the test data for the middle segment, while at ultimate the numerical predictions correspond well with the test data for the west segment.

**Table 2.3** Material properties for externally prestressed composite specimens.

| Specimen | Type | Prestressing tendons | | | | Steel beam | | Concrete |
|---|---|---|---|---|---|---|---|---|
| | | $A_p$ (mm²) | $f_{pu}$ (MPa) | $E_p$ (GPa) | $\sigma_{pe}$ (MPa) | $f_y$ (MPa) | $E_s$ (GPa) | $f_{ck}$ (MPa) |
| BS2 | Simple | 274.8 | 1860 | 195 | 819.5 | 327.7 (web) 406.5 (flange) | 200 | 30 |
| PCCB1 | Continuous | 274.8 | 1860 | 195 | 880.6 | 370 | 200 | 27.2 |

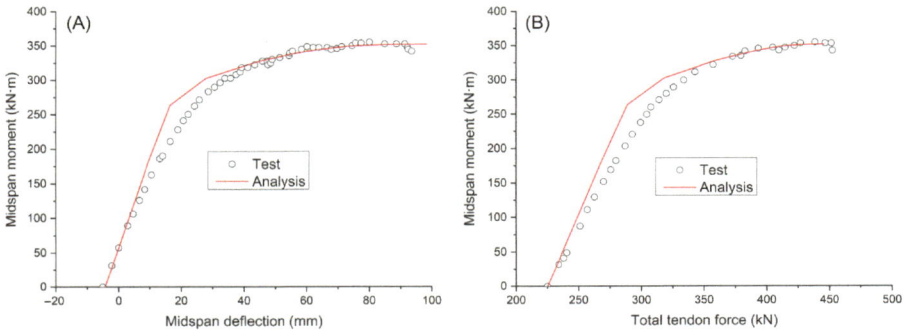

**Figure 2.15** Comparison with experimental results for simply supported composite specimen: (A) moment versus deflection; (B) moment versus total tendon force.

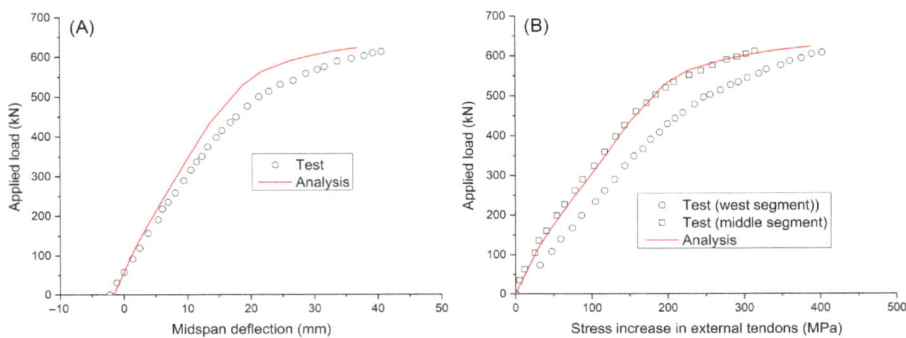

**Figure 2.16** Comparison with experimental results for continuous composite specimen: (A) load versus deflection; (B) load versus tendon stress increment.

## 2.6   Conclusions

This chapter presents a finite element model developed to predict the short-term (immediate) behavior of concrete or steel—concrete composite beams prestressed with external tendons over the entire loading process up to the ultimate limit state. Time-dependent effects are not covered in this chapter and will be discussed in Chapter 12. The finite element is formulated based on the layered Euler—Bernoulli beam theory. According to the updated Lagrangian description, the stiffness matrix consists of the material stiffness matrix representing the material nonlinear effect and the geometric stiffness matrix representing the large-displacement effect. At each step during the solution process, the geometry (length and eccentricities) of external tendons is updated in terms of the current nodal displacements of beam elements (anchorage and deviator points are set to be the element nodes). The tendon strain increment, above the reference strain, is computed from the elongation of the entire tendon between anchorages. This increment is added to the reference strain to obtain the current total strain, and thereby the tendon stress and prestressing force. The contribution of external tendons to the beam is made by transforming the current prestressing force into equivalent nodal loads acting on a finite element model.

The finite element model is verified with the experimental results of a number of specimens available in the literature, including 11 simply supported and 5 continuous prestressed concrete beams with external steel or FRP tendons as well as 2 externally prestressed steel—concrete composite beams. Comparisons between the numerical and experimental results show that the proposed model is able to capture the response characteristics of the specimens over the entire loading range up to failure. In the following, the proposed finite element model is employed to assess the short-term flexural behavior of prestressed concrete members with external FRP/steel tendons (Chapters 3—9) as well as externally prestressed steel—concrete composite girders (Chapters 10 and 11).

# References

Alkhairi, F. M., & Naaman, A. E. (1993). Analysis of beams prestressed with unbonded internal or external tendons. *Journal of Structural Engineering, 119*(9), 2680−2700. Available from https://doi.org/10.1061/(asce)0733-9445(1993)119:9(2680).

Ayyub, B. M., Sohn, Y. G., & Saadatmanesh, H. (1990). Prestressed composite girders under positive moment. *Journal of Structural Engineering, 116*(11), 2931−2951. Available from https://doi.org/10.1061/(asce)0733-9445(1990)116:11(2931).

Bennitz, A., Schmidt, J. W., Nilimaa, J., Täljsten, B., Goltermann, P., & Ravn, D. L. (2012). Reinforced concrete T-beams externally prestressed with unbonded carbon fiber-reinforced polymer tendons. *ACI Structural Journal, 109*(4), 521−530. Available from http://www.concrete.org/tempComDocs/-41246/109-s45.pdf.

Campbell, T. I., & Kodur, V. K. R. (1990). Deformation controlled nonlinear analysis of pre-stressed concrete continuous beams. *PCI Journal, 35*(5), 42−55. Available from https://doi.org/10.15554/pcij.09011990.42.55.

CEN. (2004). *Eurocode 2 (EC2): Design of concrete structures. Part 1-1: General rules and rules for buildings*. EN 1992-1-1, Brussels, Belgium.

Chen, S. M., & Gu, P. (2005). Load carrying capacity of composite beams prestressed with external tendons under positive moment. *Journal of Constructional Steel Research, 61* (4), 515−530. Available from https://doi.org/10.1016/j.jcsr.2004.09.004.

Chen, S. M., Wang, X. D., & Jia, Y. L. (2009). A comparative study of continuous steel−concrete composite beams prestressed with external tendons: Experimental investigation. *Journal of Constructional Steel Research, 65*(7), 1480−1489. Available from https://doi.org/10.1016/j.jcsr.2009.03.005.

Du, J. S., Au, F. T. K., Cheung, Y. K., & Kwan, A. K. H. (2008). Ductility analysis of pre-stressed concrete beams with unbonded tendons. *Engineering Structures, 30*(1), 13−21. Available from https://doi.org/10.1016/j.engstruct.2007.02.015.

Hognestad, E. (1951). Study of combined bending and axial load in reinforced concrete members. In *Bulletin No. 399*, Urbana, IL: University of Illinois Engineering Experiment Station.

Kang, Y. J., & Scordelis, A. C. (1980). Nonlinear analysis of prestressed concrete frames. *Journal of the Structural Division, 106*(2), 445−462. Available from https://doi.org/10.1061/jsdeag.0005367.

Lou, T., Lopes, S. M. R., & Lopes, A. V. (2016). Numerical modeling of externally pre-stressed steel-concrete composite beams. *Journal of Constructional Steel Research, 121*, 229−236. Available from https://doi.org/10.1016/j.jcsr.2016.02.008.

Lou, T., & Xiang, Y. Q. (2006). Finite element modeling of concrete beams prestressed with external tendons. *Engineering Structures, 28*(14), 1919−1926. Available from https://doi.org/10.1016/j.engstruct.2006.03.020.

Markovič, M., Krauberger, N., Saje, M., Planinc, I., & Bratina, S. (2013). Non-linear analysis of pre-tensioned concrete planar beams. *Engineering Structures, 46*, 279−293. Available from https://doi.org/10.1016/j.engstruct.2012.08.004.

Menegotto, M., & Pinto, P.E. (1973). Method of analysis for cyclically loaded reinforced concrete plane frames. *IABSE preliminary report for symposium on resistance and ulti-mate deformability of structures acted on well-defined repeated loads* (pp. 15−22).

Ng, C. K., & Tan, K. H. (2006). Flexural behaviour of externally prestressed beams. Part II: Experimental investigation. *Engineering Structures, 28*(4), 622−633. Available from https://doi.org/10.1016/j.engstruct.2005.09.016.

Ozkul, O., Nassif, H., Tanchan, P., & Harajli, M. (2008). Rational approach for predicting stress in beams with unbonded tendons. *ACI Structural Journal*, *105*(3), 338−347.

Tan, K. H., & Tjandra, R. A. (2007). Strengthening of RC continuous beams by external prestressing. *Journal of Structural Engineering*, *133*(2), 195−204. Available from https://doi.org/10.1061/(ASCE)0733-9445(2007)133:2(195).

Zhang, N., & Fu, C. C. (2009). Experimental and theoretical studies on composite steel-concrete box beams with external tendons. *Engineering Structures*, *31*(2), 275−283. Available from https://doi.org/10.1016/j.engstruct.2008.08.004.

Zona, A., Ragni, L., & Dall'Asta, A. (2009). Simplified method for the analysis of externally prestressed steel−concrete composite beams. *Journal of Constructional Steel Research*, *65*(2), 308−313. Available from https://doi.org/10.1016/j.jcsr.2008.07.015.

# Second-order effects of externally prestressed concrete members

**3**

## 3.1 Introduction

The introduction of external prestressing is widespread in engineering practice (Grace et al., 2002; Mancarti, 1984; Mutsuyoshi et al., 2010). In an externally post-tensioned member, the prestressing tendons are integrated into the member through end anchorages and deviators; thus the tendons between adjacent anchorages and/or deviators remain rectilinear throughout the whole loading history. When the member deforms under external loads, the eccentricities or effective depths of external tendons except at the points of anchorages and deviators would change, causing what is referred to as second-order effects. These effects are one of the most important features of externally prestressed members, as they make the members have different response characteristics and strength capacities from the members with internal tendons, hence bringing new issues for the analysis and design of prestressed members with external tendons. Therefore addressing the issue of the second-order effects of these members is crucial.

A few works have addressed the second-order effects of externally prestressed members. Harajli et al. (1999) carried out an analytical study to evaluate the second-order effect in externally prestressed concrete members. Their study focused on the number of deviators, tendon profile, and load type. They concluded that the second-order effect is the primary factor distinguishing the flexural behavior of externally prestressed concrete members from that of unbonded internally prestressed ones. Due to second-order effects, externally prestressed members without deviators exhibit lower flexural strength than their counterparts with deviators. The structural behaviors of externally prestressed members with one or two deviators are very similar. In addition, center-point loading develops a lower inelastic deflection, and consequently, results in less significant second-order effects, when compared to third-point loading or uniform loading. Wu and Lu (2003) developed a numerical program to predict the overall response of externally prestressed members. Their work demonstrated that a reasonable placement of deviators can effectively mitigate the second-order effects arising from the gradual change in the eccentricity of the external tendon with member deformation. It is suggested that more than one deviator be provided to minimize the second-order effects. On the other hand, the study by Ng and Tan (2006) found that the provision of a single deviator at the midspan can effectively diminish the second-order effects and maintain a high bearing capacity. However, these effects are generally observable in beams with a large span-to-depth ratio ($L/d_p$). It is recommended to place at least two deviators in long beams ($L/d_p > 30$) to minimize the impact of the second-

Prestressed Members with External Fiber-Reinforced Polymer (FRP) Tendons. DOI: https://doi.org/10.1016/B978-0-443-23877-2.00003-1

order effects. To further mitigate the second-order effects, Pisani (2005) proposed a design rule in accordance with EC2 (CEN, 1992a, 1992b). This rule aims to specify the maximum distance between deviators that is sufficiently small to minimize the second-order effects in externally prestressed structures in practice.

This chapter presents a numerical assessment on the second-order effects of externally prestressed concrete members under short-term (immediate) loads. The results of a parametric analysis performed by applying the finite element model described in Chapter 2 are illustrated. Particular emphasis is placed on the parameters closely related to second-order effects, including the number of deviators, the load type, the span-to-depth ratio, and the deviator spacing (Lou & Xiang, 2010; Lou et al., 2012a,b).

## 3.2   Influence of deviators

Prestressed concrete beams with external tendons of a straight horizontal profile, as shown in Fig. 3.1, are used. Three different deviator configurations are considered, that is, no deviators, one deviator at midspan, and two deviators at third points. Two types of loading are applied, that is, center-point loading or uniform loading. The areas of prestressing steel tendons ($A_p$), reinforcing tension steel ($A_s$), and reinforcing compressive steel ($A'_s$) are 500, 720, and 360 mm$^2$. The material properties are as follows: for prestressing steel, $\sigma_{pe} = 1120$ MPa, $f_{pu} = 1860$ MPa, $f_{py} = 1562.4$ MPa, $E_p = 195$ GPa; for reinforcing steel, $f_y = 450$ MPa, $E_s = 200$ GPa; and for concrete, $f_{ck} = 40.0$ MPa, $f_t = 3.0$ MPa, $\varepsilon_u = 0.0033$.

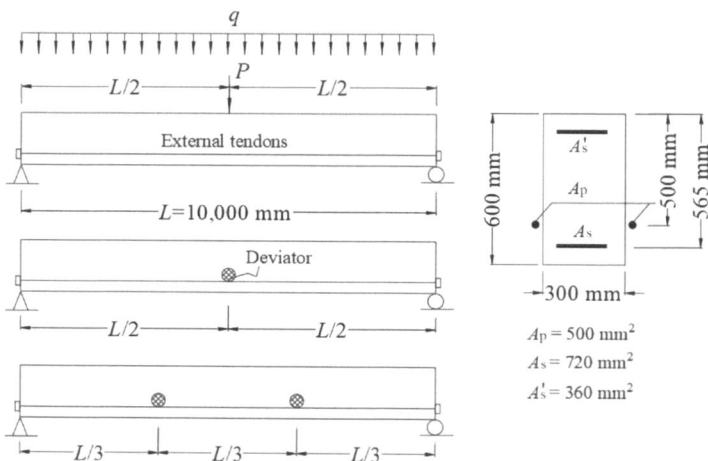

**Figure 3.1** Externally prestressed concrete members with different deviator configurations.

### 3.2.1 Load—deflection response

Fig. 3.2A and B shows the load—deflection response of externally prestressed concrete members under center-point and uniform loading, respectively. It can be seen from this figure that before the member collapse due to the crushing of the concrete at the critical section, the members experienced almost three-stage behaviors with transitions at cracking and yielding points, respectively. Once the members are loaded to be cracked, the member stiffness is reduced considerably. Thereafter the deflection increases rapidly with increasing load up to the yielding of ordinary tension reinforcement. After that, the member stiffness becomes very weak, accompanied by a sharp increase in deflection until failure.

It is seen that the load—deflection responses for three deviator configurations are largely identical up to the cracking load. Beyond that, tendons without deviators register larger member deflection at any load level than the other two deviator configurations, that is, tendons with one deviator at midspan and tendons with two deviators at third points, which exhibit almost identical response characteristics up to the yielding load. After the yielding of ordinary tension reinforcement, negative stiffness occurs in the member without any deviators, that is, the load decreases with continuously increasing deflection. Also, because two deviators are not placed at the section of the maximum deflection, the member with two deviators at third points mobilizes more significant second-order effects than the member with one deviator at midspan.

The corresponding load—deflection responses of internally unbonded prestressed concrete members are also plotted in the figure for comparisons. They are almost identical to the responses of the externally prestressed concrete member with one deviator. This indicates that for externally prestressed concrete members with a straight horizontal tendon profile, as long as one deviator is provided at the section of the maximum deflection, the second-order effects of the members can be negligible.

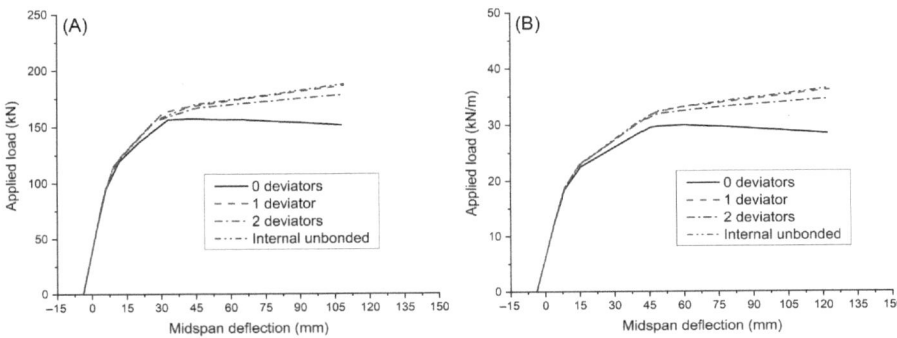

**Figure 3.2** Effect of deviator configuration on the load—deflection curves: (A) center-point loading; (B) uniform loading.

At the ultimate limit state, it can be seen that the second-order effects, associated with the reduction in the tendon eccentricities, seem to have little influence on the ultimate member deflection, but it leads to a much lower ultimate load-carrying capacity of the members. It is suggested that one deviator at midspan instead of two deviators at one-third span, as discussed earlier, be used to minimize the second-order effects. In this analysis, with the provision of a deviator at the critical section, the ultimate single concentrated load and uniform load increase by 23.2% and 27.5%, respectively, when compared to the cases of a member without deviators.

From Fig. 3.2, it can also be seen that the load type also has an important influence on the deflection development, and hence, on the degree of second-order effects. Compared to a single concentrated load, a uniform load registers a larger member deflection due to a larger length of plastic zone developed at the ultimate limit state, leading to relatively more serious second-order effects if they are not minimized by furnishing one deviator at midspan.

## 3.2.2  Eccentricities of external tendons

Fig. 3.3 presents the tendon eccentricities at the ultimate limit state. It is seen that, due to a larger ultimate deflection developed, the second-order effects for a uniform load are more serious than that for a single concentrated load. In addition, the figure shows how significant the second-order effects exist in the members without deviators: the tendon eccentricities at the midspan section reduce remarkably from an initial eccentricity of 200 to 91.7 mm for a single concentrated load and to 78.0 mm for a uniform load. It can also be seen in Fig. 3.3 that providing one or two deviators can effectively diminish the second-order effects. However, the eccentricity variation mode of tendons with one deviator is quite different from that of tendons with two deviators, resulting in different influences of second-order effects for these two deviator configurations. On the one hand, although tendons

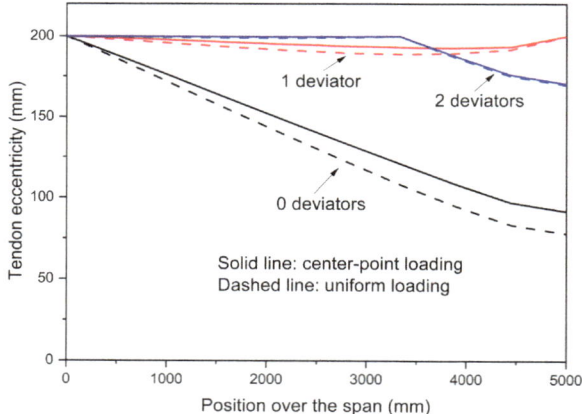

**Figure 3.3** Eccentricities of external tendons over the span.

with one deviator register a larger reduction in eccentricities along most of the span, their eccentricity variations are not serious at the critical regions around the midspan. On the other hand, for tendons with two deviators, although the eccentricity variations can be negligible at the regions between each end support and 1/3 span, the reduction in eccentricities is much more significant around the midspan section. Since the members are crushed at the midspan section, it may be concluded that, for minimizing the second-order effects of a prestressed concrete member with external straight horizontal tendons, providing one deviator at midspan is more effective than providing two deviators at third points of the span.

### 3.2.3  Stress increase in external tendons

Fig. 3.4A and B presents the stress increase in external tendons against the center-point loading and uniform loading, respectively. It is seen in this figure that, similar to the load−deflection response, the response characteristics of stress increase in tendons also consist of three different stages, attributed to that either concrete cracking or ordinary reinforcement yielding leads to much more stress increase in tendons to resist the applied load. For a given load application, before cracking, all members exhibit the same response, indicating at this stage the second-order effects are negligible. After cracking, the second-order effects of the member without deviators become more and more important, leading to a larger stress increase in tendons at a given load level as compared to the members with deviators, of which the second-order effects are still insignificant. When the ordinary tension reinforcement reaches its yield point, the stress in tendons without any deviators rapidly increases with decreasing load due to the severe second-order effects. Meanwhile, the second-order effects of the member with two deviators are no longer negligible, resulting in a larger stress increase in tendons in comparison with the member with one deviator, whose response characteristics of stress increase in tendons are basically identical to the corresponding member with internal unbonded tendons.

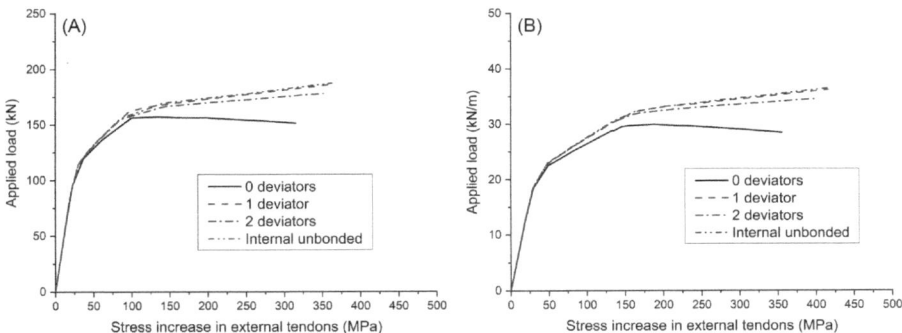

**Figure 3.4** Effect of deviator configuration on the development of tendon stress: (A) center-point loading; (B) uniform loading.

At the ultimate limit state, it can be seen in the figure that the second-order effects result in a much less ultimate stress increase in external tendons, which, together with a smaller eccentricity in tendons, would produce a much smaller resisting moment. Therefore the second-order effects cause a reduction in the ultimate moment capacity of the members. It is also seen that the provision of one deviator at midspan can effectively minimize the second-order effects. In this analysis, providing one deviator at the critical section of a member without deviators, the ultimate moment and stress increase in tendons induced by the single concentrated load increase by 24.5% and 14.3%, respectively; and those induced by the uniform load increase by 29.3% and 17.1%, respectively.

From Fig. 3.4, it can be seen that, because of a larger plastic zone length developed at ultimate, a uniform load registers a larger ultimate tendon stress increase than a single concentrated load. However, this difference is less obvious for tendons without deviators. This is attributed to that for tendons without any deviators a uniform load results in more significant second-order effects that would reduce tendon stress.

## 3.3  Influence of span-to-depth ratio

Externally prestressed concrete beams with two deviators at third points of the span, as illustrated in Fig. 3.5, are used. Third-point loads are applied. The span-to-depth ratio, $L/d_p$ (ratio of span length to initial tendon depth), varies from 8 to 32. Other geometric and material parameters of the members are the same as those used in Section 3.2.

Fig. 3.6 shows the vertical displacements over the member span at the reference state (under effective prestressing and self-weight) for various span-to-depth ratios. It is seen from the figure that the member camber increases with increasing the span-to-depth ratio up to 20. Beyond that, due to more and more significant self-weight effects, the camber diminishes gradually. When the span-to-depth ratio increases to a value of 32, a downward deflection has already existed before the application of any live loads.

Fig. 3.7 shows the effect of the span-to-depth ratio on the moment—deflection response. It is seen that the curves consist of three approximately straight segments that represent three different stages during the whole loading process. The

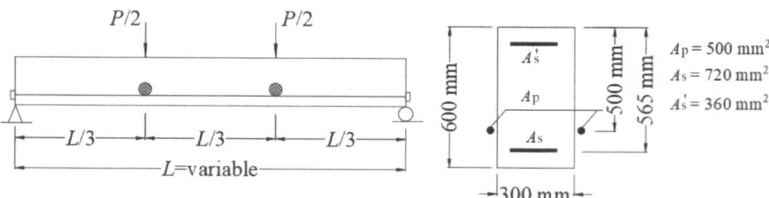

**Figure 3.5** Externally prestressed concrete members with various span-to-depth ratios.

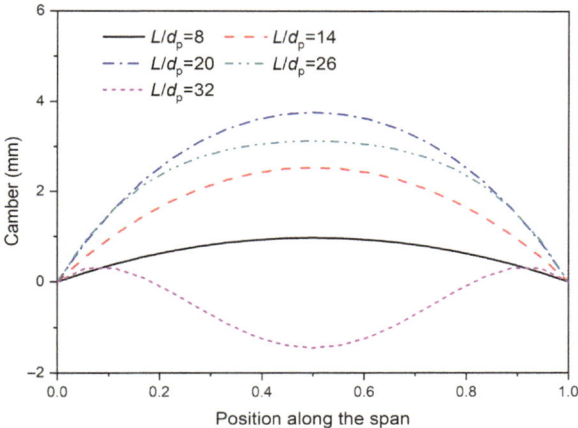

**Figure 3.6** Camber over the span at the reference state.

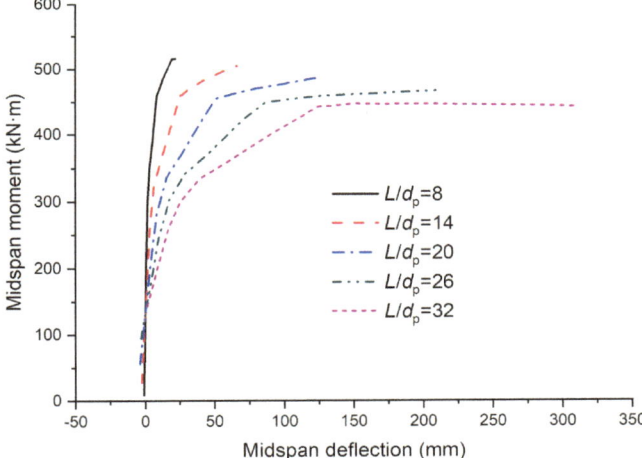

**Figure 3.7** Effect of span-to-depth ratio on the moment−deflection response.

transitions from the first to the second stage and from the second to the third stage are caused by the cracking of concrete and the yielding of ordinary tension reinforcement, respectively. It is also seen that a higher span-to-depth ratio registers a dramatically larger ultimate deflection and corresponding considerably larger reduction in the eccentricity of external tendons, leading to a lower ultimate moment resistance. In this analysis, the ultimate moment resistance decreases by 13.96% as the span-to-depth ratio increases from 8 to 32.

Fig. 3.8 shows the variations in the tendon eccentricity, at the ultimate limit state, along the member span for different levels of span-to-depth ratios. It is seen that, although the reductions in the eccentricity of external tendons are very small

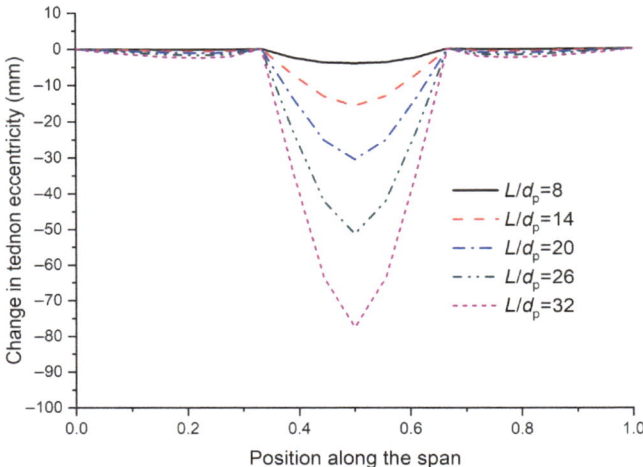

**Figure 3.8** Variation in tendon eccentricity over the span at the ultimate limit state.

between end supports and deviators, they are considerable in the critical zone between two deviator points. Therefore the second-order effects would not be minimized by the provision of two deviators at third points, and they become more and more important with the increase of the span-to-depth ratio.

Fig. 3.9 shows the stress increase in external tendons with the midspan moment for different levels of span-to-depth ratios. The entire response consists of three different stages with transitions caused by cracking and yielding, respectively. The transitions between different stages are more serious for higher span-to-depth ratios due to more significant second-order effects. At the ultimate limit state, the stresses in external tendons are below or a little bit beyond the yield point of 1562.4 MPa (corresponding stress increase in tendon, above the effective prestress, is 442.4 MPa). Also, it is seen that the ultimate stress in external tendons decreases with the increase of the span-to-depth ratio. This change in the ultimate tendon stress is partly attributed to the span-to-depth ratio parameter itself and partly attributed to the second-order effects as discussed in the following sections.

## 3.4 Second-order effects on tendon stress and ductility

Unless otherwise mentioned, the geometric and material parameters of externally prestressed concrete members used in this section are the same as those used in Sections 3.2 and 3.3. The span-to-depth ratio ranges widely from 8 to 44 except for the member without any deviators, where the span-to-depth ratio is up to 32 due to extremely significant second-order effects.

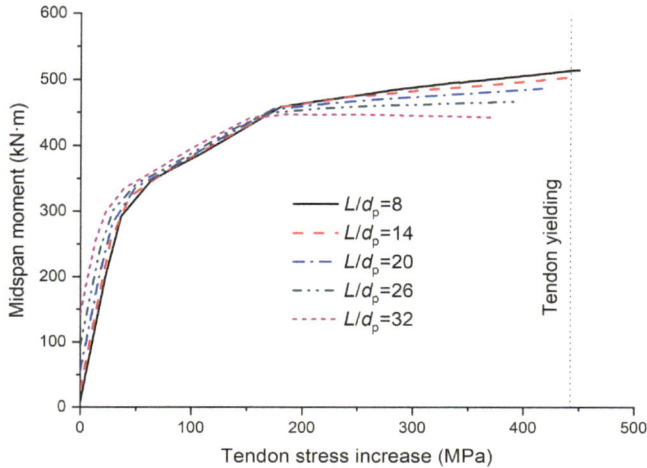

**Figure 3.9** Effect of span-to-depth ratio on the moment versus tendon stress increase response.

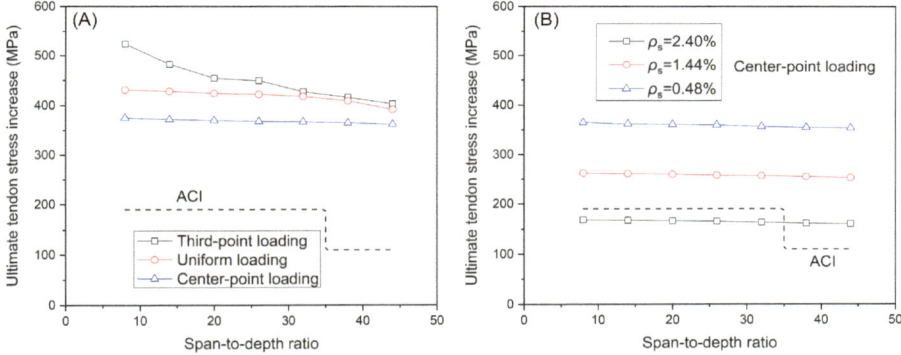

**Figure 3.10** Effect of span-to-depth ratio on the ultimate stress increase in internal unbonded tendons: (A) at different types of loading; (B) at different nonprestressed steel ratios.

### 3.4.1 Ultimate stress increase in internal unbonded tendons

Fig. 3.10A shows the effect of span-to-depth ratio and load type on the ultimate stress increase in internal unbonded tendons (assuming $\sigma_{pe} = 1024$ MPa). It is seen that the span-to-depth ratio affects the ultimate tendon stress differently according to the load applications. Under center-point loading, the span-to-depth ratio seems to have an insignificant influence on the ultimate stress in internal unbonded tendons. As the span-to-depth ratio increases from 8 to 44, the ultimate stress increase in internal unbonded tendons drops by only 3.47% at this load type. Under uniform loading, the ultimate stress increase in internal unbonded tendons decreases slightly by 3.02% as the span-to-depth ratio increases from 8 to 32 and obviously by 6.22%

from 32 to 44. On the other hand, under third-point loading, the span-to-depth ratio significantly influences the ultimate stress in internal unbonded tendons, particularly at low span-to-depth ratios. The ultimate stress increase in internal unbonded tendons decreases by 23.09% as the span-to-depth ratio increases from 8 to 44.

Depending on the length of plastic zones developed at the ultimate limit state, the type of loading also has important influences on the ultimate stress in unbonded tendons. It is seen in Fig. 3.10A that center-point loading mobilizes the lowest ultimate tendon stress increases while third-point loading exhibits the largest ones. The difference between the ultimate tendon stresses by different types of loading is obvious at a low span-to-depth ratio and gradually diminishes as the span-to-depth ratio increases. For a long beam (i.e., $L/d_p > 32$), the ultimate stress in unbonded tendons mobilized by uniform loading is close to that by third-point loading.

For comparison, Fig. 3.10A also plots results calculated from the ACI code (ACI Committee 318, 2019), which considers the effect of the span-to-depth ratio by

$$\text{for } L/d_p \leq 35, \ \Delta\sigma_p = 70 + \frac{f_{ck}}{100\rho_p} \leq 414 \tag{3.1a}$$

$$\text{for } L/d_p > 35, \ \Delta\sigma_p = 70 + \frac{f_{ck}}{300\rho_p} \leq 207 \tag{3.1b}$$

where $\rho_p$ is the ratio of prestressing tendons; and $\Delta\sigma_p$ is in MPa. This equation is subject to that the ultimate stress in unbonded tendons does not exceed its yield strength $f_{py}$.

As illustrated in Fig. 3.10A, with increasing the span-to-depth ratio from 8 to 44, the drop of the ultimate stress in unbonded tendons is 80 MPa according to the ACI code, just in the middle of those obtained from the finite element analysis for third-point and uniform loading (121 and 39 MPa). Because of a low nonprestressed steel ratio ($\rho_s$), the ultimate tendon stresses computed from the finite element analysis are much higher than those from the ACI code, in which the effect of nonprestressed steel is neglected. As the amount of nonprestressed tension steel increases, the actual ultimate tendon stresses rapidly decrease, and at a high level of nonprestressed steel ratio, the ACI code might be nonconservative, as shown in Fig. 3.10B.

### 3.4.2  Ultimate stress increase in external tendons

The parameters affecting the ultimate stress in internal unbonded tendons, such as the span-to-depth ratio, load type, and the amount of nonprestressed steel, as discussed earlier, have a similar influence on the ultimate stress in external tendons. For external tendons, however, the reduction in effective tendon depths or second-order effect is an additional factor that may influence substantially the ultimate tendon stress. The second-order effects are closely related to several parameters, including the deviator configuration, span-to-depth ratio, and the type of loading.

To determine quantitatively the contribution of second-order effects to the stress increase in external tendons at ultimate, a reduction ratio is defined as follows:

$$R_{sec} = 1 - \frac{[\Delta\sigma_p]_{ext}}{[\Delta\sigma_p]_{int}} \tag{3.2}$$

where $[\Delta\sigma_p]_{ext}$ and $[\Delta\sigma_p]_{int}$ represent the ultimate stress increases in external tendons and corresponding internal unbonded tendons, respectively.

The values of $R_{sec}$ for externally prestressed concrete members with different span-to-depth ratios, deviator configurations, and load types are listed in Table 3.1. The effect of span-to-depth ratio and load type on the ultimate stress increase in external tendons in members with 0, 1, and 2 deviators are shown in Fig. 3.11A−C, respectively. The results are produced by using $\sigma_{pe}$ = 1024 MPa.

As the span-to-depth ratio increases, the second-order effects of the members without any deviators become more and more severe, leading to a quick increase in $R_{sec}$ as illustrated in Table 3.1. As a result, with increasing the span-to-depth ratio, the ultimate stress in external tendons (see Fig. 3.11A) decreases much more rapidly than that in the corresponding internal unbonded tendons (see Fig. 3.10A). By the provision of two deviators at third points of the span, the reduction ratios due to second-order effects are effectively diminished but cannot be negligible (see Table 3.1). With the increase of the span-to-depth ratio, the reduction ratio gradually increases, causing the ultimate stress in external tendons (see Fig. 3.11C) to decrease quicker than that in the corresponding internal unbonded tendons. From Fig. 3.11A and C, it can also be seen that the rate of decrease of the ultimate tendon stress mobilized by third-point loading or uniform loading is faster than that by center-point loading. This is partly attributed to that, in comparison with center-point loading, third-point or uniform loading develops a larger ultimate deflection, and hence, a higher reduction ratio due to more important second-order effects (see Table 3.1).

**Table 3.1** Values of $R_{sec}$ for members with different deviator configurations, span-to-depth ratios, and load types.

| $L/d_p$ | $R_{sec}$ (%) | | | | | | | | |
|---|---|---|---|---|---|---|---|---|---|
| | 0 deviators | | | 1 deviator | | | 2 deviators | | |
| | TPL | UL | CPL | TPL | UL | CPL | TPL | UL | CPL |
| 8 | 2.1 | 1.9 | 1.6 | 0.8 | 0.9 | 0.5 | 1.1 | 0.9 | 0.8 |
| 14 | 11.8 | 6.8 | 5.1 | 0.8 | 0.7 | 0.3 | 6.0 | 2.6 | 1.6 |
| 20 | 18.5 | 14.2 | 12.2 | 0.4 | 0.7 | 0.0 | 8.4 | 4.7 | 3.2 |
| 26 | 28.7 | 25.1 | 20.7 | 3.8 | 0.9 | 0.3 | 11.4 | 7.8 | 6.5 |
| 32 | 39.8 | 39.2 | 33.8 | 0.7 | 0.7 | 0.3 | 13.3 | 12.7 | 9.3 |
| 38 | – | – | – | 3.4 | 2.9 | 0.0 | 17.1 | 16.9 | 13.2 |
| 44 | – | – | – | 6.0 | 4.6 | 0.3 | 22.8 | 21.7 | 17.7 |

TPL, Third-point loading; UL, uniform loading; CPL, center-point loading.

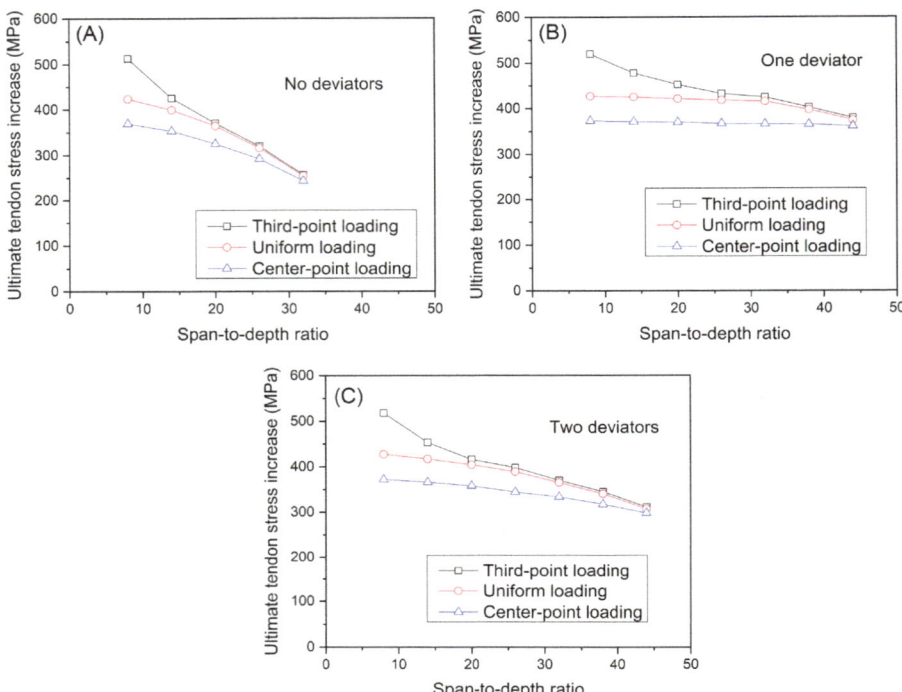

**Figure 3.11** Effect of span-to-depth ratio and load type on the ultimate stress increase in external tendons for members with different numbers of deviators: (A) 0 deviators; (B) 1 deviator at midspan; (C) 2 deviators at third points.

Comparing Fig. 3.11B with Fig. 3.10A, it is seen that the variations in ultimate tendon stress with the span-to-depth ratio for the members with one deviator at midspan are similar to those for the members with internal unbonded tendons, indicating that for the members with one deviator, the reduction in the ultimate tendon stress due to second-order effects is negligible at low and medium span-to-depth ratios or not significant at a high span-to-depth ratio (see Table 3.1). It is noted that for long members with one deviator under third-point loading, the ultimate failure occurs at a section where the effective tendon depth is the smallest. Therefore in this case, the drop in the ultimate load and strength capacities of the members due to second-order effects is important.

### 3.4.3 Deflection ductility

The ductility of a structure may be expressed by the deflection ductility factor, which is defined by

$$\mu_\Delta = \frac{\Delta_u}{\Delta_y} \tag{3.3}$$

where $\Delta_u$ and $\Delta_y$ are midspan deflections at the ultimate limit state and at first yielding of nonprestressed tension steel, respectively.

For a member with a low nonprestressed steel ratio, the deflection at yielding ($\Delta_y$) is much smaller than the deflection at ultimate ($\Delta_u$). In this case, the value of $\mu_\Delta$ is very sensitive to $\Delta_y$. Therefore the value of $\Delta_y$ must be reached as accurately as possible. To achieve the exact value of $\Delta_y$, a very small load increment is used in the numerical analysis when the applied load is near the yielding load.

Fig. 3.12A–C shows the influence of span-to-depth ratio and load type on the deflection ductility factor for members with 0, 2 deviators, and internal unbonded tendons, respectively. It is seen that for a member under a certain type of loading, the deflection ductility factor remains almost constant as the span-to-depth ratio increases. In addition, the members with different deviator configurations and with internal unbonded tendons exhibit almost the same ductile behavior. Since both the span-to-depth ratio and deviator configuration are closely related to the second-order effects, it can be deduced that the ductility of externally prestressed concrete members is independent of the second-order effects. It is also seen that the flexural

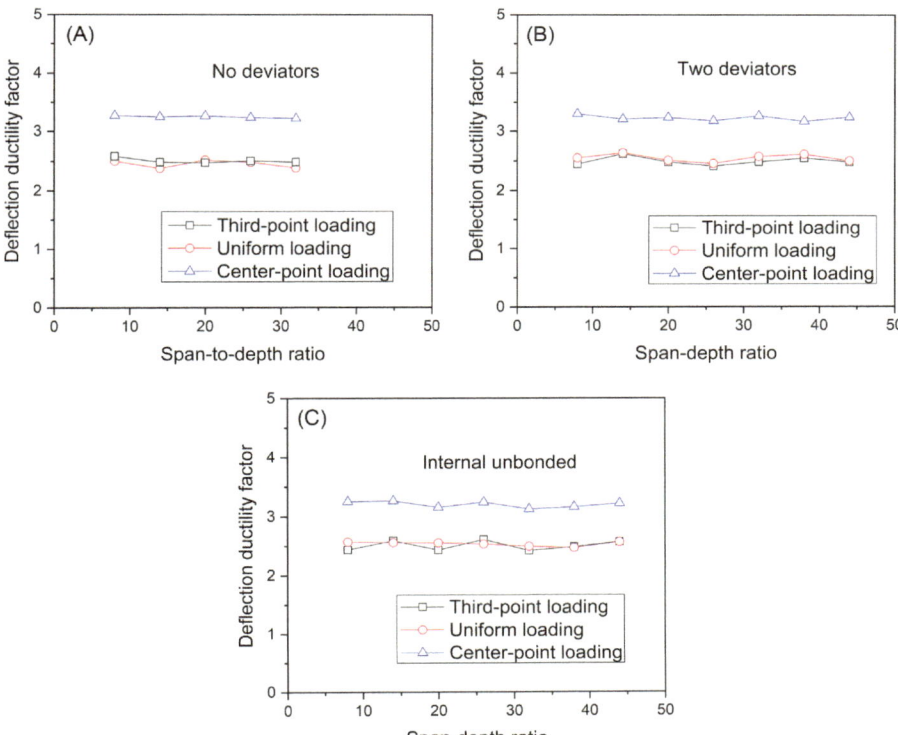

**Figure 3.12** Effect of span-to-depth ratio and load type on the ductility: (A) members without deviators; (B) members with 2 deviators at third points; (C) members with internal unbonded tendons.

ductility is affected by the type of loading. On the one hand, the deflection ductility factors of members under third-point and uniform loading are almost identical (around 2.5). On the other hand, although center-point loading mobilizes a lower ultimate deflection, it exhibits better ductile behavior, with a deflection ductility factor of around 3.2, when compared to third-point or uniform loading.

## 3.5  Optimum deviator position

For simply supported members, tendons draped at two deviators that are symmetrical with respect to the midspan section, as illustrated in Fig. 3.13, are fairly common in the actual engineering of retrofits and new constructions using external prestressing. The distance between two deviators is a concern for designers because this parameter is directly related to second-order effects. In addition, this parameter may affect the location of the failure section of members under point loads that are away from midspan. It is important to adopt appropriate deviator spacing to achieve the desirable flexural strength and behavior of the members.

Fig. 3.13 shows some details of externally prestressed concrete beams that are used to illustrate the effect of deviator spacing represented by $S_d/L$ (ratio of the deviator spacing to the span length). It is noted that $S_d/L$ equal to 0 represents a single-draped tendon profile, while $S_d/L$ equal to 1 indicates a straight tendon profile without any deviators. The tendon depths at the anchorage and deviator points are 300 and 500 mm, respectively.

Fig. 3.14 shows the variations in eccentricities of external tendons with the applied load for various values of $S_d/L$. It is seen that, when $S_d/L$ is equal to or below 1/6, the reduction in tendon eccentricity is negligible. As $S_d/L$ increases to 1/3 or above, the reduction in tendon eccentricity is unimportant before cracking load, while it becomes more and more important with a continuing increase of loading, particularly after the member is yielded. At the ultimate limit state, as $S_d/L$ increases from 1/3 to 1, the reductions in the eccentricity of external tendons increase from 41.3 to 171.7 mm.

For draped profile under third-point loading, the members with a small $S_d/L$ may fail prematurely at a section near one of the loading points where the effective tendon depth is relatively small. However, this is not a desirable failure mode because

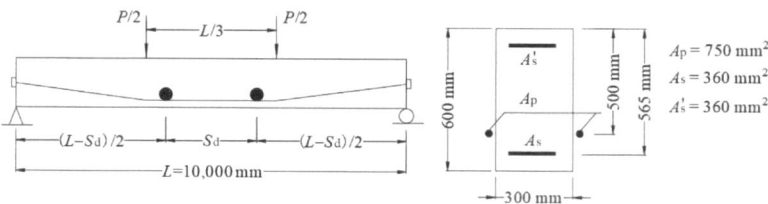

**Figure 3.13** Externally prestressed concrete members with different layouts of external tendons.

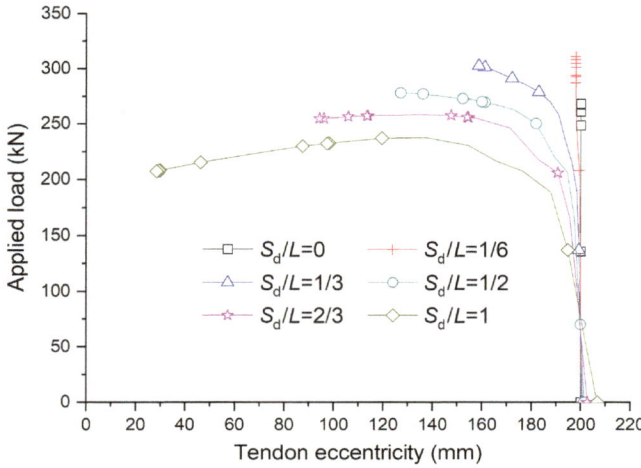

**Figure 3.14** Variation of tendon eccentricity with applied load.

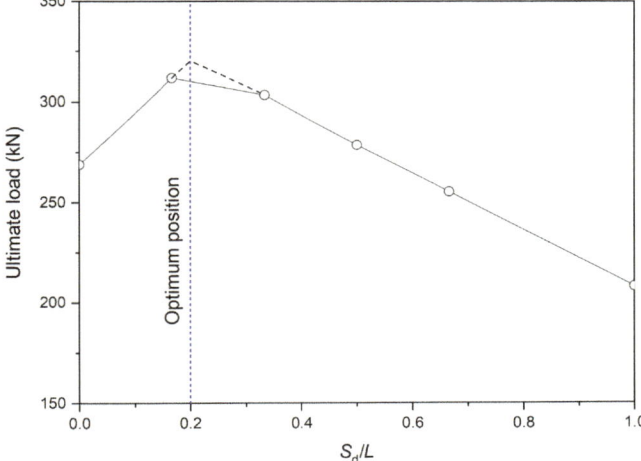

**Figure 3.15** Variation of ultimate load with $S_d/L$.

both the flexural strength and ductility have not been fully developed yet. With increasing $S_d/L$, failure would happen at midspan that is desirable, but the second-order effects become more and more significant. The optimum distance between deviators would be reached when a member that fails at midspan has the lowest reduction in tendon eccentricity.

With increasing $S_d/L$, the variation of the ultimate load-carrying capacity is plotted in Fig. 3.15. Due to the premature failure as mentioned earlier, the ultimate load or moment of the member with $S_d/L$ of 0 is undesirable although the second-order effects are minimal. When $S_d/L$ increases to 1/6, the ultimate load is significantly enhanced but the member still collapses at a section near the loading point rather than at the midspan. As $S_d/L$ increases to 1/3 or above, failure occurs at midspan. Also, it is interesting to find that the decrease in the ultimate load due to second-order effects shows a linear manner with the increase of $S_d/L$. Based on this finding, it could be inferred that the optimum value of $S_d/L$ is located at the point where the ascending and descending lines meet, as illustrated in Fig. 3.15. The optimum value of $S_d/L$ in this analysis is 0.2. The corresponding ultimate load is 320.2 kN.

## 3.6 Conclusions

The second-order effects are the most important characteristic that distinguishes an external tendon system from an internal unbonded one. If the second-order effects are negligible by proper measures, the response of externally prestressed concrete members is basically identical to that of unbonded internally prestressed concrete members. The analysis shows that the second-order effects have a significant influence on the response characteristics of externally prestressed concrete members. At a particular level of applied loads, these effects lead to a larger member deflection as well as a larger stress in external tendons. Due to these effects, the stiffness of members without deviators may become negative after the yielding of ordinary tension reinforcement. At failure, the second-order effects lead to much lower ultimate load and moment capacities of the members and a less ultimate stress increase in external tendons, but they appear to have little influence on the ultimate member deflection. For simply supported prestressed concrete members with external straight horizontal tendons, the response of a member with one deviator at midspan is nearly identical to that of a corresponding unbonded internally prestressed member, indicating that the provision of one deviator at the section of the maximum member deflection can effectively minimize the second-order effects.

The second-order effects of externally prestressed concrete members with two deviators at third points of the span cannot be negligible, and these effects become increasingly important with the increase of the span-to-depth ratio. Due to the second-order effects, a higher span-to-depth ratio results in a lower ultimate moment capacity of externally prestressed concrete members. The reduction ratio $R_{sec}$ as a result of second-order effects of members with no deviators or two deviators at third points quickly or gradually increases with increasing the span-to-depth ratio, and the reduction ratio by third-point or uniform loading is higher than that by center-point loading. On the other hand, the second-order effects on the ultimate stress in external tendons with one deviator at midspan are insignificant. The span-to-depth ratio and deviator configuration have null effects on the ductility of externally prestressed concrete members. Since both the span-to-depth ratio and deviator configuration are closely related to the second-order effects, it can be deduced that the second-order effects have no influence on the ductility of externally prestressed concrete members.

For external tendons draped at two deviators, a proper distance between deviators is important to gain the desirable structural performance. It is indicated that the reduction in the ultimate load-carrying capacity due to second-order effects displays a linear manner with the deviator spacing, provided that failure happens at midspan. A method for determining the optimum deviator spacing is proposed in this chapter.

# References

ACI Committee 318. (2019). *Building code requirements for structural concrete (ACI 318-19) and commentary (ACI 318R-19)*. Farmington Hills, MI.

CEN. (1992a). *Eurocode 2 (EC2): Design of concrete structures. Part 1-1: General rules and rules for buildings*. EN 1992-1-1. Brussels, Belgium.

CEN. (1992b). *Eurocode 2 (EC2): Design of concrete structures. Part 1—5: Structures with unbonded and external prestressing tendons*. EN 1992-1-5. Brussels, Belgium.

Grace, N. F., Navarre, F. C., Nacey, R. B., Bonus, W., & Collavino, L. (2002). Design-construction of bridge street bridge-first CFRP bridge in the United States. *PCI Journal*, *47*(5), 20—35. Available from https://doi.org/10.15554/pcij.09012002.20.35.

Harajli, M., Khairallah, N., & Nassif, H. (1999). Externally prestressed members: Evaluation of second-order effects. *Journal of Structural Engineering*, *125*(10), 1151—1161. Available from https://doi.org/10.1061/(ASCE)0733-9445(1999)125:10(1151).

Lou, T., Lopes, A. V., & Lopes, S. M. R. (2012a). Influence of span-depth ratio on behavior of externally prestressed concrete beams. *ACI Structural Journal*, *109*(5), 687—695. Available from https://doi.org/10.14359/51684046.

Lou, T., Lopes, S. M. R., & Lopes, A. V. (2012b). Numerical analysis of behaviour of concrete beams with external FRP tendons. *Construction and Building Materials*, *35*, 970—978. Available from https://doi.org/10.1016/j.conbuildmat.2012.04.055.

Lou, T., & Xiang, Y. (2010). Numerical analysis of second-order effects of externally prestressed concrete beams. *Structural Engineering and Mechanics*, *35*(5), 631—643. Available from https://doi.org/10.12989/sem.2010.35.5.631.

Mancarti, G. D. (1984). Strengthening California steel bridges by prestressing. *Transportation Research Record*, *950*, 183—187.

Mutsuyoshi, H., Hai, N.D., & Kasuga, A. (2010). Recent technology of prestressed concrete bridges in Japan. In *IABSE-JSCE joint conference on advances in bridge engineering-II* (pp. 46—55).

Ng, C. K., & Tan, K. H. (2006). Flexural behaviour of externally prestressed beams. Part II: Experimental investigation. *Engineering Structures*, *28*(4), 622—633. Available from https://doi.org/10.1016/j.engstruct.2005.09.016.

Pisani, M. A. (2005). Geometrical nonlinearity and length of external tendons. *Journal of Bridge Engineering*, *10*(3), 302—311. Available from https://doi.org/10.1061/(ASCE)1084-0702(2005)10:3(302).

Wu, X. H., & Lu, X. (2003). Tendon model for nonlinear analysis of externally prestressed concrete structures. *Journal of Structural Engineering*, *129*(1), 96—104. Available from https://doi.org/10.1061/(ASCE)0733-9445(2003)129:1(96).

# Simply supported prestressed concrete members with external FRP tendons

## 4.1 Introduction

External prestressing is considered as a powerful technique not only in the construction of new bridges (Brockmann and Rogenhofer, 2000) but also in the strengthening of damaged structures (Nordin, 2005). As external tendons are exposed to a complicated environment, high anticorrosion protection for their use is needed. The conventional prestressing steel, however, is naturally prone to corrosion. Because of this problem, external prestressing had ever been dormant during the 1960s and 1970s. Adopting nonmetal materials seems to be the best strategy to solve completely the corrosion problem. In recent years, the use of fiber-reinforced polymer (FRP) composite materials has been becoming popular in prestressed structures for their merits such as high corrosion resistance, high strength, and low density (Kim, 2010; Pisani, 1998). The adoption of FRP composites instead of prestressing steel as external tendons is expected to be increasingly popular (Bennitz et al., 2012). These composite materials include carbon FRP (CFRP), glass FRP (GFRP), aramid FRP (AFRP), and basalt FRP (BFRP). The mechanical properties of FRPs are quite different from those of steel tendons. FRP is brittle in nature, with linearly elastic behavior up to failure, and usually a low modulus of elasticity (ACI Committee 440, 2004). As a result, the external FRP tendon members may exhibit different response characteristics from the external steel tendon members. Accordingly, new design concerns for this type of system would arise.

This chapter introduces the use of external FRP tendons for flexural strengthening of simply supported concrete members (Lou, Hu, & Pang, 2023; Lou, Lopes, & Lopes, 2012). First, a comparative analysis on simply supported concrete beams with external FRP or steel tendons is performed, and the results of comprehensive aspects of structural behavior are presented. Second, the flexural performance of simply supported prestressed concrete beams with external CFRP tendons is examined, focusing on the effects of various variables related to CFRP tendons. Based on the results of the numerical parametric analysis, a simplified model is developed to predict the ultimate tendon stress and the flexural strength of prestressed concrete members with external FRP tendons.

## 4.2 Using external fiber-reinforced polymer tendons instead of external steel tendons

A rectangular prestressed concrete beam with external tendons draped at two deviators at third points, as shown in Fig. 4.1, is used. The tendon depth $d_p$ at the

Prestressed Members with External Fiber-Reinforced Polymer (FRP) Tendons. DOI: https://doi.org/10.1016/B978-0-443-23877-2.00004-3

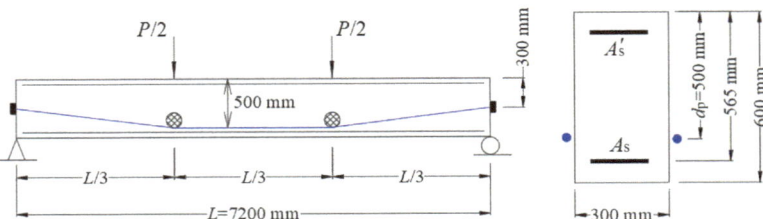

**Figure 4.1** Simply supported concrete beams with external FRP/steel tendons. *FRP*, Fiber-reinforced polymer.

**Table 4.1** Mechanical properties of prestressing tendons.

| Tendons | Ultimate strength (MPa) | Yield stress (MPa) | Elastic modulus (GPa) | Rupture strain (%) |
|---|---|---|---|---|
| Steel | 1860 | 1674 | 195 | >3.5 |
| CFRP | 1840 | – | 147 | 1.25 |
| GFRP | 1500 | – | 50 | 3.0 |

*CFRP*, Carbon fiber-reinforced polymer; *GFRP*, glass fiber-reinforced polymer.

anchorage and deviator points are 300 and 500 mm, respectively. Third-point loading is used. Three types of external prestressing tendons are selected, namely, GFRP, CFRP, and steel tendons. GFRP and CFRP are chosen because they represent two extreme values of the elastic modulus among FRP composites that have been widely used in civil engineering. The mechanical properties of the steel and FRP tendons are presented in Table 4.1. The effective prestress $\sigma_{\mathrm{pe}}$ for all tendons is taken as the same value of 1024 MPa for comparative purposes. This is a common value both for prestressing steel and for CFRP (values between 40% and 60% of the ultimate strength are quite normal in practical applications), and when only the short-term loading condition is considered, such value is also acceptable for GFRP (the ultimate tendon stress developed is far below the ultimate strength of the tendons). The concrete strength $f_{\mathrm{ck}}$ is 40 MPa. The yield strengths of nonprestressed tension and compression steels are 530 and 450 MPa, respectively. The areas of both nonprestressed tension and compression steel are 360 mm². Various tendon areas $A_{\mathrm{p}}$ are used to produce different levels of prestressing reinforcement index $\omega_{\mathrm{p}}$, which is defined as

$$\omega_{\mathrm{p}} = \frac{A_{\mathrm{p}}\sigma_{\mathrm{pe}}}{bd_{\mathrm{p}}f_{\mathrm{ck}}} \tag{4.1}$$

where $b$ is the section width and $d_{\mathrm{p}}$ is the maximum initial tendon depth along the span. Three levels of $\omega_{\mathrm{p}}$ are used, namely, $\omega_{\mathrm{p}} = 0.023$, $0.085$, and $0.145$.

## 4.2.1  Load–deformation characteristics

The moment–curvature responses and load–deflection responses of the members are shown in Figs. 4.2 and 4.3, respectively. At the initial state, the members have midspan moments of about 29 kN m, which are produced by their self-weight. After the application of live loading, the members sequentially experience three

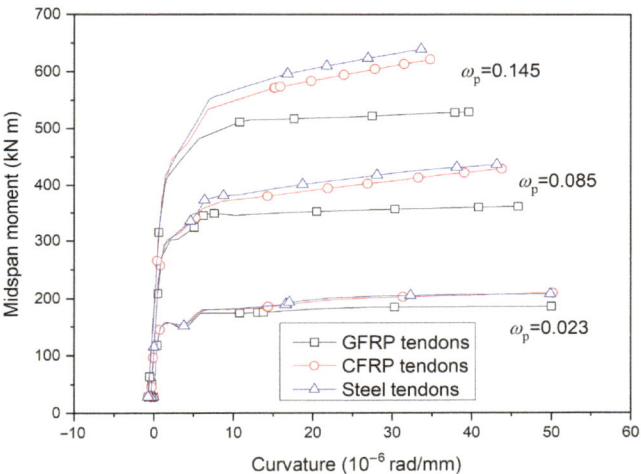

**Figure 4.2** Moment–curvature responses for members with different tendon types and $\omega_p$ levels.

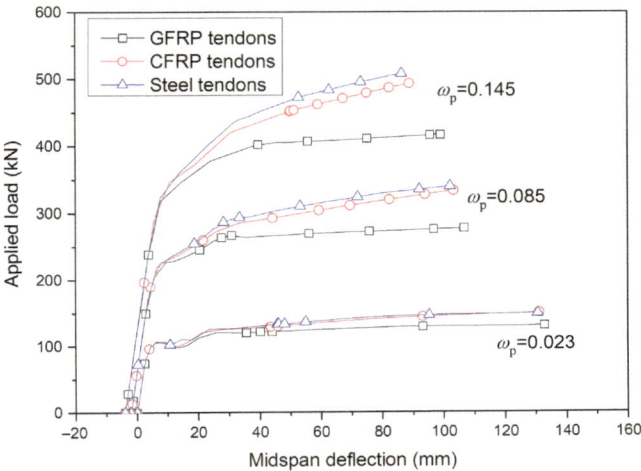

**Figure 4.3** Load–deflection responses for members with different tendon types and $\omega_p$ levels.

different stages before collapse happens. The first stage is characterized by elastic behavior and is finished by the appearance of cracks at the bottom fiber of the members. This is followed by the second stage up to the yielding of nonprestressed tension steel and subsequently by the third stage until failure. Each transition of the stages is accompanied by a significant reduction of the member stiffness due to cracking or yielding. In the second stage, the members with $\omega_p = 0.085$ or 0.145 also exhibit approximately linear moment−curvature or load−deflection behavior, that is, cracked-elastic behavior. This is not the case, however, for the members with $\omega_p = 0.023$. The moment or load of these significantly under-reinforced members suddenly drops on cracking due to the loss of the prevailing contribution by concrete and gradually recovers following the increasing contribution by the tension reinforcement. It is also seen that, as $\omega_p$ increases, the loads (moments) at cracking, yielding, and ultimate increase significantly, while the ultimate deformations gradually decrease.

Because the elastic behavior before cracking is mostly governed by the concrete, the members with different types of external tendons exhibit almost identical moment−curvature and load−deflection responses at this stage and have nearly the same cracking loads and moments. After cracking, the discrepancy in the responses appears. It is observed that the discrepancy between external CFRP and steel tendon members is really small up to failure load. On the other hand, due to a much smaller elastic modulus, external GFRP tendon members exhibit more significant stiffness reductions on cracking and yielding than external CFRP or steel tendon members, particularly for high levels of $\omega_p$. Owing to smaller tendon stress developed at ultimate, external GFRP tendon members mobilize lower ultimate loads and moments than external CFRP or steel tendon members.

Table 4.2 gives the values of the deflection ductility factor $\mu_\Delta$, which is defined as follows:

$$\mu_\Delta = \Delta_u / \Delta_y \tag{4.2}$$

where $\Delta_u$ and $\Delta_y$ are the midspan deflections at failure and at first yielding of nonprestressed tension steel, respectively. When $\omega_p$ increases, the ductility factor decreases, and the rate of decrease of ductility of external GFRP tendon members is slower than that of external CFRP or steel tendon members. At a given level of $\omega_p$,

Table 4.2 Ductility factor of members with external FRP or steel tendons.

| Tendons | Deflection ductility | | |
|---------|---------------------|---|---|
| | $\omega_p = 0.023$ | $\omega_p = 0.085$ | $\omega_p = 0.145$ |
| GFRP | 5.69 | 3.71 | 3.18 |
| CFRP | 5.60 | 3.53 | 2.74 |
| Steel | 5.57 | 3.45 | 2.58 |

CFRP, Carbon fiber-reinforced polymer; GFRP, glass fiber-reinforced polymer.

external GFRP tendon members have higher ductility than external CFRP or steel tendon members. The discrepancy is insignificant for a low level of $\omega_p$, but it turns out to be more noticeable when $\omega_p$ increases.

### 4.2.2 Neutral axis depth

Under effective prestressing force and self-weight load, the initial neutral axis lies either inside or outside the cross-section, depending on the level of $\omega_p$. For a low level of $\omega_p$, the effect of self-weight load may override the prestressing effect, hereby causing a slightly sagging curvature before the application of live loading. In this situation, the neutral axis initially lies far below the bottom fiber of the section (in this analysis, the initial neutral depth at midspan is about 3200 mm for $\omega_p = 0.023$), and gradually rises after the application of live loading. On the other hand, when the prestressing effect prevails over the self-weight effect, which causes a camber of the member, the initial neutral axis may lie someplace at the upper half or above the top fiber of the section, depending on whether there is a tensile strain at the top fiber or not. In this analysis, the initial neutral depth at midspan is about 116 mm for $\omega_p = 0.145$. With the application and increase of the live load, the camber gradually diminishes and, correspondingly, the neutral axis gradually moves upward. When the camber vanishes and the sag begins to appear, the neutral axis jumps to a place far below the bottom fiber of the section, and hereafter it gradually rises with the continuing increase of the applied load.

Fig. 4.4 shows the load versus neutral axis depth responses of the members after the neutral axis rises to the bottom fiber of the section. It is seen that the responses for members with different types of external tendons deviate after apparent cracking. The term "apparent cracking" indicates a stabilizing development of cracks.

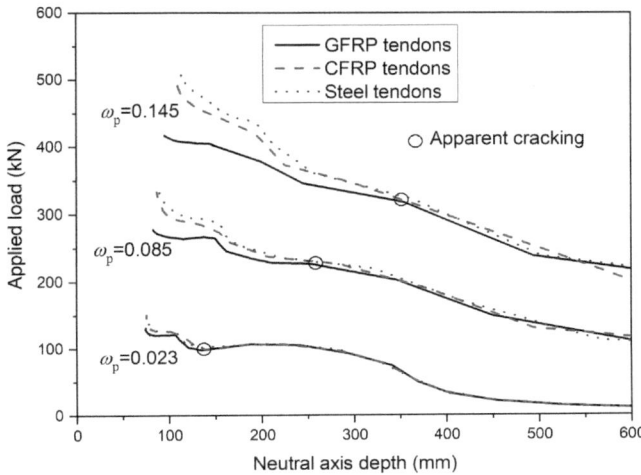

**Figure 4.4** Variations of neutral axis depth with applied load for members with different tendon types and $\omega_p$ levels.

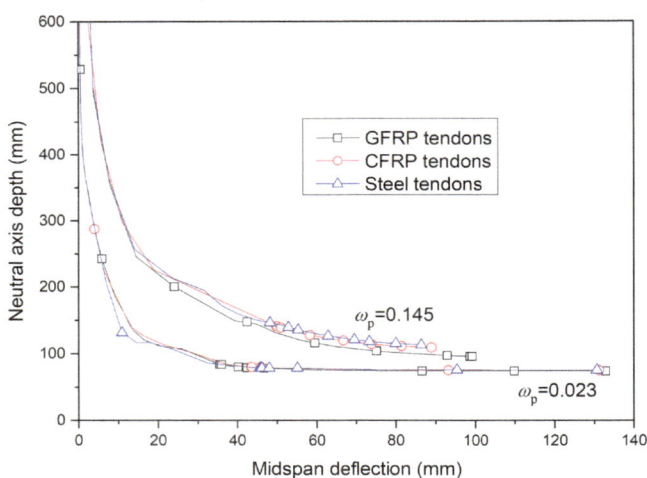

**Figure 4.5** Variations of neutral axis depth with midspan deflection for members with different tendon types and $\omega_p$ levels.

With increasing load, the neutral axis of external GFRP tendon members rises quicker than that of external CFRP or steel tendon members. At ultimate, the external GFRP tendon members have smaller neutral axis depths than the external CFRP or steel tendon members. For $\omega_p = 0.023$, the rise of the neutral axis displays a notably nonlinear manner with applied load, and the neutral axis depth at apparent cracking is close to the ultimate depth. As $\omega_p$ increases, the nonlinear manner is obviously mitigated, and the difference between neutral axis depths at apparent cracking and ultimate enlarges.

Fig. 4.5 shows the variation of the neutral axis, after it rises to the bottom fiber of the section, with the midspan deflection of the members. The neutral axis rapidly rises until the member is apparently cracked. Thereafter, the neutral axis slows its movement and, after nonprestressed tension steel has apparently yielded, the rise of the neutral axis becomes very slow, particularly for low levels of $\omega_p$. At a given deflection, members with a higher level of $\omega_p$ exhibit larger neutral axis depths than members with a lower level of $\omega_p$, particularly notable at a low deflection. It can be also observed that a higher level of $\omega_p$ leads to larger neutral axis depths at the cracking, yielding, and ultimate states, as compared to a lower level of $\omega_p$. The discrepancy is most remarkable at cracking while it is not so important at ultimate. In addition, regardless of the level of $\omega_p$, the influence of tendon type on the deflection versus neutral axis depth response seems to be insignificant, as shown in Fig. 4.5.

### 4.2.3 Stress in external tendons

Fig. 4.6 shows the stress increase in external tendons with the midspan moment of the members. The entire response of external CFRP tendon members is similar to

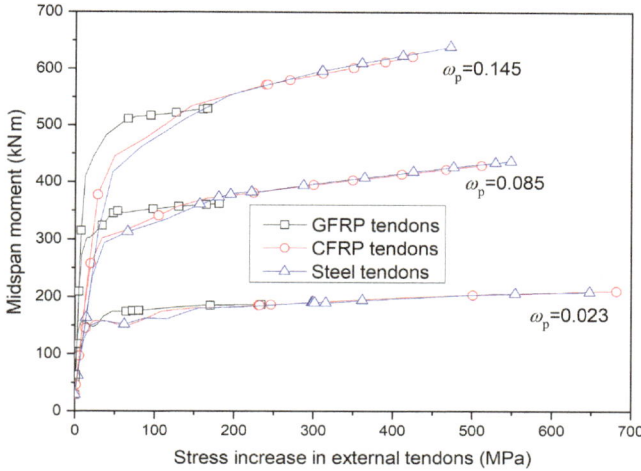

**Figure 4.6** Stress increase in external tendons with midspan moment for members with different tendon types and $\omega_p$ levels.

that of external steel tendon members. The difference between the stresses in external CFRP and steel tendons at a given moment is small and finally vanishes after both members have yielded. The response of external GFRP tendon members differs from that of external CFRP or steel tendon members at the beginning, and this difference becomes more and more obvious with the increase of the applied load. The maximum difference appears when the applied load reaches the yielding load of the external GFRP tendon member. After that, the difference gradually diminishes. Due to a much smaller elastic modulus, GFRP tendons develop a significantly lower stress at ultimate in comparison to CFRP or steel tendons. On the other hand, CFRP tendons may have either a little bit higher or lower ultimate stress than steel tendons, depending on the ductility deformation of members that can be developed at ultimate.

The increase in stress of external tendons with midspan deflection is shown in Fig. 4.7. The responses are dependent on the stress—strain laws of tendon materials. The stress in FRP tendons increases almost linearly with the development of the member deflection throughout the loading process, attributed to the elastic behavior of FRP composites. The rate of the stress increase is determined by the elastic modulus of FRP composites. The slopes for external CFRP and GFRP tendon members are about 5.20 and 1.76 MPa/mm, respectively. The slope ratio is 2.95, which is almost identical to the elastic modulus ratio of 2.94. A trivial difference may be attributed to slightly different levels of second-order effects between external CFRP and GFRP tendon members. For steel tendons, on the other hand, the response displays an obviously nonlinear manner when the member deformation is significant

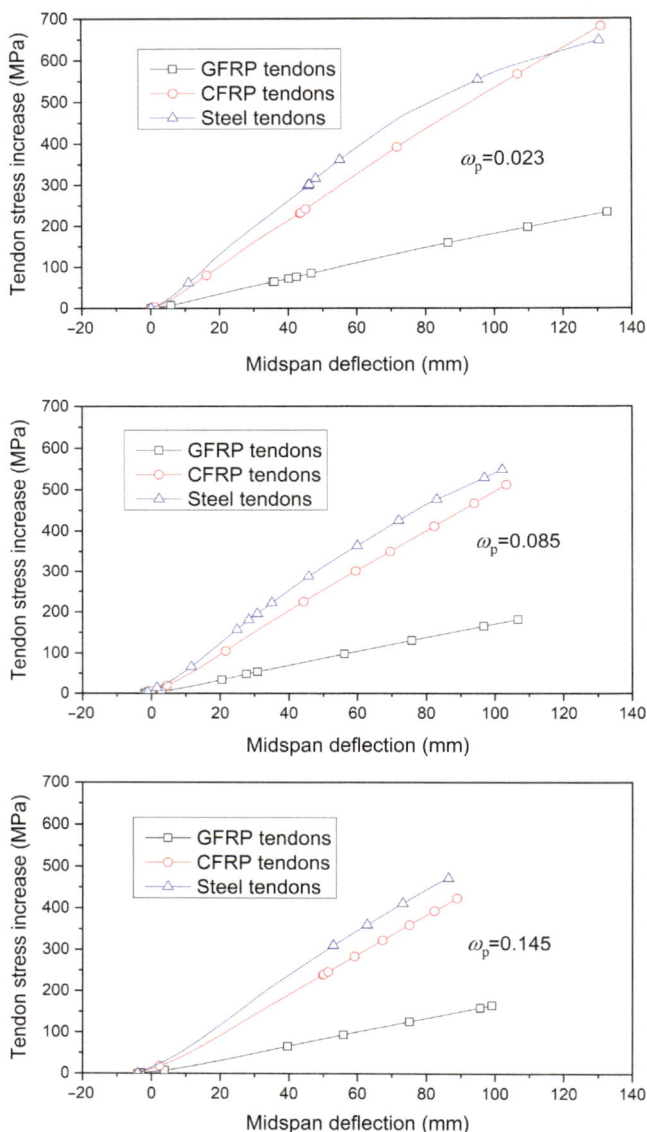

**Figure 4.7** Stress increase in external tendons with midspan deflection for members with different tendon types and $\omega_p$ levels.

(for low reinforcing index) and exhibits almost a linear manner when the member deformation is not serious (for high reinforcing index), following the stress–strain path of prestressing steel.

### 4.2.4 Stress in nonprestressed steel

Fig. 4.8 shows the response of the stress in the bottom and top nonprestressed steel at midspan with the applied load. It is seen that the tensile or compressive stress increases linearly with the applied load up to cracking. At this stage, the rates of stress increase in the bottom and top nonprestressed steel are identical. After cracking, the stress increase is much quicker and, when the development of cracks stabilizes, it resumes a linear manner up to yielding of bottom nonprestressed steel. Because cracks occur and develop in the tension region of the members, at this stage, the rate of the tensile stress increase is greater than that of the compressive stress increase. After yielding, the tensile stress remains at a constant value of 530 MPa, following the elasto-perfectly plastic stress−strain relationship adopted in this study. Meanwhile, the increase in compressive stress becomes very quickly. It is also observed that a higher level of $\omega_p$ leads to a higher compressive strain in top nonprestressed steel at ultimate. The theoretical mechanism of this observation can be described by

$$\varepsilon_s' = \kappa(c - d_s') = \frac{\varepsilon_u}{c}(c - d_s') = \varepsilon_u\left(1 - \frac{d_s'}{c}\right) \tag{4.3}$$

where $\varepsilon_s'$ and $d_s'$ are the strain and effective depth in top nonprestressed steel, respectively; $\kappa$ and $c$ correspond to curvature and neutral axis depth, respectively; $\varepsilon_u$ is the ultimate concrete compressive strain at the top fiber of the critical section. Since the neutral axis depth $c$ at ultimate is larger for a higher level of $\omega_p$ as mentioned previously, the strain $\varepsilon_s'$ is, therefore, higher according to Eq. (4.3). At failure load, the top nonprestressed steel has usually not yielded (for low level of $\omega_p$) or

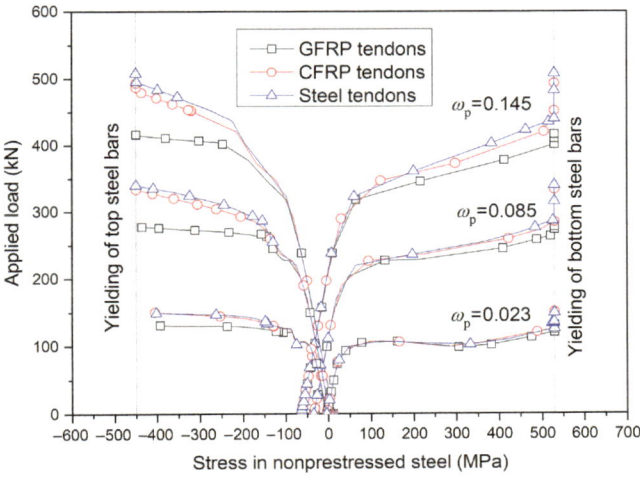

**Figure 4.8** Stresses in nonprestressed steel with applied load for members with different tendon types and $\omega_p$ levels.

slightly yielded (for high level of $\omega_p$), as shown in Fig. 4.8. It is also seen that the responses of members with different types of external tendons differ after cracking. The nonprestressed steel strain of external GFRP tendon members increases quicker than that of external CFRP or steel tendon members.

### 4.2.5 Curvature distribution and crack pattern at failure

The influence of tendon type on the curvature distribution at the ultimate limit state for $\omega_p = 0.145$ is shown in Fig. 4.9. The curvature is large around the midspan zone but small at other zones, indicating that there are a few main cracks with large widths around the midspan, while at other cracking zones the crack widths are small. It is seen that the type of external tendons has limited influence on the crack pattern. The members with different types of external tendons exhibit the same cracking zone. In addition, the zones where the main cracks exist are also identical, while the external GFRP tendon member exhibits a larger main crack width than the CFRP or steel tendon member.

The internal bonded tension reinforcement is recognized as the most important parameter that influences the crack pattern of unbonded prestressed concrete members. A minimum amount of bonded reinforcement in unbonded prestressed concrete members is required by various codes to guarantee a favorable crack pattern and failure mode. In the ACI building code (ACI Committee 318, 2019), the minimum amount is $0.004\,A$, in which $A$ is the area between the centroidal axis and the flexural tension face of a concrete section. To examine the influence of bonded reinforcement on the crack pattern, externally prestressed concrete members with different amounts of nonprestressed tension steel are analyzed using the proposed nonlinear analysis. The curvature distribution over the span for external GFRP

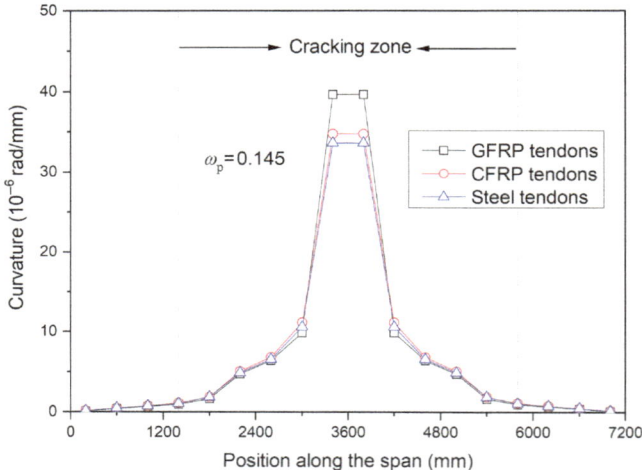

**Figure 4.9** Effect of tendon type on the curvature distribution.

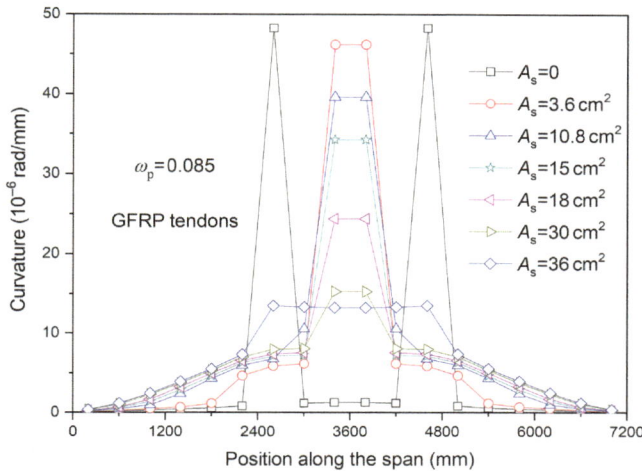

**Figure 4.10** Effect of nonprestressed tension steel on the curvature distribution.

tendon members is shown in Fig. 4.10. The results are produced using $\omega_p = 0.085$ and third-point loading.

It is seen that the member without any nonprestressed tension steel displays an extremely irregular curvature distribution, indicating a completely undesirable crack pattern. In this member, there are two main cracks with large width concentrating at the failure sections near the loading points, while the other zone of the member remains almost uncracked. By the provision of the minimum amount of nonprestressed tension steel (360 mm$^2$) required by the ACI building code (ACI Committee 318, 2019), the crack pattern and failure mode are greatly improved. The member is crushed at the midspan section. The maximum crack width is reduced and more cracks appear in the member. With a continuously increasing amount of the bonded reinforcement, the cracking zone gradually extends to the end sides of the member, and meanwhile, the crack width diminishes. When the bonded reinforcement amounts to 3600 mm$^2$, the cracking zone almost extends to the whole member, and the maximum crack width becomes tiny, as can be observed in Fig. 4.10.

## 4.3   Effects of critical parameters related to carbon fiber-reinforced polymer tendons

Fig. 4.11 shows a prestressed concrete reference beam with external CFRP tendons for numerical parametric analysis. The concrete strength $f_{ck}$ is 60 MPa. The tendon area is 1100 mm$^2$. CFRP tendons have a rupture strength of 1840 MPa, an elastic modulus of 150 GPa, and an initial prestress of 1104 MPa. Either the bottom or top steel rebars have a cross-sectional area of 360 mm$^2$, a yield strength of 450 MPa, and an elastic modulus of 200 GPa.

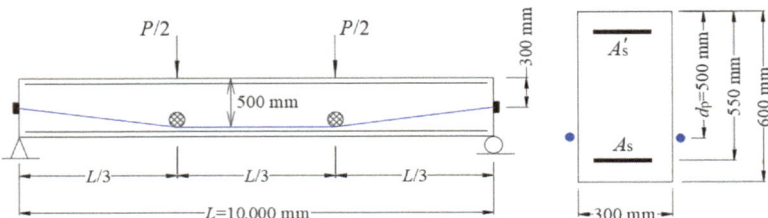

**Figure 4.11** Reference beam with external CFRP tendons for parametric analysis. *CFRP*, Carbon fiber-reinforced polymer.

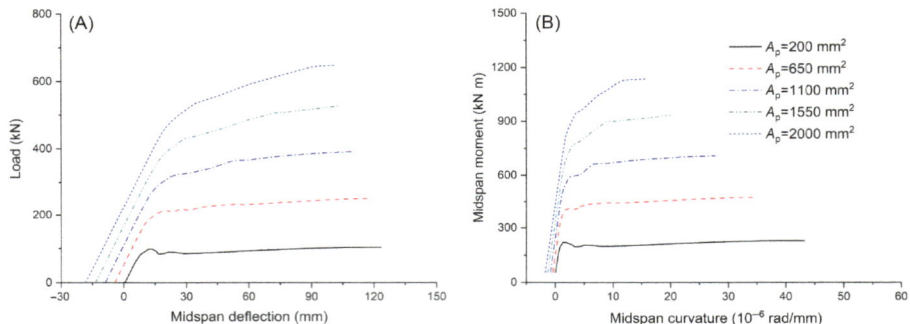

**Figure 4.12** Effect of tendon area on the deformation development: (A) load−deflection; (B) moment−curvature.

## 4.3.1 Effect of tendon area

The tendon area, $A_p$, varies between 200 and 2000 mm$^2$ to examine its effect on the structural response. Fig. 4.12A shows that at a small area of 200 mm$^2$, the prestressing effect is not prominent, just counteracting the effect of self-weight, that is, the initial deflection before the live load is around zero. At $A_p = 650$ mm$^2$ or above, the prestressing effect becomes pronounced, leading to an upward deflection at the initial state. However, at high areas of 1550 and 2000 mm$^2$, there occur excessive prestressing effects. In these cases, the top face of the critical section has cracked before load application and the upward deflection is rather big (13.5 and 18.3 mm). It is observed in Fig. 4.12A that as the tendon area increases, the cracking and ultimate loads significantly increase while the ultimate deflection decreases obviously. In this analysis, increasing the tendon area from 200 to 2000 mm$^2$ leads to an increase in ultimate load by 528.5% and a decrease in ultimate deflection by 18.3%. Fig. 4.12B shows that at the initial state, the self-weight moment is 56 kN m. A higher tendon area causes substantially higher flexural stiffness and flexural strength but a markedly smaller ultimate curvature.

The flexural ductility can be expressed by either deflection or curvature ductility, defined as the ratio of deflection or curvature at ultimate to that at yielding. Table 4.3 presents the data for flexural ductility of members with different tendon areas. It is seen that the flexural ductility quickly decreases as the tendon area

**Table 4.3** Effect of tendon area on flexural ductility.

| $A_p$ (mm$^2$) | Deflection (mm) | | Curvature ($10^{-6}$ rad/mm) | | Deflection ductility | Curvature ductility |
|---|---|---|---|---|---|---|
| | Yielding | Ultimate | Yielding | Ultimate | | |
| 200 | 21.9 | 123.4 | 5.3 | 43.3 | 5.63 | 8.16 |
| 650 | 43.2 | 119.1 | 5.9 | 34.3 | 2.75 | 5.78 |
| 1100 | 52.7 | 110.0 | 6.6 | 27.8 | 2.09 | 4.24 |
| 1550 | 57.3 | 102.7 | 7.0 | 20.0 | 1.79 | 2.86 |
| 2000 | 61.4 | 100.9 | 7.5 | 15.9 | 1.64 | 2.13 |

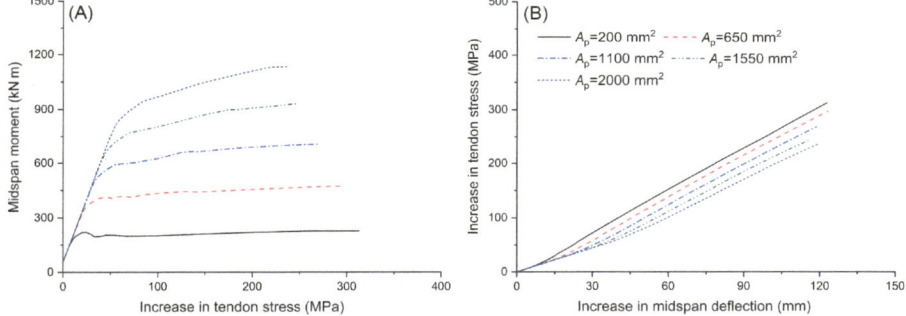

**Figure 4.13** Effect of tendon area on tendon stress development: (A) moment versus increase in tendon stress; (B) increase in deflection versus increase in tendon stress.

increases, that is, increasing the tendon area from 200 to 2000 mm$^2$ leads to a decrease in deflection ductility by 70.8% and curvature ductility by 73.9%.

Fig. 4.13A and B illustrates the tendon stress versus midspan moment and deflection for members with various tendon areas, respectively. In the initial range of loading up to tensile cracking, the rate of increase in tendon stress with the moment is independent of the tendon area, as demonstrated in Fig. 4.13A. A higher tendon area causes a higher cracking moment and correspondingly a larger increase in tendon stress at cracking. Moreover, the ultimate stress increase in tendons decreases as the tendon area increases. In this analysis, increasing the tendon area from 200 to 2000 mm$^2$ reduces the ultimate stress increase in external tendons by 24.4%. Fig. 4.13B demonstrates that for the same deflection, a high tendon area causes a small increase in tendon stress. Slopes of the stress—deflection relationship for members with $A_p$ = 200, 650, 1100, 1550, and 2000 mm$^2$ are 2.54, 2.50, 2.45, 2.39, and 2.35 MPa/mm, respectively.

### 4.3.2 Effect of prestress level

Five initial prestress levels are considered, that is, 0%, 20%, 40%, 60%, and 80% of the rupture strength, or equivalently, the initial prestress, $\sigma_{p0}$, corresponds to 0,

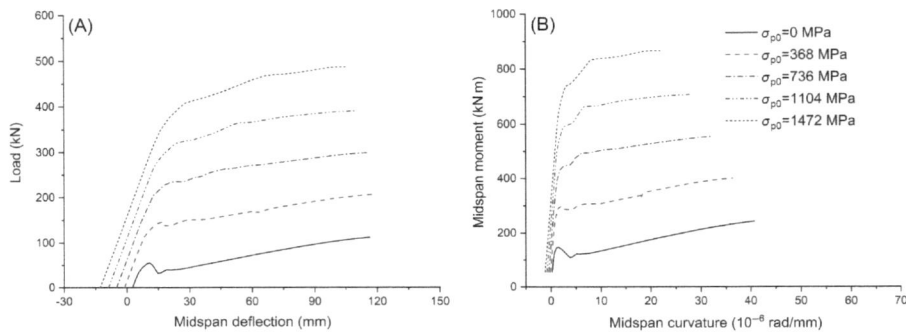

**Figure 4.14** Effect of initial prestress on the deformation development: (A) load–deflection; (B) moment–curvature.

368, 736, 1104, and 1472 MPa. Fig. 4.14A and B shows the load–deflection and moment–curvature behavior of members with various prestress levels.

The initial deflection is highly dependent on the prestress level, that is, a higher prestress level corresponds to a larger upward deflection. At zero prestress level, there is a downward deflection due to the self-weight effect, as expected. In addition, a higher prestress level improves effectively not only the cracking load and moment but also the ultimate load and flexural strength. It is worth mentioning that this observation is different from the case of using bonded tendons. Generally, bonded FRP tendons have ruptured at failure, and therefore, the prestress level has practically no influence on the ultimate load or flexural strength of the members (Kim, 2010). On the other hand, the stress increase in external tendons is much slower than that in bonded tendons due to strain incompatibility between external tendons and the adjacent concrete and also to the second-order effects. Therefore the ultimate stress in external tendons is generally below their rupture strength. The ultimate load or flexural strength is controlled by the tendon stress, and therefore, is strongly dependent on the prestress level in external tendons. Moreover, a higher prestress level results in substantially higher flexural stiffness and lower ultimate curvature, while this variable appears to have a marginal effect on the ultimate deflection. Table 4.4 presents the data for flexural ductility for members with different prestress levels. It is seen that as the prestress level increases, the flexural ductility quickly decreases. Increasing the initial prestress from 0 to 1472 MPa leads to a decrease in deflection ductility by 70.3% and curvature ductility by 60.9%.

Fig. 4.15A and B illustrates the effect of initial prestress on tendon stress versus midspan moment and deflection, respectively. Over the elastic range, the initial prestress has no impact on the increase in tendon stress as the moment develops, as shown in Fig. 4.15A. A higher initial prestress corresponds to a higher increment in tendon stress at cracking, but generally a lower one at ultimate. In this analysis, increasing the prestress level from 20% to 80% causes a reduction in ultimate stress increase in external tendons by 10.7%. It is seen in Fig. 4.15B that when deflecting, the greater the initial prestress, the slower the increase in tendon stress.

**Table 4.4** Effect of initial prestress on flexural ductility.

| $\sigma_{p0}$ (MPa) | Deflection (mm) | | Curvature ($10^{-6}$ rad/mm) | | Deflection ductility | Curvature ductility |
|---|---|---|---|---|---|---|
| | Yielding | Ultimate | Yielding | Ultimate | | |
| 0 | 18.6 | 116.5 | 5.0 | 40.8 | 6.25 | 8.13 |
| 368 | 25.8 | 118.7 | 5.6 | 36.3 | 4.61 | 6.53 |
| 736 | 45.8 | 115.2 | 6.1 | 32.0 | 2.51 | 5.25 |
| 1104 | 52.7 | 110.0 | 6.6 | 27.8 | 2.09 | 4.24 |
| 1472 | 56.5 | 105.0 | 6.9 | 22.0 | 1.86 | 3.18 |

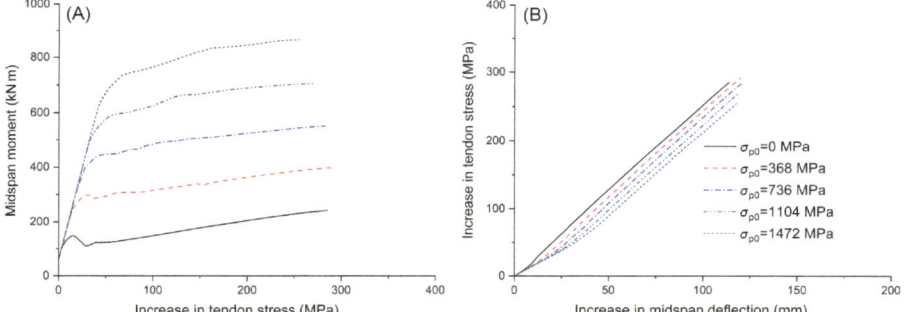

**Figure 4.15** Effect of initial prestress on tendon stress development: (A) moment versus increase in tendon stress; (B) increase in deflection versus increase in tendon stress.

### 4.3.3 Effect of tendon depth

The initial tendon depth (maximum depth), $d_p$, ranging from 400 to 600 mm is used to evaluate its effect on the structural response. The load−deflection and moment−curvature behavior are shown in Fig. 4.16A and B, respectively. From Fig. 4.16, it is seen that a larger tendon depth registers a larger upward deflection before load application. In addition, the cracking and ultimate loads or moments substantially increase as the tendon depth increases. In this analysis, increasing the tendon depth from 400 to 600 mm causes an increase in ultimate load by 74.2%. However, the ultimate deflection, curvatures, and flexural ductility are insensitive to the tendon depth, as presented in Table 4.5.

Fig. 4.17A and B illustrates the tendon stress versus midspan moment and deflection for members with various tendon depths, respectively. As the tendon depth increases, the ultimate tendon stress increases substantially. As mentioned previously, the members with different tendon depths exhibit approximately the same deformation at ultimate, implying there are approximately the same concrete strain distributions along the depth of the sections of these members. Therefore the larger the tendon depth, the larger the average change in concrete strain at the same

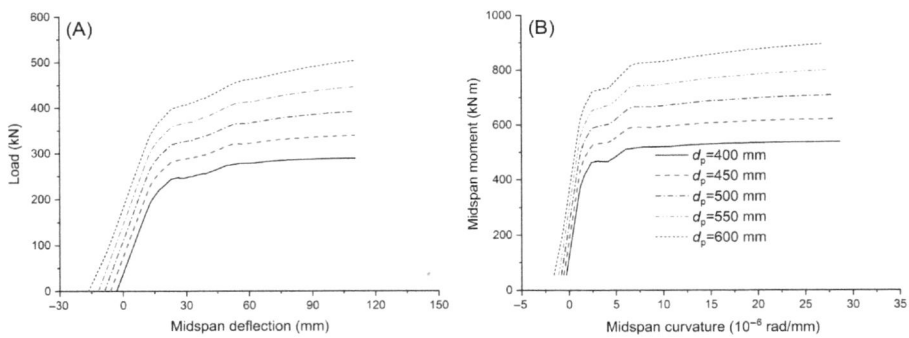

**Figure 4.16** Effect of tendon depth on the deformation development: (A) load—deflection; (B) moment—curvature.

**Table 4.5** Effect of tendon depth on flexural ductility.

| $d_p$ (mm) | Deflection (mm) | | Curvature ($10^{-6}$ rad/mm) | | Deflection ductility | Curvature ductility |
|---|---|---|---|---|---|---|
| | Yielding | Ultimate | Yielding | Ultimate | | |
| 400 | 50.4 | 110.4 | 6.5 | 28.6 | 2.19 | 4.42 |
| 450 | 52.0 | 110.1 | 6.5 | 28.2 | 2.12 | 4.32 |
| 500 | 52.7 | 110.0 | 6.6 | 27.8 | 2.09 | 4.24 |
| 550 | 53.6 | 109.7 | 6.6 | 27.2 | 2.05 | 4.13 |
| 600 | 53.8 | 110.1 | 6.6 | 26.7 | 2.04 | 4.06 |

level of the external tendons, and thereby the larger the tendon stress increase. Besides, due to a lower reduction in effective tendon depth, a larger tendon depth assumes relatively less significant second-order effects, which would reduce relatively less tendon stresses. In this analysis, increasing the tendon depth from 400 to 600 mm causes an increase in ultimate tendon stress increment by 93.2%. For the same deflection, the greater the tendon depth, the greater the tendon stress increase. The slopes of the stress—deflection relationship for members with $d_p = 400$, 450, 500, 550, and 600 mm are 1.70, 1.96, 2.45, 2.86, and 3.30 MPa/mm, respectively.

## 4.3.4 Effect of tendon elastic modulus

The elastic modulus of CFRP tendons, $E_p$, ranges from 80 to 500 GPa (FIB, 2012). Fig. 4.18A and B shows the effect of tendon elastic modulus on the load—deflection and moment—curvature behavior, respectively. The members exhibit identical behavior up to cracking. After that, the behaviors differ because the contribution of external tendons becomes increasingly important. A higher tendon modulus of elasticity mobilizes the stiffer behavior of the members. As the tendon modulus of elasticity increases, the ultimate load and flexural strength increase while the ultimate

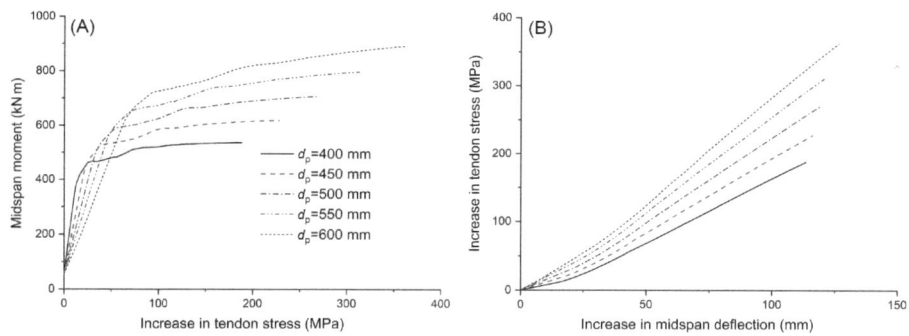

**Figure 4.17** Effect of tendon depth on tendon stress development: (A) moment versus increase in tendon stress; (B) increase in deflection versus increase in tendon stress.

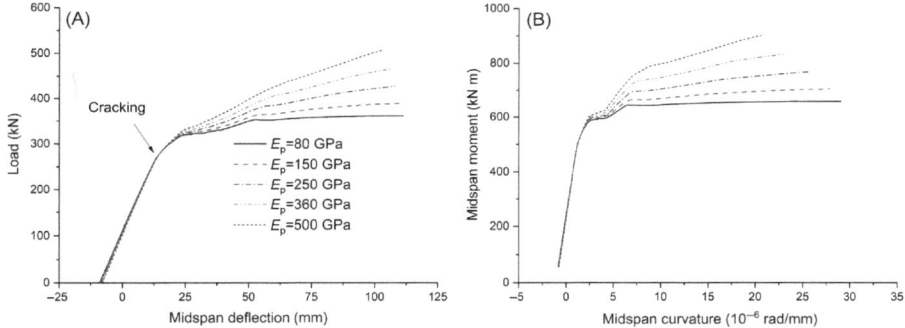

**Figure 4.18** Effect of tendon elastic modulus on the deformation development: (A) load−deflection; (B) moment−curvature.

**Table 4.6** Effect of tendon elastic modulus on flexural ductility.

| $E_p$ (GPa) | Deflection (mm) | | Curvature ($10^{-6}$ rad/mm) | | Deflection ductility | Curvature ductility |
|---|---|---|---|---|---|---|
| | Yielding | Ultimate | Yielding | Ultimate | | |
| 80 | 52.6 | 111.2 | 6.5 | 29.0 | 2.12 | 4.45 |
| 150 | 52.7 | 110.0 | 6.6 | 27.8 | 2.09 | 4.24 |
| 250 | 53.5 | 107.8 | 6.6 | 25.6 | 2.01 | 3.87 |
| 360 | 54.6 | 105.9 | 6.7 | 23.2 | 1.94 | 3.47 |
| 500 | 54.6 | 102.7 | 6.8 | 20.6 | 1.88 | 3.05 |

deflection and curvature tend to decrease. As presented in Table 4.6, increasing the tendon elastic modulus from 80 to 500 GPa leads to a decrease in deflection ductility by 11.1% and curvature ductility by 31.4%.

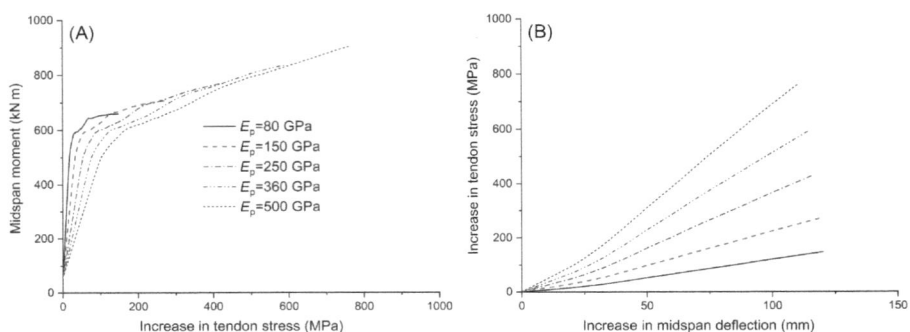

**Figure 4.19** Effect of tendon elastic modulus on tendon stress development: (A) moment versus increase in tendon stress; (B) increase in deflection versus increase in tendon stress.

Fig. 4.19A and B shows the effect of tendon elastic modulus on the tendon stress versus midspan moment and deflection, respectively. At a given moment or deflection level, a high tendon elastic modulus corresponds to a significantly higher tendon stress. The ultimate stress increase in external tendons with an elastic modulus of 500 GPa is 5.17 times that of 80 GPa. It should be noted that the stress ratio of 5.17 is smaller than the modulus ratio of 6.25 as a higher tendon elastic modulus would lead to a smaller ultimate tendon strain. The slopes of the stress−deflection relationship for members with $E_p = 80$, 150, 250, 360, and 500 GPa are 1.33, 2.45, 3.98, 5.57, and 7.43 MPa/mm, respectively.

## 4.4 Simplified model for flexural strength prediction

In externally posttensioned members, the ultimate tendon stress, $\sigma_{pu}$, depends on the whole member deformation, and therefore, its accurate predictions are complicated. The calculation of $\sigma_{pu}$ is essential for the flexural strength design of externally prestressed members (Roberts-Wollmann et al., 2005). The stress $\sigma_{pu}$ is commonly expressed by

$$\sigma_{pu} = \sigma_{pe} + \Delta\sigma_p \tag{4.4}$$

where $\Delta\sigma_p$ is the ultimate stress increment; $\sigma_{pe}$ is effective prestress.

JGJ 92−2016 (JGJ 92−2016, 2016) recommended the following equation for predicting $\Delta\sigma_p$ in simply supported unbonded prestressed members:

$$\Delta\sigma_p = (240 - 335\omega_0)(0.45 + 5.5h/L) \tag{4.5}$$

where $\omega_0$ is the combined reinforcement index; $h$ is section height; $L$ is the span length. The $\omega_0$ is expressed by

$$\omega_0 = \frac{A_p\sigma_{pe} + A_sf_y}{bd_pf_{ck}} \tag{4.6}$$

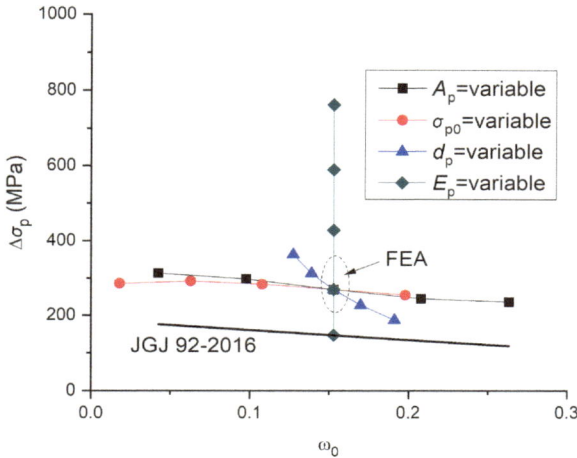

**Figure 4.20** Relationships between $\Delta\sigma_p$ and $\omega_0$ according to FEA and JGJ 92−2016. *FEA*, Finite element analysis.

Fig. 4.20 shows the $\Delta\sigma_{p-}\omega_0$ relationships for the members with different tendon-related variables obtained by the finite element analysis (FEA) along with the JGJ 92−2016 curve. By adopting the key parameter $\omega_0$, JGJ 92−2016 takes into account the effect of tendon area, prestress level, and tendon depth but neglects the effect of tendon modulus of elasticity. In addition, JGJ 92−2016 significantly underestimates the value of $\Delta\sigma_p$, except for the member with $E_p = 80$ GPa.

The numerical analysis demonstrates that the ultimate tendon stress is affected by all four tendon-related variables, that is, the tendon area, prestress level, tendon depth, and modulus of elasticity. As the parameter $\omega_0$ involves three of the tendon-related variables (i.e., $A_p$, $d_p$, and $\sigma_{pe}$), this parameter is adopted in a new equation to be developed herein for the prediction of $\Delta\sigma_p$. As illustrated in Fig. 4.21, fitting to the FEA data about $\Delta\sigma_p$-$\omega_0$ relationship of members with various tendon areas leads to the following expression:

$$\Delta\sigma_p = 330 - 372\omega_0 \tag{4.7}$$

Note the above equation does not consider the tendon modulus of elasticity, which has been demonstrated to be crucial for $\Delta\sigma_p$. To include the effect of tendon elastic modulus, Eq. (4.7) is modified by introducing a coefficient as follows:

$$\Delta\sigma_p = \lambda_E(330 - 372\omega_0) \tag{4.8}$$

where $\lambda_E$ is a coefficient related with the tendon elastic modulus $E_p$. The FEA data regarding the variation of $\lambda_E$ against $E_p/E_{ps}$ is presented in Fig. 4.22, where $E_{ps}$ is

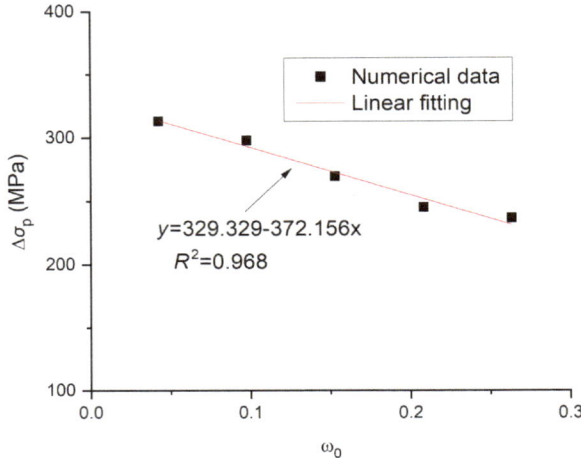

**Figure 4.21** Fitting to FEA data about $\Delta\sigma_p - \omega_0$ relationship. *FEA*, Finite element analysis.

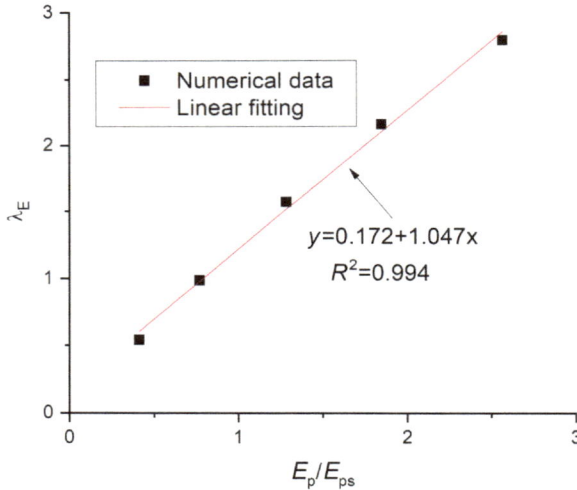

**Figure 4.22** Fitting to FEA data about $\lambda_E$ versus $E_p/E_{ps}$ relationship. *FEA*, Finite element analysis.

the elastic modulus of prestressing steel, taken equal to 195 GPa. According to the fit curve, $\lambda_E$ is expressed by

$$\lambda_E = 0.172 + 1.047\left(\frac{E_p}{E_{ps}}\right) \tag{4.9}$$

**Table 4.7** Comparison of $\Delta\sigma_p$ and $M_u$ by simplified models with FEA data.

| $A_p$ (mm²) | $\sigma_{p0}$ (MPa) | $d_p$ (mm) | $E_p$ (GPa) | $\Delta\sigma_p$ (MPa) | | | $M_u$ (kN m) | | | $(\Delta\sigma_p)_{sim}/(\Delta\sigma_p)_{fea}$ | | $(M_u)_{sim}/(M_u)_{fea}$ | |
|---|---|---|---|---|---|---|---|---|---|---|---|---|---|
| | | | | JGJ | Pro | FEA | JGJ | Pro | FEA | JGJ | Pro | JGJ | Pro |
| 200 | 1104 | 500 | 150 | 176 | 307 | 313 | 197 | 209 | 228 | 0.56 | 0.98 | 0.86 | 0.92 |
| 650 | | | | 162 | 287 | 298 | 439 | 472 | 474 | 0.54 | 0.96 | 0.93 | 1.00 |
| 1100 | | | | 147 | 267 | 270 | 655 | 703 | 707 | 0.55 | 0.99 | 0.93 | 0.99 |
| 1550 | | | | 133 | 247 | 245 | 846 | 904 | 932 | 0.54 | 1.01 | 0.91 | 0.97 |
| 2000 | | | | 118 | 227 | 237 | 1014 | 1078 | 1135 | 0.50 | 0.96 | 0.89 | 0.95 |
| 1100 | 0 | 500 | 150 | 182 | 316 | 285 | 172 | 238 | 243 | 0.64 | 1.11 | 0.71 | 0.98 |
| | 368 | | | 171 | 300 | 292 | 343 | 402 | 400 | 0.59 | 1.03 | 0.86 | 1.01 |
| | 736 | | | 159 | 283 | 283 | 504 | 558 | 554 | 0.56 | 1.00 | 0.91 | 1.01 |
| | 1104 | | | 147 | 267 | 270 | 655 | 703 | 707 | 0.55 | 0.99 | 0.93 | 0.99 |
| | 1472 | | | 135 | 251 | 255 | 795 | 838 | 867 | 0.53 | 0.98 | 0.92 | 0.97 |
| 1100 | 1104 | 400 | 150 | 137 | 253 | 188 | 497 | 530 | 537 | 0.73 | 1.35 | 0.93 | 0.99 |
| | | 450 | | 143 | 261 | 228 | 576 | 616 | 620 | 0.63 | 1.14 | 0.93 | 0.99 |
| | | 500 | | 147 | 267 | 270 | 655 | 703 | 707 | 0.55 | 0.99 | 0.93 | 0.99 |
| | | 550 | | 151 | 272 | 313 | 734 | 790 | 798 | 0.48 | 0.87 | 0.92 | 0.99 |
| | | 600 | | 154 | 276 | 363 | 813 | 877 | 894 | 0.42 | 0.76 | 0.91 | 0.98 |
| 1100 | 1104 | 500 | 80 | 147 | 164 | 148 | 655 | 661 | 660 | 1.00 | 1.11 | 0.99 | 1.00 |
| | | | 150 | 147 | 267 | 270 | 655 | 703 | 707 | 0.55 | 0.99 | 0.93 | 0.99 |
| | | | 250 | 147 | 414 | 429 | 655 | 761 | 770 | 0.34 | 0.96 | 0.85 | 0.99 |
| | | | 360 | 147 | 575 | 590 | 655 | 822 | 835 | 0.25 | 0.97 | 0.78 | 0.98 |
| | | | 500 | 147 | 780 | 763 | 655 | 897 | 902 | 0.19 | 1.02 | 0.73 | 0.99 |

*Note*: JGJ, JGJ 92–2016; Pro, proposed simplified model; $(\Delta\sigma_p)_{sim}$ and $(\Delta\sigma_p)_{fea}$, ultimate stress increase in external tendons predicted by simplified models and by FEA, respectively; $(M_u)_{sim}$ and $(M_u)_{fea}$, flexural strength by simplified models and FEA, respectively.

The axial equilibrium of members is given by

$$0.85f_{ck}b\beta_1 c_u = A_p\sigma_{pu} + A_sf_y - A'_sf'_y \tag{4.10}$$

where $\beta_1 = 0.85$; $c_u$ is neutral axis depth at ultimate; $A'_s$ and $f'_y$ are area and yield strength of nonprestressed compression steel, respectively.

According to Eq. (4.10), $c_u$ is calculated by

$$c_u = \frac{A_p\sigma_{pu} + A_sf_y - A'_sf'_y}{0.85f_{ck}b\beta_1} \tag{4.11}$$

The flexural strength is determined by

$$M_u = A_p\sigma_{pu}d_e + A_sf_yd_s - A'_sf'_yd'_s - 0.85f_{ck}b(\beta_1 c_u)^2/2 \tag{4.12}$$

where $d_s$ and $d'_s$ are effective depths of nonprestressed tension and compression steel, respectively; $d_e$ is the effective depth of external tendons, which is given by

$$d_e = R_dd_p \tag{4.13}$$

where $R_d$ is a reduction coefficient due to second-order effects. According to ACI 440.4R-04 (ACI Committee 440, 2004), for third-point loading, the value of $R_d$ is calculated from

$$R_d = 1.25 - 0.01(L/d_p) - 0.38(S_d/L) \leq 1.0 \tag{4.14}$$

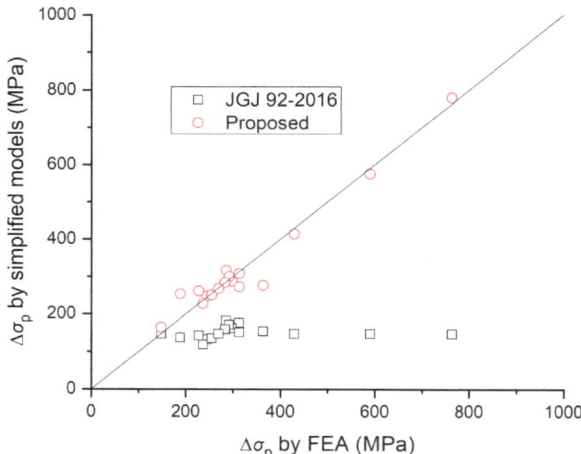

**Figure 4.23** Correlation of $\Delta\sigma_p$ by simplified models to FEA data. *FEA, Finite element analysis.*

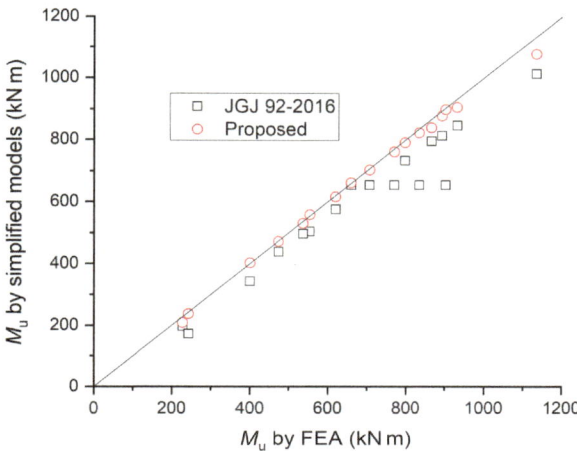

**Figure 4.24** Correlation of $M_u$ by simplified models to FEA data. *FEA*, Finite element analysis.

A comparison of $\Delta\sigma_p$ and $M_u$ for members with different tendon-related variables predicted by the simplified models and FEA is presented in Table 4.7 and Figs. 4.23 and 4.24. It is seen that JGJ 92−2016 leads to a significant underestimate in $\Delta\sigma_p$ and thereby this code underestimates the flexural strength of the members. The predicted $\Delta\sigma_p$ is 53.5% of the FEA data on average with a standard deviation of 16.2%, while the predicted $M_u$ is 88.6% of the FEA data on average with a standard deviation of 6.9%. The proposed simplified model shows much better predictions than JGJ 92−2016. According to the proposed model, the mean discrepancy for $\Delta\sigma_p$ is 0.9% with a standard deviation of 11.1%, while the mean discrepancy for $M_u$ is −1.6% with a standard deviation of 2.1%.

## 4.5 Conclusions

All aspects of the structural behavior of external CFRP tendon members are similar to those of external steel tendon members, indicating that the traditional steel tendons can be well replaced by CFRP tendons. Due to the much lower elastic modulus of the GFRP composite, external GFRP tendon members register lower flexural strengths but exhibit better ductility behavior in comparison with external steel tendon members. With the increase of the prestressing reinforcement index, the cracking, yielding, and ultimate loads (moments) of the members are significantly enhanced, while the ductility diminishes gradually. A higher prestressing reinforcement index mobilizes a larger neutral axis depth than a lower prestressing reinforcement index. The discrepancy is most remarkable at cracking while it is not so important at ultimate. A member without any bonded tension reinforcement displays an extremely irregular curvature distribution and correspondingly significant crack concentration at failure load. By the provision of a minimum amount of bonded reinforcement, the curvature distribution and crack pattern are greatly improved.

The tendon-related variables play a crucial role in the structural performance of prestressed concrete members with external FRP tendons, including flexural stiffness, ultimate load-carrying capacity, stress increase in external tendons, deformation, and ductility. A higher tendon area, initial prestress, or elastic modulus causes lower flexural ductility. The ultimate tendon stress decreases as the tendon area or initial prestress level increases, or as the tendon depth or elastic modulus decreases.

JGJ 92−2016 significantly underestimates the ultimate tendon stress, and hence, this code is overconservative for flexural strength predictions of prestressed concrete members with external FRP tendons. The predicted $\Delta\sigma_p$ and $M_u$ by JGJ 92−2016 are 53.5% and 88.6% of the FEA data on average, respectively. An equation is proposed to calculate the ultimate tendon stress in this type of members, considering the influence of tendon area, effective prestress, tendon depth, and modulus of elasticity. The proposed simplified model shows satisfactory predictions of the flexural strength of prestressed concrete members with external FRP tendons.

# References

ACI Committee 318. (2019). *Building code requirements for structural concrete (ACI 318-19) and commentary (ACI 318R-19)*. Farmington Hills, MI.

ACI Committee 440. (2004). *Prestressing concrete structures with FRP tendons*. ACI 440.4R-04. Farmington Hills, MI.

Bennitz, A., Schmidt, J. W., Nilimaa, J., Täljsten, B., Goltermann, P., & Ravn, D. L. (2012). Reinforced concrete T-beams externally prestressed with unbonded carbon fiber-reinforced polymer tendons. *ACI Structural Journal*, *109*(4), 521−530. Available from http://www.concrete.org/tempComDocs/-41246/109-s45.pdf.

Brockmann, C., & Rogenhofer, H. (2000). Bang Na expressway, Bangkok, Thailand - World's longest bridge and largest precasting operation. *PCI Journal*, *45*(1), 26−38. Available from https://doi.org/10.15554/pcij.01012000.26.38.

FIB. (2012). Model Code 2010 (MC10). Bulletins 55 and 56, Lausanne, Switzerland.

JGJ 92-2016. (2016). *Technical specification for concrete structures prestressed with unbonded tendons*. China Architecture & Building Press, Beijing.

Kim, Y. J. (2010). Flexural response of concrete beams prestressed with AFRP tendons: Numerical investigation. *Journal of Composites for Construction*, *14*(6), 647−658. Available from https://doi.org/10.1061/(ASCE)CC.1943-5614.0000128.

Lou, T., Hu, H., & Pang, M. (2023). Effect of tendon-related variables on the behavior of externally CFRP prestressed concrete beams. *Materials*, *16*(14), 5197. Available from https://doi.org/10.3390/ma16145197.

Lou, T., Lopes, S. M. R., & Lopes, A. V. (2012). Numerical analysis of behaviour of concrete beams with external FRP tendons. *Construction and Building Materials*, *35*, 970−978. Available from https://doi.org/10.1016/j.conbuildmat.2012.04.055.

Nordin, H. (2005). Technical report. *Strengthening structures with externally prestressed tendons*. Lulea University of Technology, Sweden.

Pisani, M. A. (1998). Numerical survey on the behaviour of beams pre-stressed with FRP cables. *Construction and Building Materials*, *12*(4), 221−232. Available from https://doi.org/10.1016/S0950-0618(97)00081-0.

Roberts-Wollmann, C. L., Kreger, M. E., Rogowsky, D. M., & Breen, J. E. (2005). Stresses in external tendons at ultimate. *ACI Structural Journal*, *102*(2), 206−213.

# Moment redistribution in continuous prestressed concrete members with external CFRP tendons

## 5.1 Introduction

Of different fiber-reinforced polymer (FRP) composite materials, carbon FRP (CFRP) is most resistant to creep rupture (i.e., sustaining about 80% of the ultimate strength) (FIB, 2012), and therefore, this material appears to be the best choice for prestressing applications (ACI committee 440, 2004). In external prestressing systems, the tendons are usually subject to harsh environments, thereby resulting in corrosive issues for conventional prestressing steel. Replacement of steel tendons by CFRP composites is an effective way to overcome the corrosive problem (Grace & Abdel-Sayed, 1998). The use of CFRP tendons as a replacement for steel alternatives on the behavior of prestressed concrete beams with external tendons has been experimentally investigated (Bennitz et al., 2012; Tan & Tjandra, 2007). The studies showed that CFRP tendons lead to similar structural behavior of both simply supported (Bennitz et al., 2012) and continuous beams (Tan & Tjandra, 2007) in comparison with steel tendons.

In statically indeterminate members such as continuous beams, redistribution of bending moments occurs once the members exhibit nonlinear behavior. Moment redistribution is associated with flexural ductility. As FRP composites are brittle material, the ability of FRP-reinforced or strengthened concrete beams to redistribute moments is often limited (Kara & Ashour, 2013; Santos et al., 2013; Tajaddini et al., 2017). Such limitation can be effectively overcome by providing longitudinal non-prestressed steel in prestressed concrete beams with external FRP tendons. The current design codes suggest an elastic analysis with limited redistribution for exploiting the ductile capacity of continuous members. According to the current design codes worldwide, the redistribution of moments is heavily dependent on the ductile behavior of the critical section. The neutral axis depth is used by many codes of practice, for example, CSA (CSA, 2004), BSI (BSI, 2007), and EC2 (CEN, 2004), as a key parameter to quantify the permissible moment redistribution. On the other hand, the ACI code (ACI Committee 318, 2019) adopts the net strain in extreme tension steel as a key parameter for predicting the moment redistribution in reinforced and prestressed concrete members. Both the neutral axis depth and net strain in extreme tension steel are parameters that can well reflect the ductility of a section. However, moment redistribution is not only a sectional aspect but also a

Prestressed Members with External Fiber-Reinforced Polymer (FRP) Tendons. DOI: https://doi.org/10.1016/B978-0-443-23877-2.00005-5

structural aspect of behavior since the redistribution is mainly a consequence of structural redundancy (Oehlers et al., 2004). Therefore the current design codes may not be reasonable for calculating the amount of moment redistribution since the codes take into account sectional ductility only, neglecting the structural aspect which is supposed to be very important to the redistribution of moments.

This chapter presents the redistribution of moments in two-span continuous pre-stressed concrete members with external CFRP tendons (Lou, Lopes, et al., 2014; Lou, Peng, et al., 2020). A wide range of factors affecting moment redistribution are examined, including the content of non-prestressed tension steel, eccentricities of external tendons at midspan and center support, amount of external tendons, effective prestress, span-to-height ratio, concrete strength, CFRP elastic modulus and type of loading. Based on the results of the parametric analysis, a modified ACI equation for the calculation of the amount of moment redistribution at ultimate is proposed. In addition, moment redistribution against neutral axis depth in two-span prestressed concrete members with external CFRP tendons is examined, focusing on the influence of the content of non-prestressed reinforcement at either the positive or negative moment zone. This variable is selected to produce varying stiffness differences between the midspan and inner support. Typical results in relation to global redistribution and neutral axis behavior are presented. Several codes of practice that use the neutral axis depth for redistribution quantification are assessed. Reasonable modifications of these code equations are proposed by introducing a parameter describing the impact of stiffness difference.

## 5.2   Measurement of moment redistribution and codes of practice

Several approaches have been used to measure quantitatively the amount of moment redistribution in a statically indeterminate structure. One of the approaches was based on a plastic adaption ratio (PAR) defined by Tichy and Rakosnik (1977):

$$PAR = P_{col}/P_{pl} \tag{5.1}$$

where $P_{col}$ is the actual ultimate load; and $P_{pl}$ is the ultimate load calculated by a plastic analysis. PAR = 1 indicates full redistribution of moments.

Some investigators (Campbell & Moucessian, 1988) defined the plastic adaption ratio using three ultimate loads as follows:

$$PAR1 = (P_{col} - P_{el})/(P_{pl} - P_{el}) \tag{5.2}$$

where $P_{el}$ is the ultimate load calculated by an elastic analysis. PAR1 = 0 ($P_{col} = P_{el}$) corresponds to zero redistribution, while PAR1 = 1 ($P_{col} = P_{pl}$) corresponds to full redistribution.

Cohn (1986) defined the degree of moment redistribution by

$$\beta = 1 - (M/M_e) \tag{5.3}$$

where $M$ is the actual moment; $M_e$ is the elastic moment calculated based on the theory of elasticity. $\beta = 0$ indicates nil redistribution. This definition is adopted by various codes. In calculating the design moments in continuous flexural members, the codes allow designers to take advantage of a linear analysis with an adjustment of the elastic moments through the use of the degree of moment redistribution $\beta$. However, the empirical equations for the calculation of $\beta$ in various codes are quite different.

In the ACI code (ACI Committee 318, 2019), the degree of moment redistribution for prestressed concrete members with sufficient bonded reinforcement is calculated using the net strain in extreme tension steel $\varepsilon_t$ by

$$\beta(\%) \leq 1000\varepsilon_t \tag{5.4}$$

with a maximum of 20%. Also, the moment redistribution can be done only when $\varepsilon_t$ is not less than 0.0075 at the section where the moment is reduced.

The CSA code (CSA, 2004) indicates that the negative moment calculated by an elastic analysis can be increased or decreased by

$$\beta(\%) \leq 30 - 50c/d \tag{5.5}$$

with a maximum of 20%. In Eq. (5.5), $c/d$ is the ratio of the neutral axis depth to the effective depth of a cross-section at the ultimate limit state.

In Europe, EC2 (CEN, 2004) and MC10 (FIB, 2012) also use the parameter $c/d$ to calculate the degree of moment redistribution:

for $f_{ck} \leq 50$ MPa,

$$\beta \leq 0.56 - 1.25(0.6 + 0.0014/\varepsilon_{cu2})c/d \tag{5.6a}$$

for $f_{ck} > 50$ MPa,

$$\beta \leq 0.46 - 1.25(0.6 + 0.0014/\varepsilon_{cu2})c/d \tag{5.6b}$$

with a maximum of 30% for high- and normal-ductility steel and of 20% for low-ductility steel. In Eq. (5.6), $\varepsilon_{cu2}$ is the ultimate strain which is determined according to EC2 by

$$\varepsilon_{cu2}(\%_o) = \begin{cases} 3.5 \text{ for } f_{ck} \leq 50 \text{ MPa} \\ 2.6 + 35[(90 - f_{ck})/100]^4 \text{ for } f_{ck} > 50 \text{ MPa} \end{cases} \tag{5.7}$$

## 5.3  Parametric study

A two-span continuous prestressed concrete rectangular beam with external CFRP tendons, as shown in Fig. 5.1, is used as a reference beam for the parametric analysis.

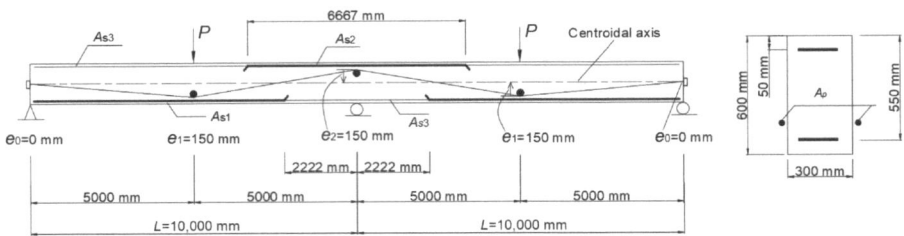

**Figure 5.1** Details of reference beam used for parametric analysis of moment redistribution.

The material parameters are as follows: unless otherwise stated, the areas of non-prestressed tension steel reinforcement over positive moment region $A_{s1}$ and negative moment region $A_{s2}$ are 720 and 360 mm², respectively; the area of non-prestressed compression steel reinforcement $A_{s3}$ is 360 mm²; the yield strength $f_y$ and elastic modulus $E_s$ of non-prestressed steel are 450 MPa and 200 GPa, respectively; the area of external tendons $A_p$ is 450 mm², and the modulus of elasticity $E_f$ and tensile strength $f_f$ of CFRP tendons are 147 GPa and 1840 MPa, respectively; the effective prestress $\sigma_{pe}$ for the CFRP tendons is considered to be 930 MPa; the concrete compressive strength $f_{ck}$ is 40 MPa.

In the finite element idealization, the concrete beam is divided into 36 identical beam elements, and the cross-section of a beam element is subdivided into 10 concrete layers and two steel layers each of which represents the bottom or top non-prestressed steel reinforcement. The external tendon is also divided into 36 tendon segments corresponding to the beam elements. Using the finite element analysis (FEA), the influence of various factors on the redistribution of moments is evaluated. These factors include the non-prestressed steel area, $A_{s2}/A_{s1}$ ratio, tendon eccentricities, tendon area, effective prestress, span-to-height ratio, concrete strength, CFRP elastic modulus, and load type. Unless otherwise stated, the results ($\beta$, $c/d$, $\varepsilon_t$) presented in the following sections of this chapter are for the critical negative moment (center support) section of the members at the ultimate limit state.

## 5.3.1  Effect of non-prestressed steel area

The effect of non-prestressed steel area is examined by varying $A_{s2}$ from 360 to 2280 mm² and maintaining the $A_{s2}/A_{s1}$ ratio at 0.8. Fig. 5.2 shows the variation of $\beta$ with the amount of non-prestressed steel. Both the FEA results and code predictions are presented. The FEA results are obtained using Eq. (5.3) where the actual moment capacity $M$ and elastic moment $M_e$ are computed by FEA. In the calculation of $M$, both geometric and material nonlinearities are considered. On the other hand, in the calculation of $M_e$, all the materials are assumed to be linear elastic while the geometric nonlinearity is taken into account. To obtain the elastic moment $M_e$, the ultimate load corresponding to the actual moment capacity $M$ is applied and the incremental load method is employed to solve the equilibrium equations.

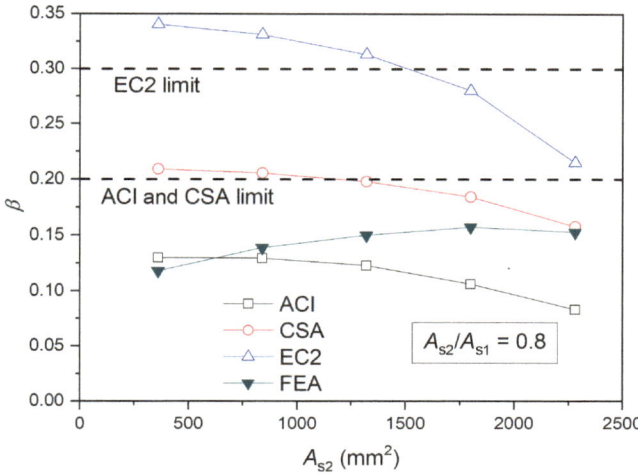

**Figure 5.2** Effect of non-prestressed steel area on moment redistribution according to FEA and code predictions.

A summary of results in relation to moment redistribution for different amounts of non-prestressed steel is given in Table 5.1.

From Fig. 5.2 and Table 5.1, it is seen that, according to the FEA predictions, the $\beta$ value increases with increasing $A_{s2}$ up to 1800 mm$^2$ and then gradually decreases with a continuing increase of the steel area. This observation can be attributed to the combined effects of ductility and stiffness differences between critical sections. When $A_{s2}$ ($A_{s2}/A_{s1} = 0.8$) increases, the flexural ductility tends to decrease (the less the ductility, the lower the moment redistribution) while the stiffness difference between critical sections enlarges (the larger the stiffness difference, the higher the moment redistribution). Therefore if the effect of stiffness difference transcends the effect of ductility (for $A_{s2}$ increased up to 1800 mm$^2$), the moment redistribution increases; on the other hand, if the effect of ductility prevails against the effect of stiffness difference (for $A_{s2}$ increased beyond 1800 mm$^2$), the moment redistribution decreases.

It is also observed that, according to the predictions by various codes, the moment redistribution consistently decreases as the amount of non-prestressed steel increases. This implies that the codes account for the section ductility only, neglecting the stiffness difference between critical sections. As a consequence, the code predictions fail to reflect accurately the actual trend of the variation of $\beta$ with the amount of non-prestressed steel. In this analysis, it is seen that EC2 and the CSA code are nonconservative particularly at a low amount of non-prestressed steel, while the ACI code is generally conservative.

### 5.3.2  Effect of $A_{s2}/A_{s1}$

The effect of $A_{s2}/A_{s1}$ is examined assuming a minimum non-prestressed steel ($A_{s1} = A_{s2} = 360$ mm$^2$) and varying $A_{s1}$ or $A_{s2}$ from 360 to 1800 mm$^2$. Fig. 5.3A

**Table 5.1** Moment redistribution in beams with different contents of nonprestressed steel.

| Beam | $A_{s1}$ (mm$^2$) | $A_{s2}$ (mm$^2$) | $A_{s2}/A_{s1}$ | $\varepsilon_t$ (%) | $c/d$ (%) | $M$ (kN m) | $M_e$ (kN m) | $\beta$ (%) | | | |
|------|------|------|------|------|------|------|------|------|------|------|------|
| | | | | | | | | ACI | CSA | EC2 | FEA |
| B01 | 450 | 360 | 0.8 | 1.297 | 18.11 | −327.31 | −371.03 | 12.97 | 20.94 | 34.07 | 11.78 |
| B02 | 1050 | 840 | | 1.295 | 18.90 | −444.76 | −516.40 | 12.95 | 20.55 | 33.12 | 13.87 |
| B03 | 1650 | 1320 | | 1.230 | 20.40 | −559.45 | −658.23 | 12.30 | 19.80 | 31.31 | 15.01 |
| B04 | 2250 | 1800 | | 1.059 | 23.12 | −671.36 | −796.61 | 10.59 | 18.44 | 28.02 | 15.72 |
| B05 | 2850 | 2280 | | 0.833 | 28.50 | −781.44 | −922.29 | 8.33 | 15.75 | 21.50 | 15.27 |
| B10 | 360 | 360 | 1 | 1.226 | 18.20 | −324.37 | −353.75 | 12.26 | 20.90 | 33.97 | 8.31 |
| B11 | 720 | | 0.5 | 1.420 | 17.96 | −332.51 | −418.69 | 14.20 | 21.02 | 34.26 | 20.58 |
| B12 | 1080 | | 0.33 | 1.512 | 17.85 | −338.38 | −480.26 | 15.12 | 21.07 | 34.39 | 29.54 |
| B13 | 1440 | | 0.25 | 1.577 | 17.77 | −342.93 | −539.64 | 15.77 | 21.12 | 34.49 | 36.45 |
| B14 | 1800 | | 0.2 | 1.612 | 17.71 | −346.60 | −597.30 | 16.12 | 21.15 | 34.56 | 41.97 |
| B15 | 360 | 360 | 1 | 1.226 | 18.20 | −324.37 | −353.75 | 12.26 | 20.90 | 33.97 | 8.31 |
| B16 | | 720 | 2 | 1.076 | 19.25 | −405.41 | −385.15 | 10.76 | 20.38 | 32.70 | −5.26 |
| B17 | | 1080 | 3 | 0.936 | 21.13 | −486.52 | −416.56 | 9.36 | 19.44 | 30.43 | −16.79 |
| B18 | | 1440 | 4 | 0.768 | 24.18 | −567.80 | −447.57 | 7.68 | 17.91 | 26.73 | −26.86 |
| B19 | | 1800 | 5 | 0.625 | 28.51 | −648.27 | −478.97 | 6.25 | 15.75 | 21.49 | −35.35 |

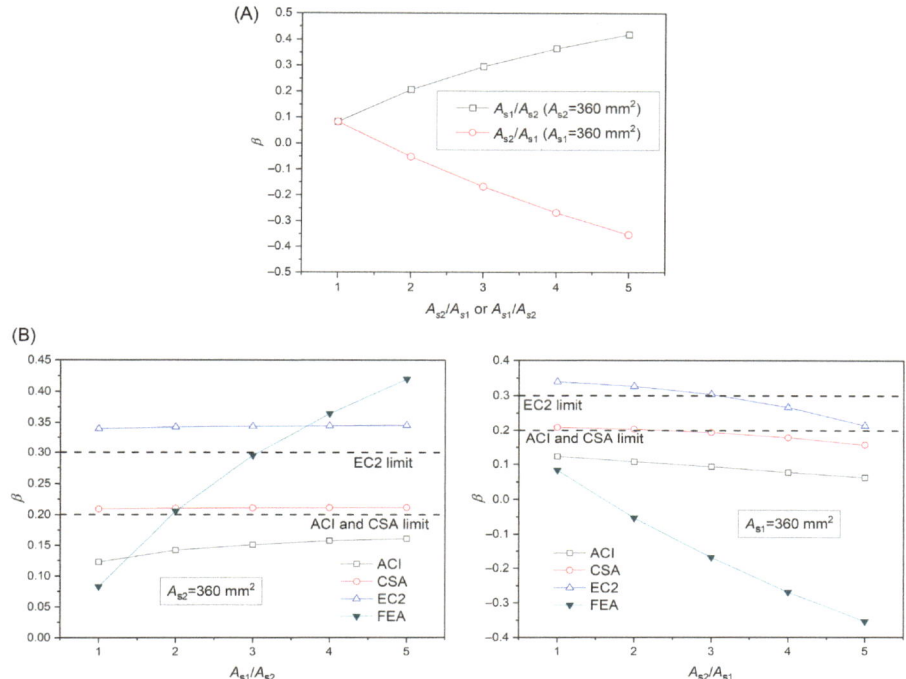

**Figure 5.3** Effect of As2/As1 or As1/As2 on moment redistribution: (A) FEA results; (B) comparison between FEA and code predictions.

shows the variation of $\beta$ with the $A_{s2}/A_{s1}$ or $A_{s1}/A_{s2}$ ratio according to FEA predictions. A comparison between the $\beta$ values predicted by FEA and various code equations is illustrated in Fig. 5.3B and Table 5.1.

It is observed from Fig. 5.3A that the $A_{s2}/A_{s1}$ (or $A_{s1}/A_{s2}$) ratio strongly affects the degree of moment redistribution, attributed primarily to the change in the stiffness difference between the critical midspan and center support sections. The $\beta$ value increases significantly with the increase of $A_{s1}/A_{s2}$ or decreases significantly with the increase of $A_{s2}/A_{s1}$. When $A_{s2}/A_{s1}$ increases to a level of about 1.6, the positive redistribution at the center support disappears and the negative redistribution begins to appear. The negative redistribution, which indicates that the actual moment is greater than the elastic value, becomes more and more significant with the continuing increase of $A_{s2}/A_{s1}$.

From Fig. 5.3B and Table 5.1, it is observed that for a fixed value of $A_{s2}$, the effect of the $A_{s2}/A_{s1}$ ratio is slightly reflected in the ACI code but neglected in other codes where the parameter $c/d$ is used. On the other hand, for a fixed value of $A_{s1}$, the $\beta$ values predicted by various codes gradually decrease as $A_{s2}/A_{s1}$ increases, but the importance of the parameter $A_{s2}/A_{s1}$ is significantly underestimated. It should be noted that this observation is attributed to the change in the ductility of the center support section rather than the change in the stiffness difference between critical sections. For a minimum amount of non-prestressed steel over the center support ($A_{s2} = 360$ mm$^2$), the ACI code is generally conservative except when the non-prestressed steel over midspan is close to the minimum amount. The CSA code is nonconservative for $A_{s1}/A_{s2}$ less than about 2, while EC2 is nonconservative for $A_{s1}/A_{s2}$ less than about 3. On the other hand, for minimum non-prestressed steel over midspan ($A_{s1} = 360$ mm$^2$), all the codes are nonconservative, particularly at high values of $A_{s2}/A_{s1}$.

### 5.3.3  Effect of midspan and center support tendon eccentricities

To study the influence of tendon eccentricities on the degree of moment redistribution, four levels of the midspan eccentricity $e_1$ or center support eccentricity $e_2$ are selected: 0, 100, 200, and 300 mm. The variation of $\beta$ with the midspan or center support tendon eccentricity is shown in Fig. 5.4A. A comparison between the $\beta$ values predicted by FEA and various code equations is illustrated in Fig. 5.4B and Table 5.2.

It is observed from Fig. 5.4A that the $\beta$ value increases with the increase of $e_1$ but decreases with increasing $e_2$. The decreasing rate is much more significant than the increasing rate. The $\beta$ value increases by 18.5% as $e_1$ increases from 0 to 300 mm,

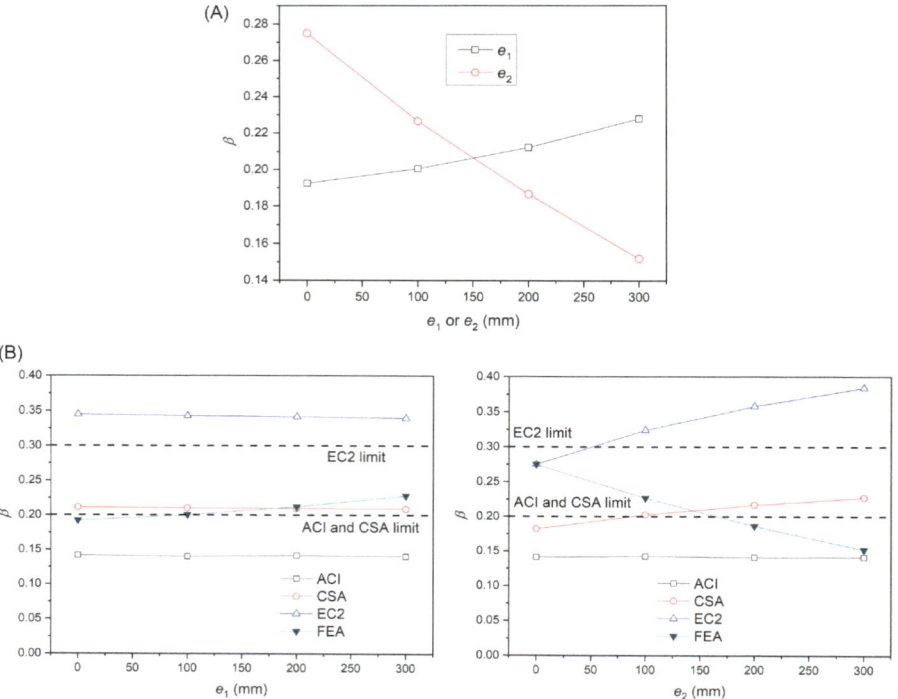

**Figure 5.4** Effect of midspan and center support tendon eccentricities on moment redistribution: (A) FEA results; (B) comparison between FEA and code predictions.

**Table 5.2** Moment redistribution in beams with different levels of $e_1$ and $e_2$.

| Beam | $e_1$ (mm) | $e_2$ (mm) | $\varepsilon_t$ (%) | $c/d$ (%) | $M$ (kN m) | $M_e$ (kN m) | $\beta$ (%) ACI | CSA | EC2 | FEA |
|------|-----------|-----------|--------|--------|---------|----------|------|------|------|------|
| B21 | 0 | 150 | 1.416 | 17.75 | −320.71 | −397.10 | 14.16 | 21.13 | 34.52 | 19.24 |
| B22 | 100 | | 1.401 | 17.91 | −328.35 | −410.72 | 14.01 | 21.05 | 34.33 | 20.06 |
| B23 | 200 | | 1.414 | 18.03 | −336.84 | −427.61 | 14.14 | 20.99 | 34.18 | 21.23 |
| B24 | 300 | | 1.403 | 18.18 | −344.55 | −446.29 | 14.03 | 20.91 | 33.99 | 22.80 |
| B25 | 150 | 0 | 1.413 | 23.53 | −249.52 | −344.16 | 14.13 | 18.23 | 27.52 | 27.50 |
| B26 | | 100 | 1.425 | 19.50 | −304.77 | −394.10 | 14.25 | 20.25 | 32.40 | 22.67 |
| B27 | | 200 | 1.409 | 16.66 | −361.26 | −444.11 | 14.09 | 21.67 | 35.84 | 18.66 |
| B28 | | 300 | 1.413 | 14.52 | −420.20 | −495.53 | 14.13 | 22.74 | 38.42 | 15.20 |

while decreases by 44.72% as $e_2$ increases from 0 to 300 mm. The important influence of the tendon eccentricity is partly attributed to the change in the stiffness difference between critical midspan and center support sections and partly attributed to the change in secondary moments, which is mainly controlled by the profile of the prestressing tendons.

From Table 5.2, it can be observed that the change in the values of $\varepsilon_t$ and $c/d$ with varying $e_1$ is negligible, indicating that the effect of the variable $e_1$ is not included in all code equations, as can be seen in Fig. 5.4B. On the other hand, as $e_2$ increases, the value of $\varepsilon_t$ remains almost unchanged while the value of $c/d$ quickly decreases. In fact, the variable $e_2$ ($e_1$ as well) does not affect the neutral axis depth $c$. The significant variation in the value of $c/d$ with $e_2$ is due to the change of the effective depth, $d$, of the center support section. From Fig. 5.4B and Table 5.2, it can also be observed that the effect of the variable $e_2$ is neglected in the ACI code, while it is incorrectly included in the CSA code and EC2 because the trend predicted by these code equations is opposite to the actual trend by FEA. In addition, the ACI code is conservative while EC2 is nonconservative. The CSA code may be nonconservative for low levels of $e_1$ or high levels of $e_2$.

### 5.3.4 Effect of tendon area and effective prestress

The tendon area $A_p$ and effective prestress $\sigma_{pe}$ are two variables that determine the effective prestressing force $N_{pe}$ ($=A_p\sigma_{pe}$), which is a fundamental parameter in the design of prestressing. To study the effect of $N_{pe}$ on the moment redistribution, either $A_p$ varies from 0 to 600 mm$^2$ ($\sigma_{pe} = 930$ MPa) or $\sigma_{pe}$ varies from 0 to 1240 MPa ($A_p = 450$ mm$^2$) to produce $N_{pe}$ from 0 to 558 kN.

Fig. 5.5A shows the variation of $\beta$ with the effective prestressing force. It is observed that the $\beta$ value quickly decreases as the effective prestressing force increases. The phenomenon is particularly obvious when the amount of external tendons varies. When $A_p = 0$ mm$^2$, namely, in the case of a reinforced concrete continuous member, the $\beta$ value is as high as 41.55%. The value is significantly reduced to 28.95% when the reinforced concrete member is slightly prestressed with external tendons of 150 mm$^2$. On the other hand, when $\sigma_{pe} = 0$ MPa ($A_p = 450$ mm$^2$), the $\beta$ value is 31.43%, which is much lower than that for $A_p = 0$ mm$^2$.

For different levels of $A_p$ and $\sigma_{pe}$, a comparison between the $\beta$ values by FEA and various code equations is illustrated in Fig. 5.5B and Table 5.3. It is observed in Table 5.3 that as $A_p$ or $\sigma_{pe}$ increases, the value of $\varepsilon_t$ gradually decreases while the value of $c/d$ increases gradually. As a consequence, all the codes take into account the effect of these variables, as shown in Fig. 5.5B. It can also be seen that the ACI code is conservative, while EC2 is nonconservative except at a very low level of $N_{pe}$. The CSA code may be nonconservative at high levels of $N_{pe}$.

### 5.3.5 Effect of span-to-height ratio and concrete strength

Fig. 5.6A shows the variation of $\beta$ with the span-to-height ratio $L/h$ (ratio of span to the overall height of a cross-section). The results are produced using concrete

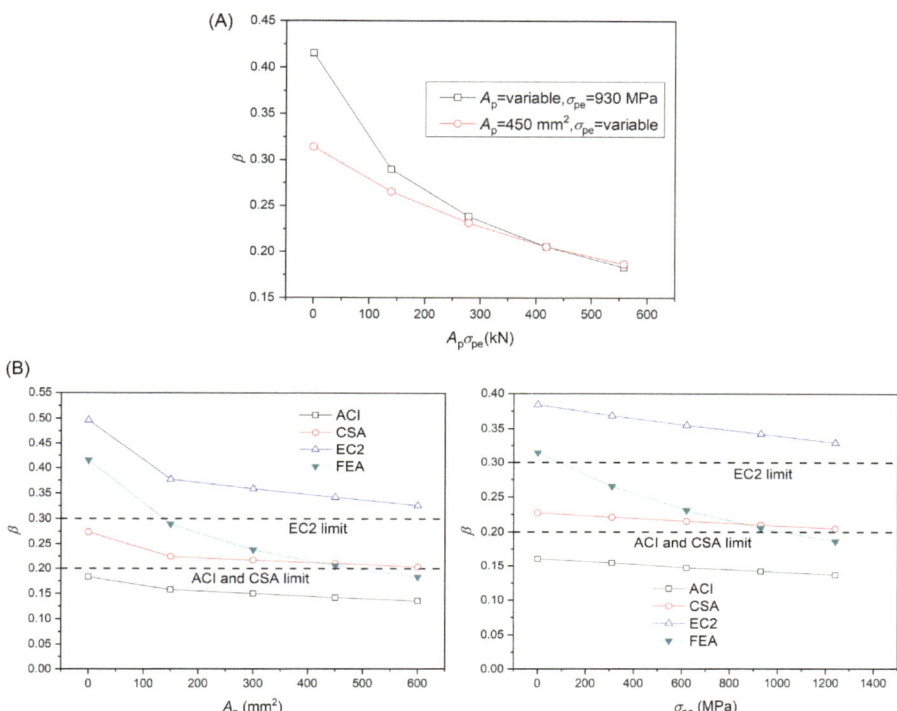

**Figure 5.5** Effect of tendon area and effective prestress on moment redistribution: (A) FEA results; (B) comparison between FEA and code predictions.

strengths $f_{ck}$ of 30 and 50 MPa. For a concrete strength of 50 MPa, the maximum redistribution of moments in the members appears at the ultimate limit state. For concrete strength of 30 MPa, on the other hand, the maximum redistribution of a very slender beam may not take place at ultimate due to softening load-deformation behavior during the loading process. For example, for $L/h$ of 33.33, the maximum redistribution of moments, occurred at the maximum load, is 17% higher than the redistribution at ultimate, as shown in Fig. 5.6A. Provided that there is no softening behavior, a higher span-to-height ratio produces an obviously higher redistribution at ultimate, while a lower concrete strength leads to a slightly higher redistribution. However, a long beam with a lower concrete strength may exhibit softening load-deformation behavior, hereby causing a lower redistribution at ultimate compared to that with a higher concrete strength, as shown in Fig. 5.6A.

For different levels of $L/h$ and $f_{ck}$, a comparison between the $\beta$ values by FEA and various code equations is illustrated in Fig. 5.6B and Table 5.4. It is observed from Table 5.4 that as $L/h$ increases, the value of $\varepsilon_t$ gradually increases while the decrease in $c/d$ is negligible, provided that there is no softening load-deformation behavior. Therefore the effect of $L/h$ is reflected in the ACI code but neglected in the CSA code and EC2, as shown in Fig. 5.6B. It is also observed that a higher

**Table 5.3** Moment redistribution in beams with different levels of $A_p$ and $\sigma_{pe}$.

| Beam | $A_p$ (mm²) | $\sigma_{pe}$ (MPa) | $\varepsilon_t$ (%) | $c/d$ (%) | $M$ (kN m) | $M_e$ (kN m) | $\beta$ (%) ACI | $\beta$ (%) CSA | $\beta$ (%) EC2 | $\beta$ (%) FEA |
|---|---|---|---|---|---|---|---|---|---|---|
| B30 | 0 | 930 | 1.831 | 5.23 | −112.18 | −191.93 | 18.31 | 27.39 | 49.67 | 41.55 |
| B31 | 150 | | 1.577 | 15.08 | −192.96 | −271.56 | 15.77 | 22.46 | 37.74 | 28.95 |
| B32 | 300 | | 1.500 | 16.60 | −263.79 | −346.26 | 15.00 | 21.70 | 35.91 | 23.82 |
| B33 | 450 | | 1.420 | 17.96 | −332.87 | −419.12 | 14.20 | 21.02 | 34.26 | 20.58 |
| B34 | 600 | | 1.354 | 19.32 | −400.78 | −490.69 | 13.54 | 20.34 | 32.61 | 18.32 |
| B35 | 450 | 0 | 1.605 | 14.53 | −171.98 | −250.82 | 16.05 | 22.73 | 38.41 | 31.43 |
| B36 | | 310 | 1.545 | 15.81 | −225.48 | −306.75 | 15.45 | 22.10 | 36.86 | 26.50 |
| B37 | | 620 | 1.471 | 16.91 | −278.69 | −362.42 | 14.71 | 21.55 | 35.54 | 23.10 |
| B38 | | 930 | 1.420 | 17.96 | −332.87 | −419.12 | 14.20 | 21.02 | 34.26 | 20.58 |
| B39 | | 1240 | 1.367 | 19.03 | −386.94 | −475.71 | 13.67 | 20.48 | 32.96 | 18.66 |

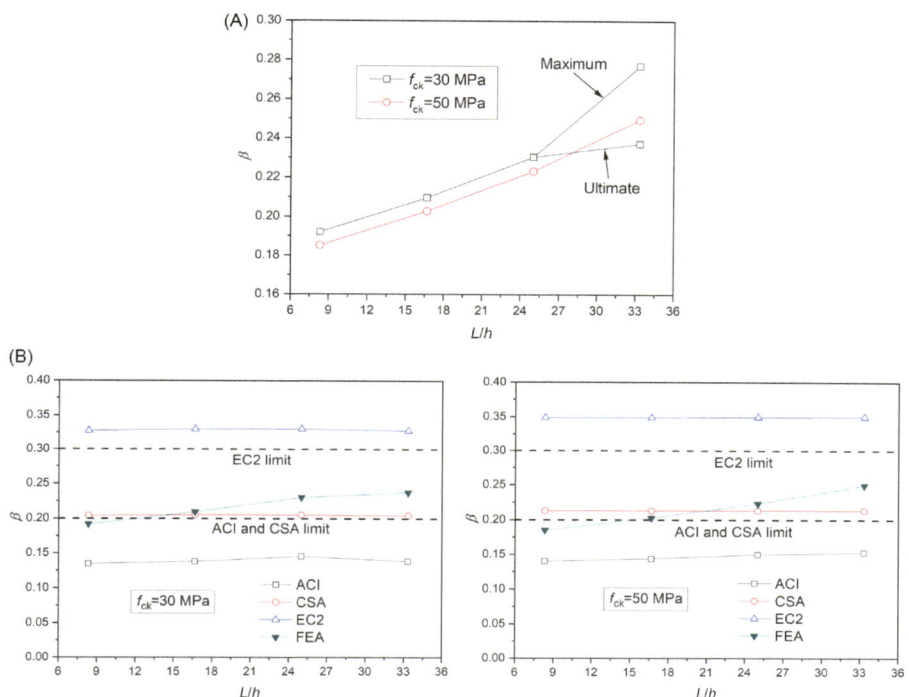

**Figure 5.6** Effect of span-to-height ratio and concrete strength on moment redistribution: (A) FEA results; (B) comparison between FEA and code predictions.

concrete strength leads to a higher value of $\varepsilon_t$ and a lower value of $c/d$, hereby causing higher redistribution according to the code equations. However, this is opposite to the fact that a higher concrete strength produces a lower redistribution as discussed previously. In EC2, the effect of concrete strength is considered using Eq. (5.6a) for normal-strength concrete and Eq. (5.6b) for high-strength concrete. It is also seen that the ACI code is conservative while EC2 is nonconservative. The CSA code may be nonconservative for a low level of $L/h$.

## 5.3.6  Effect of CFRP elastic modulus and load type

The CFRP composites cover a wide range of modulus of elasticity which may vary from 80 to 500 GPa (FIB, 2012). Four levels of the CFRP tendon elastic modulus $E_f$ are selected, namely, 80, 147, 270, and 500 GPa. The corresponding tensile strengths $f_f$ are 1440, 1840, 2160, and 2500 MPa, respectively. Fig. 5.7A shows the variation of $\beta$ with the CFRP modulus of elasticity for center-point loading and uniform loading. It is observed that the $\beta$ value decreases slightly as $E_f$ increases. In addition, uniform loading mobilizes an obviously higher redistribution compared to

**Table 5.4** Moment redistribution in beams with different levels of $L/h$ and $f_{ck}$.

| Beam | $L/h$ | $f_{ck}$ (MPa) | $\varepsilon_t$ (%) | $c/d$ (%) | $M$ (kN m) | $M_e$ (kN m) | $\beta$ (%) | | | |
|------|-------|----------------|---------------------|-----------|------------|--------------|------|------|------|------|
| | | | | | | | ACI | CSA | EC2 | FEA |
| B41 | 8.33 | 30 | 1.346 | 19.20 | −322.16 | −398.69 | 13.46 | 20.40 | 32.75 | 19.19 |
| B42 | 16.67 | | 1.380 | 19.01 | −324.58 | −410.58 | 13.80 | 20.49 | 32.99 | 20.95 |
| B43 | 25.00 | | 1.453 | 18.97 | −327.56 | −425.66 | 14.53 | 20.52 | 33.04 | 23.05 |
| B44 | 33.33 | | 1.383 | 19.16 | −325.00 | −426.11 | 13.83 | 20.42 | 32.80 | 23.73 |
| B45 | 8.33 | 50 | 1.399 | 17.45 | −337.12 | −413.73 | 13.99 | 21.27 | 34.87 | 18.52 |
| B46 | 16.67 | | 1.438 | 17.44 | −339.61 | −425.93 | 14.38 | 21.28 | 34.89 | 20.27 |
| B47 | 25.00 | | 1.500 | 17.39 | −342.26 | −440.60 | 15.00 | 21.30 | 34.95 | 22.32 |
| B48 | 33.33 | | 1.529 | 17.36 | −343.76 | −458.01 | 15.29 | 21.32 | 34.99 | 24.95 |

center-point loading. In this analysis, the $\beta$ value for uniform loading is about 1.4 times that for center-point loading.

For different levels of $E_f$ and different types of loading, a comparison between the $\beta$ values predicted by FEA and various code equations is illustrated in Fig. 5.7B and Table 5.5. It is observed in Table 5.5 that the variable $E_f$ affects the values of $\varepsilon_t$ and $c/d$. The load type also influences the value of $\varepsilon_t$ but has a null effect on the value of $c/d$. As a consequence, the effect of the variable $E_f$ is considered in all the code equations, while the effect of the load type is considered in the ACI code but neglected in the CSA code and EC2, as illustrated in Fig. 5.7B. In this analysis, the ACI code is conservative but may be overconservative when uniform loading is used. EC2 is nonconservative, particularly for center-point loading. The CSA code may be nonconservative in the case of low CFRP modulus of elasticity and center-point loading.

## 5.4    Proposed modification of ACI equation

Among various factors examined in the parametric study, the $A_{s2}/A_{s1}$ ratio is found to be a leading parameter affecting the moment redistribution. The results presented

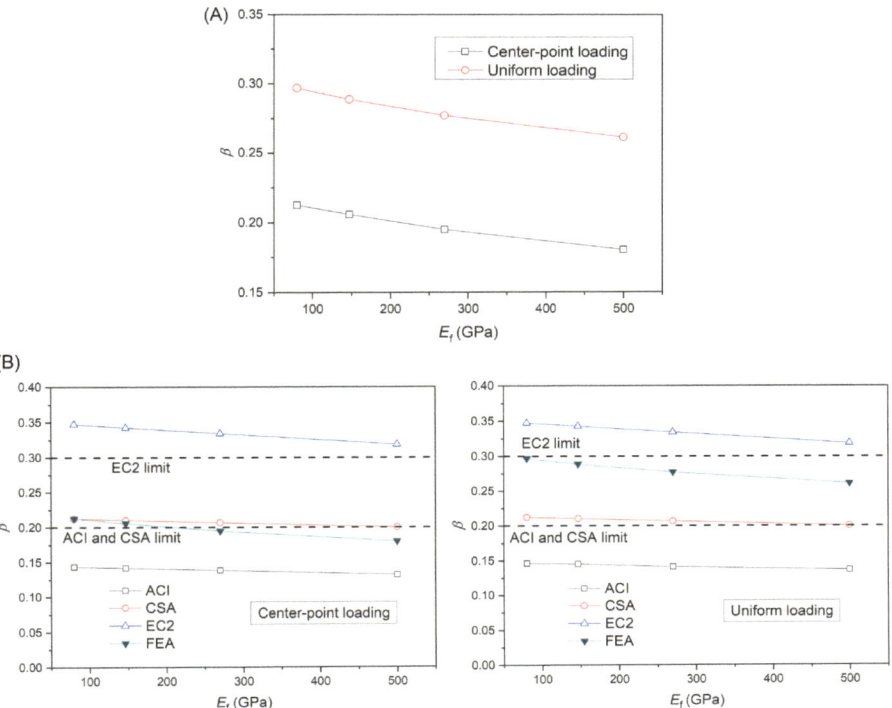

**Figure 5.7** Effect of CFRP elastic modulus and load type on moment redistribution: (A) FEA results; (B) comparison between FEA and code predictions.

**Table 5.5** Moment redistribution in beams with different levels of $E_f$ and types of loading.

| Beam | $E_f$ (GPa) | Load type | $\varepsilon_t$ (%) | c/d (%) | M (kN m) | $M_e$ (kN m) | β (%) | | | |
|------|------|------|------|------|------|------|------|------|------|------|
| | | | | | | | ACI | CSA | EC2 | FEA |
| B51 | 80 | CPL | 1.435 | 17.56 | −312.67 | −397.13 | 14.35 | 21.22 | 34.74 | 21.27 |
| B52 | 147 | | 1.420 | 17.96 | −332.87 | −419.12 | 14.20 | 21.02 | 34.26 | 20.58 |
| B53 | 270 | | 1.388 | 18.65 | −368.38 | −457.54 | 13.88 | 20.67 | 33.42 | 19.49 |
| B54 | 500 | | 1.323 | 19.93 | −428.19 | −522.31 | 13.23 | 20.03 | 31.87 | 18.02 |
| B55 | 80 | UL | 1.466 | 17.54 | −330.15 | −469.56 | 14.66 | 21.23 | 34.76 | 29.69 |
| B56 | 147 | | 1.454 | 17.94 | −352.27 | −495.23 | 14.54 | 21.03 | 34.29 | 28.87 |
| B57 | 270 | | 1.413 | 18.63 | −389.90 | −539.52 | 14.13 | 20.68 | 33.45 | 27.73 |
| B58 | 500 | | 1.370 | 19.93 | −455.16 | −616.21 | 13.70 | 20.04 | 31.88 | 26.14 |

*Note*: CPL, center-point loading; UL, uniform loading.

in Section 5.3.2 show that the degree of moment redistribution decreases remarkably from 41.97% to $-35.35\%$ when $A_{s2}/A_{s1}$ increases from 0.2 to 5 (see Table 5.1). This indicates that the moment redistribution depends on not only the ductility of one critical section as reflected in the code equations, but also on the structural characteristics of the whole member. In addition, the parameter $\varepsilon_t$ (adopted by the ACI code) seems to be better than the parameter $c/d$ (adopted by the CSA code and EC2) when used to calculate the degree of moment redistribution in continuous external tendon members, because $\varepsilon_t$ can reflect more important factors affecting the moment redistribution. Therefore a simplified equation including the two parameters, $A_{s2}/A_{s1}$ and $\varepsilon_t$, may be reasonable to calculate the degree of moment redistribution since this equation can take into account both the structural characteristics and the section ductility. Based on the earlier discussion, the ACI code equation indicated by Eq. (5.4) can be modified as follows:

$$\beta(\%) = \lambda_{aci}(1000\varepsilon_t) \tag{5.8}$$

in which $\lambda_{aci}$ is a coefficient related to the parameter $A_{s2}/A_{s1}$. To get the form of $\lambda_{aci}$, the relationship between $\beta/(1000\varepsilon_t)$ and $\ln(A_{s2}/A_{s1})$ for the beams analyzed in Section 5.3.2 is plotted in Fig. 5.8. According to the fit curves, $\lambda_{aci}$ is related to $A_{s2}/A_{s1}$ by

$$\lambda_{aci} = 0.65 - 1.2\ln\left(\frac{A_{s2}}{A_{s1}}\right) \text{ for } A_{s2}/A_{s1} \leq 1 \tag{5.9a}$$

$$\lambda_{aci} = 0.65 + 0.67\ln\left(\frac{A_{s2}}{A_{s1}}\right) - 2.76\ln^2\left(\frac{A_{s2}}{A_{s1}}\right) \text{ for } A_{s2}/A_{s1} > 1 \tag{5.9b}$$

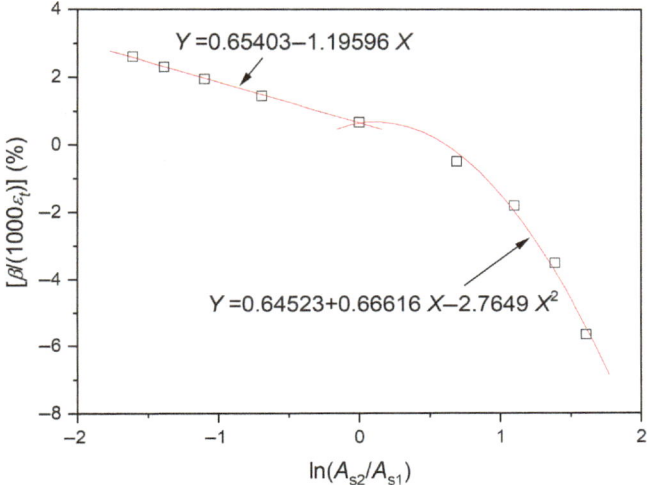

**Figure 5.8** Relationship between $\beta/(1000\varepsilon_t)$ and $\ln(A_{s2}/A_{s1})$.

Fig. 5.9 illustrates the correlation of simplified equations with the actual $\beta$ values. In addition to the specimens of the present numerical test, 16 two-span unbonded prestressed concrete beam specimens tested by Zhou and Zheng (2010) are also used for the correlation. The actual values of $\beta$ are obtained from FEA (for numerical test specimens) or experiment (for laboratory test specimens). It can be seen from Fig. 5.9A that the data that the ACI code equation is fitted to the actual values are rather scattered. By introducing the coefficient $\lambda_{aci}$, the modified equation proposed in this section correlates well with the actual values, as shown in Fig. 5.9B. In addition, most of the data shown in Fig. 5.9B are on the safe side, indicating that the proposed equation is generally conservative in predicting the degree of moment redistribution at ultimate in such members.

## 5.5 Effect of relative stiffness on global redistribution behavior

The structure and section of the reference beam are illustrated in Fig. 5.10. The prestressed concrete beam with external CFRP tendons has two identical spans with a total length of 20,000 mm. The reinforcement arrangement is as follows: $\rho_p = 0.34\%$, $\rho_{s1} = \rho_{s2} = 1.41\%$, $\rho_{s3} = 0.27\%$, where $\rho_p = A_p/(bd_p)$, $\rho_s = A_s/(bd_p)$. The subscripts

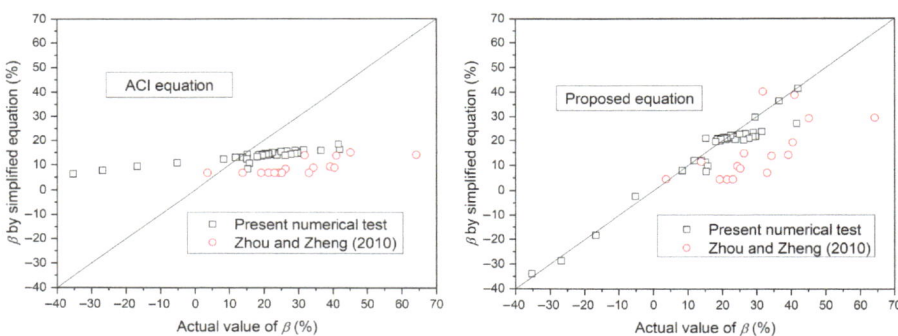

**Figure 5.9** Correlation of simplified equations with actual $\beta$ values: (A) ACI code equation; (B) proposed equation.

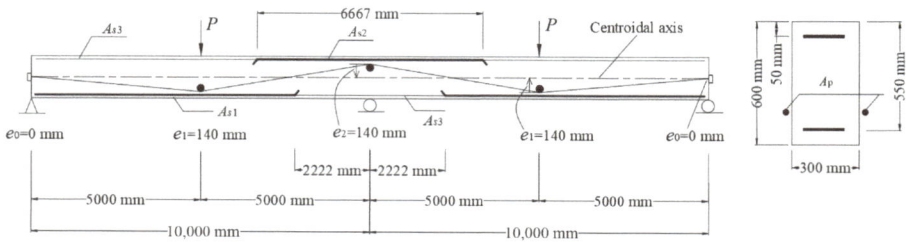

**Figure 5.10** Details of reference prestressed concrete beam with external CFRP tendons.

p, s1, s2 and s3 represent the tendon, non-prestressed tension steel at the midspan and inner support and the non-prestressed compression steel, respectively; $b$ denotes the cross-sectional width; $d_p$ denotes the tendon depth. The material properties of CFRP tendons and non-prestressed steel are the same as those of the reference beam used in Section 5.3. The concrete strength $f_{ck}$ is 60 MPa in this section.

The main variable is $\rho_{s1}$ or $\rho_{s2}$ to produce varying stiffness differences between the midspan and inner support. In this section, the results for varied $\rho_{s1}$ (0.27%− 4.03%) and fixed $\rho_{s2}$ (1.41%) are presented and discussed.

### 5.5.1    Failure and crack mode

Before failure that is caused by concrete crushing, the following phases are experienced sequentially in the members with $\rho_{s1}$ equal to or less than 1.03%: cracks occurring at the inner support and then at the midspan; yielding of non-prestressed tension steel at the midspan and then at the inner support. For $\rho_{s1}$ equal to or greater than 1.41%, the order of yielding at the critical sections is reversed, namely, yielding occurs first at the inner support, followed by yielding at the midspan.

Fig. 5.11 shows the strain distribution at ultimate at the bottom and top fibers of the members with different $\rho_{s1}$ values. The beam concrete is crushed when the specified ultimate compressive strain of 0.003 is reached at the midspan. Meanwhile, the compressive strain at the inner support may be far below (for $\rho_{s1} = 0.27\%$), below (for $\rho_{s1} = 1.41\%$) or very close to its ultimate capacity (for $\rho_{s1} = 4.03\%$). In addition, when $\rho_{s1} = 0.27\%$, there occurs a large tensile strain at the midspan against slight strains over other positive moment regions, indicating strain or crack concentration. As $\rho_{s1}$ increases, the crack width at the midspan is significantly reduced and the crack zone over the positive moment region is extended. On the other hand, increasing $\rho_{s1}$ leads to increased crack width at the inner support and reduced crack zone over the negative moment region.

### 5.5.2    Deformation behavior

Fig. 5.12 shows that the members with different $\rho_{s1}$ values exhibit the same load−deflection responses until cracking. Thereafter, a higher $\rho_{s1}$ leads to stiffer structural behavior, attributed to the higher cross-sectional stiffness over the positive moment region. The member with $\rho_{s1}$ of 0.27% exhibits the smallest deformation capacity. This can be explained by the fact the inner support section is well below its strength capacity when the concrete at the midspan is crushed. As $\rho_{s1}$ increases up to 1.41%, the exploitation of the inner support section is improved and, therefore, the deformation capacity is increased. The change in deformation capacity with varying $\rho_{s1}$ appears to be not apparent when $\rho_{s1}$ is greater than 1.41%. This is caused by the combined effect of the exploitation of the inner support section and the change of flexural ductility of the midspan section.

Fig. 5.13 shows the deflection development against the stress increase in external tendons for different $\rho_{s1}$ values. There is a nearly linear relationship between the tendon stress increase and the deflection. For a given tendon stress, a higher $\rho_{s1}$ leads

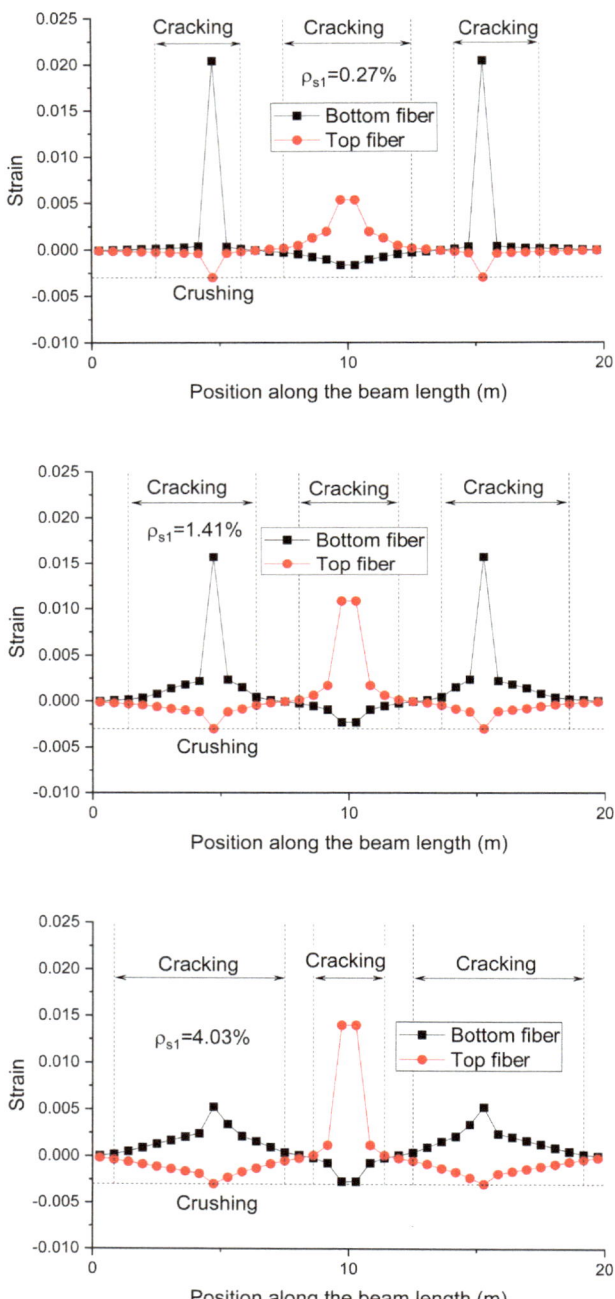

**Figure 5.11** Strain distribution at ultimate at the bottom and top fibers of the beams with different $\rho_{s1}$ values.

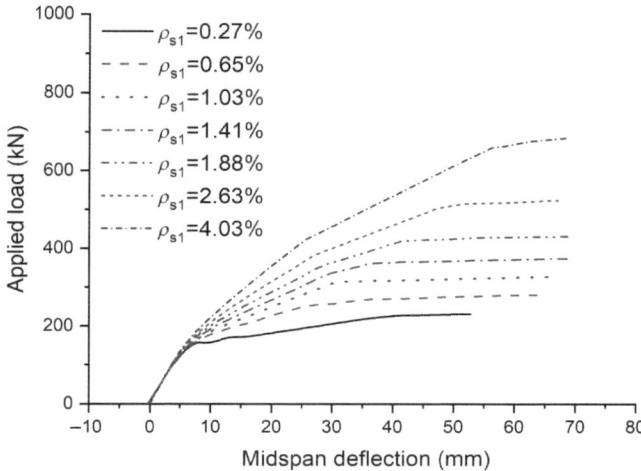

**Figure 5.12** Load−deflection curves for different $\rho_{s1}$ values.

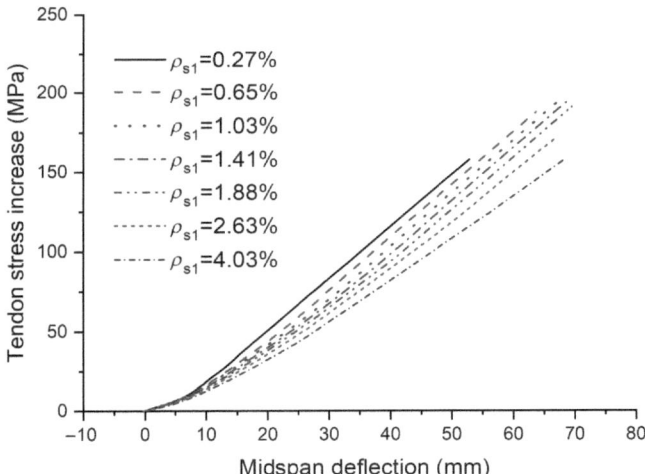

**Figure 5.13** Deflection development with the tendon stress increase for different $\rho_{s1}$ values.

to a higher deflection. Since the stress in external tendons is evenly distributed over the full length, external tendons are well below their tensile capacity at failure.

### 5.5.3 Neutral axis evolution with the moment

The neutral axis depth $c$ is recognized to have a critically important relationship with moment redistribution. This is reflected in various design codes such as CSA, BSI, and EC2. A good understanding of the neutral axis evolution during the loading process is therefore essential. For the members under self-weight and external prestressing, there is a slight hogging curvature at the midspan and a slight sagging curvature at the inner support. Accordingly, the initial value of $c$ is negative, namely, the neutral axis situated above (below) the top (bottom) of the midspan (inner support). The initial curvature gradually vanishes after loading. Then, the sagging (hogging) curvature at the midspan (inner support) appears and increases. Accordingly, the value of $c$ is quickly reduced to the negative infinity and then suddenly changed to the positive infinity and rapidly reduced thereafter.

The evolution of $c$ (for $c \leq 600$ mm) with the bending moment $M$ for different $\rho_{s1}$ values is presented in Fig. 5.14. The $M-c$ curve is characterized by four stages. The first stage is featured by a dramatic decrease in $c$ until the cracking moment is reached. The second stage is initiated by the appearance of flexural cracks and ended by the stabilization of the crack development. This stage shows a reduced decrease rate of $c$ with increasing moment. The decrease rate is significantly further reduced in the third stage until the yielding of non-prestressed tension steel. This is followed by the fourth stage, which shows a resume of a quick decrease of $c$ until failure. At the midspan, the $M-c$ curves for different $\rho_{s1}$ values differ after cracking, that is, a higher $\rho_{s1}$ value leads to a slower decrease in $c$. For the inner support, the $M-c$ curves for different $\rho_{s1}$ values are almost identical. In other words, the neutral axis evolution at the inner support is independent of the reinforcement content at the midspan. This observation confirms that the neutral axis depth is only a section-related parameter, rather than a structure-related parameter. Therefore it is not adequate to use this parameter to quantify the moment redistribution, which mainly results from structural redundancy.

### 5.5.4 Load−reaction relationship

Fig. 5.15 shows the load−reaction curves for different $\rho_{s1}$ values. Both the actual reactions generated by a nonlinear analysis and the elastic reactions generated by

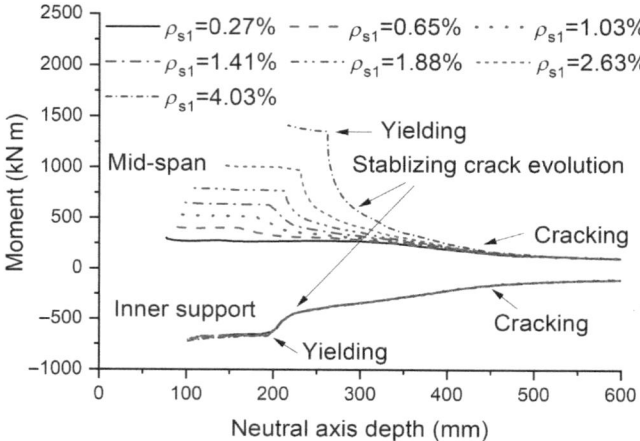

**Figure 5.14** Neutral axis evolution with the moment for different $\rho_{s1}$ values.

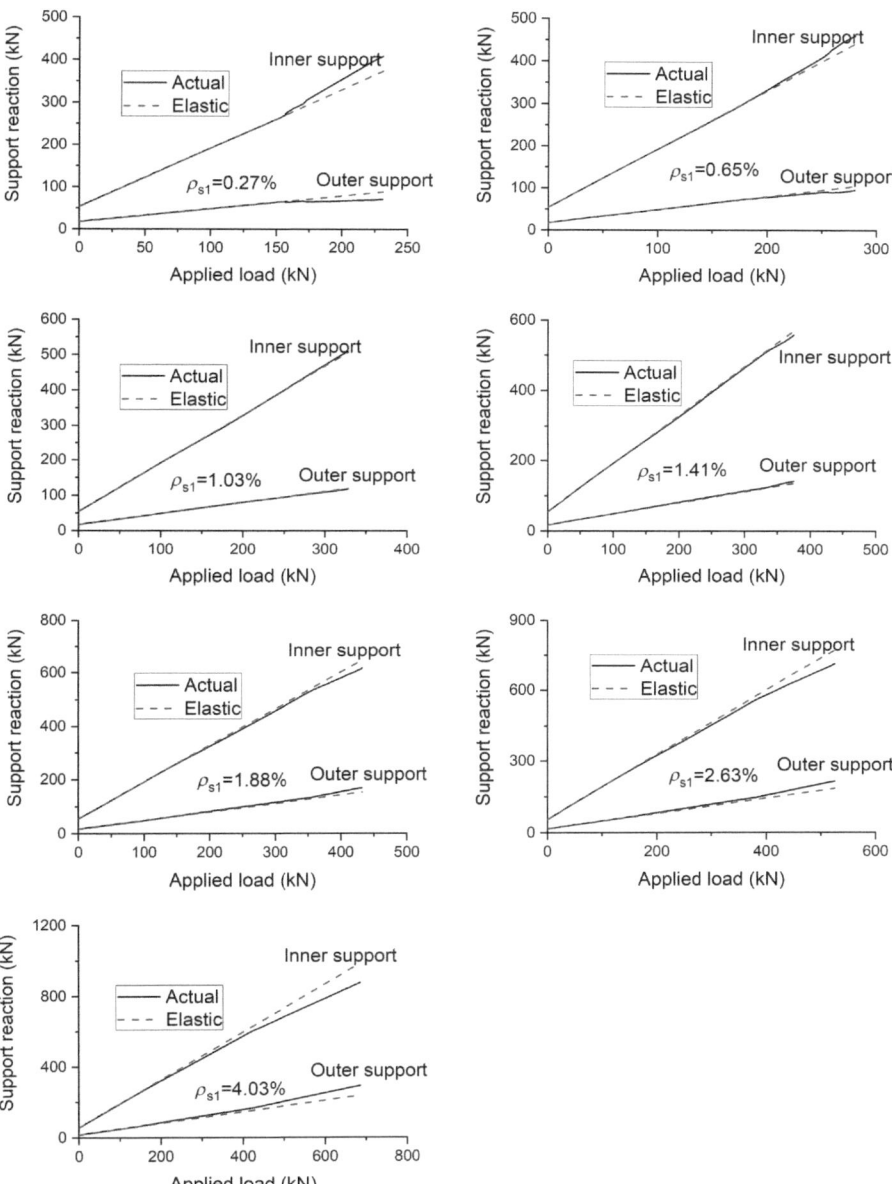

**Figure 5.15** Load—reaction curves for different $\rho_{s1}$ values.

an elastic analysis are demonstrated. The reaction comprises the load-induced reaction and prestressing-induced reaction (secondary reaction). The secondary reaction for the members is positive at the outer support and negative at the inner support. Therefore if the tendons are linearly transformed to be concordant to eliminate the secondary reaction, the outer support reaction would be smaller and the inner support reaction would be bigger.

According to the linear-elastic theory, the support reaction develops linearly throughout the loading process. Before cracking, the actual reaction development is identical to the elastic one. On cracking, the actual reaction begins to deviate from the elastic one due to the redistribution of moments. The deviation varies significantly according to the $\rho_{s1}$ value. For $\rho_{s1} = 0.27\%$, the cracking-induced reduction in flexural stiffness at the midspan is significantly more pronounced than that at the inner support, leading to a redistribution of moments from the midspan to the inner support. Consequently, the actual reaction at the outer support tends to be smaller than the elastic one while the actual reaction at the inner support tends to be larger than the elastic one. The deviation tends to be increasingly apparent as the load increases. As $\rho_{s1}$ gradually increases, the difference between the actual and elastic reactions diminishes, indicating reduced moment redistribution. When $\rho_{s1}$ is equal to 1.03%, the difference is negligible over the inelastic loading range, indicating that moment redistribution is negligible. When $\rho_{s1}$ increases to 1.41%, the actual reaction at the outer (inner) support after cracking turns out to be slightly larger (smaller) than the elastic one. This indicates that bending moments are redistributed slightly from inner support toward midspan. The deviation of the actual reaction from the elastic reaction is more and more apparent with the continuing increase of $\rho_{s1}$. When $\rho_{s1}$ is equal to 4.03%, the actual reaction at the outer (inner) support after cracking is significantly larger (smaller) than the elastic one, indicating a significant redistribution of moments from inner support toward midspan.

### 5.5.5 Evolution of bending moments and moment ratio with the load

Fig. 5.16 shows the evolution of bending moments and moment ratio for different $\rho_{s1}$ values. In Fig. 5.16, $M_1$ and $M_2$ refer to the actual moments, induced by the applied load, at the midspan and inner support, respectively; $(M_1)_{\text{ela}}$ and $(M_2)_{\text{ela}}$ refer to elastic moments, induced by the applied load, at the midspan and inner support, respectively. The elastic moment development shows a linear behavior and, therefore, the elastic moment ratio remains constant over the loading process. Because of the influence of reinforcement content, the value of $(M_1)_{\text{ela}}/(M_2)_{\text{ela}}$ varies from 0.8 for $\rho_{s1} = 0.27\%$ to 0.91 for $\rho_{s1} = 4.03\%$.

Similar to the behavior of reaction development, the actual moment development differs from the elastic one after the cracking load is reached. Correspondingly, the actual moment ratio is no longer a constant. The actual moment ratio is associated with moment redistribution: an increase in $M_1/M_2$ represents a redistribution of moments from inner support toward midspan; a decrease in $M_1/M_2$ indicates moment redistribution from midspan toward inner support; a stabilization of $M_1/M_2$ implies stabilizing moment redistribution; the larger the deviation between the actual moment ratio and elastic one, the higher the amount of moment redistribution. The evolution of the actual moment ratio strongly depends on $\rho_{s1}$, as can be seen in Fig. 5.16.

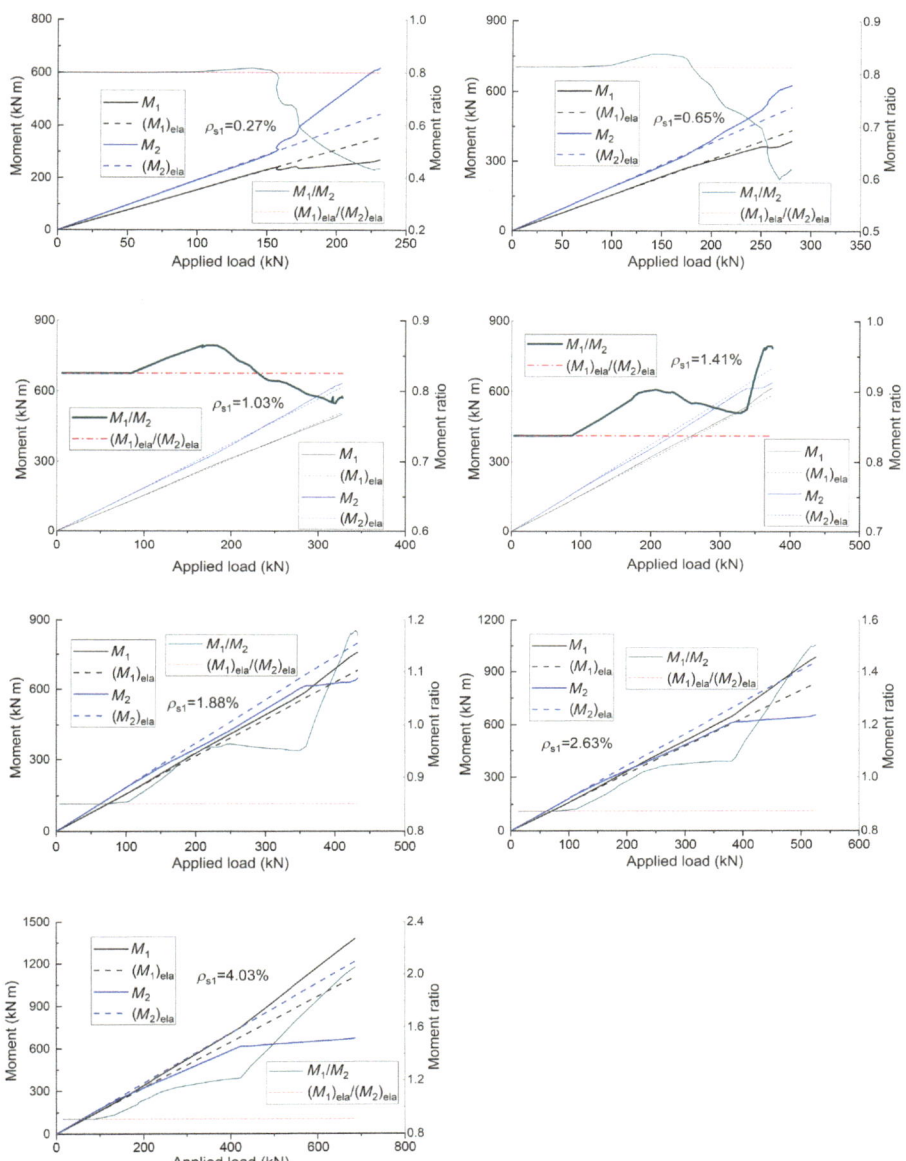

**Figure 5.16** Evolution of moments and moment ratio for different $\rho_{s1}$ values.

## 5.5.6  *Neutral axis evolution against moment redistribution*

Fig. 5.17 presents the evolution of $\beta$ against $c/d$ (for $c/d \leq 1$) for different $\rho_{s1}$ values. The effective depth $d$ is calculated by

$$d = \frac{A_p \sigma_{pe} d_p + A_s f_y d_s}{A_p \sigma_{pe} + A_s f_y} \tag{5.10}$$

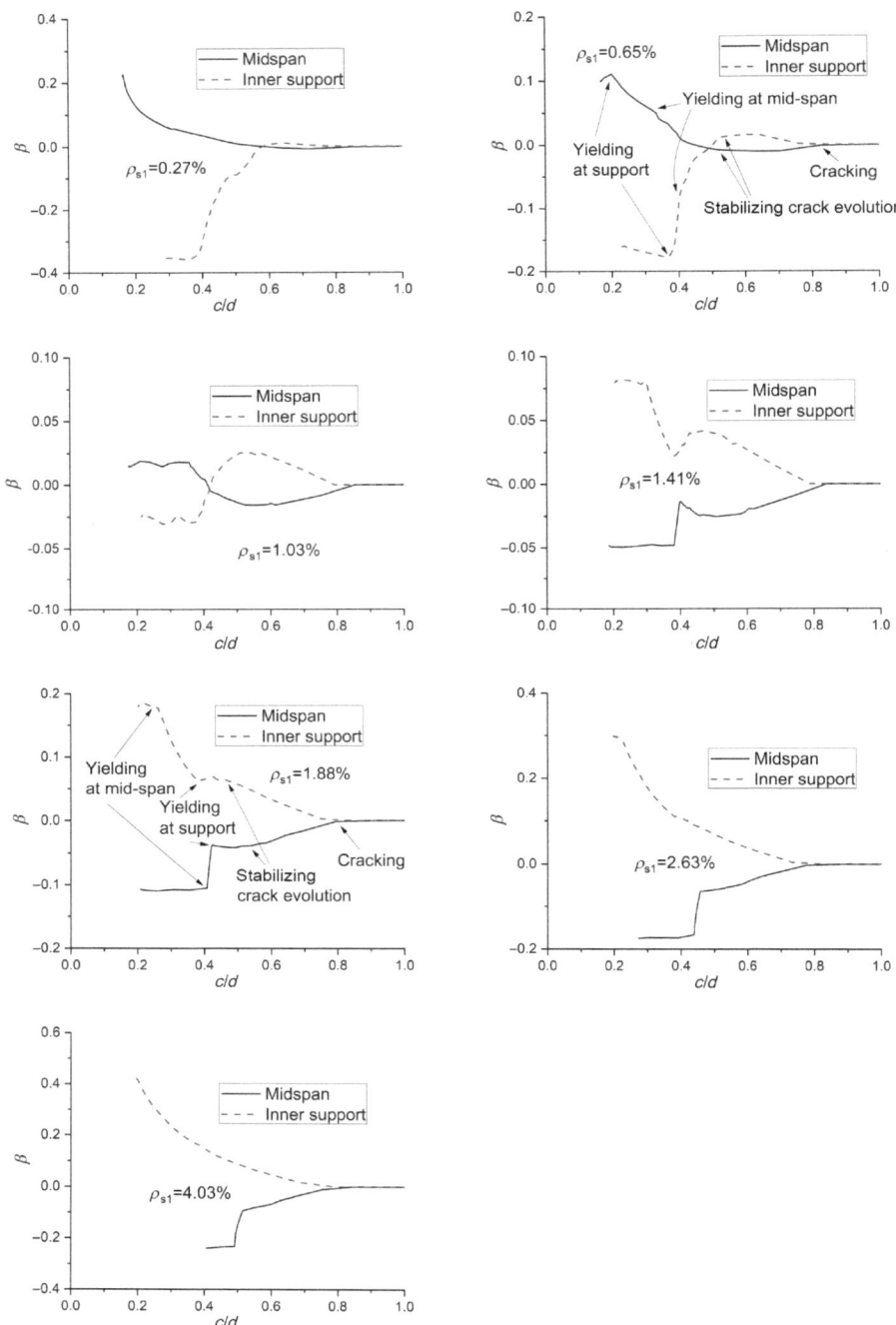

**Figure 5.17** Neutral axis evolution against moment redistribution for different $\rho_{s1}$ values.

where $d_p$ and $d_s$ are the depths of prestressed and non-prestressed reinforcement, respectively. The redistribution is zero with the rapid decrease in $c/d$ until cracking. Moment redistribution occurs afterward, and the behavior is affected typically by the stabilizing of crack evolution and yielding of non-prestressed tension steel at the critical sections. The members having $\rho_{s1} = 0.27\%$ and $0.65\%$ show similar behavior. Immediately after cracking, there occurs a slightly positive moment redistribution over inner support and a negative one over midspan. This is attributed to the occurrence of the first crack at the inner support, leading to moment redistribution from inner support toward midspan. Cracking also occurs at the midspan very soon. When the crack evolution at both the critical sections stabilizes, moments start to be redistributed from midspan (weaker section) toward inner support (stronger section). Consequently, the positive (negative) redistribution at the inner support (midspan) decreases (increases) quickly to a negative (positive) value, accompanied by a slower decrease in $c/d$. When first yielding occurs at the midspan, further moments are redistributed from midspan to inner support, resulting in a faster decrease (increase) of moment redistribution over inner support (midspan). Meanwhile, the variation in $c/d$ is limited. Such behavior continues until yielding at the inner support. Thereafter, the evolution of moment redistribution is insignificant with varying $c/d$. For $\rho_{s1} = 1.03\%$, yielding at critical sections occurs almost simultaneously. Therefore the decrease (increase) in positive (negative) redistribution over inner support (midspan) after stabilizing of crack evolution is not so important. For $\rho_{s1} = 1.41\%$ or above, moment redistribution tends to slightly decrease, stabilize, or slightly increase, after stabilizing of crack evolution, until first yielding at the inner support. This is followed by a quick increase (decrease) in positive (negative) moment redistribution over inner support (midspan) until the second yielding at the midspan, accompanied by limited variation in $c/d$. Afterward, the variation of $\beta$ is negligible while the variation of $c/d$ is significant at the midspan and not so significant at the inner support.

## 5.6 Proposed equations based on neutral axis depth

The parameter $c/d$ (at ultimate) is mostly adopted in current design codes around the world for calculating the moment redistribution in statically indeterminate structures. The results presented in the previous section show that the parameter $c/d$ is only section-related while moment redistribution is remarkably affected by varying $\rho_{s1}$ ($\rho_{s2}$ is fixed) and consequently by the stiffness difference between the critical sections. In this section, typical codes of practice are assessed and possible modifications of the code equations are suggested to take the stiffness difference into consideration. Unless otherwise stated, the numerical results presented herein are generated for $f_{ck} = 60$ MPa. The flexural stiffness of an externally prestressed concrete section could be described by the combined reinforcement index $\omega$, which consists of the prestressed reinforcement index $\omega_p$ and non-prestressed reinforcement index $\omega_s$.

$$\omega = \omega_p + \omega_s \tag{5.11}$$

$$\omega_p = \frac{\rho_p \sigma_{pe}}{f_{ck}}, \quad \omega_s = \frac{\rho_s f_y}{f_{ck}} \tag{5.12}$$

Three typical design codes are investigated herein, that is, CSA (CSA, 2004), BSI (BSI, 2007), and EC2 (CEN, 2004). A single parameter $c/d$ is used in CSA (Eq. (5.13)) and BSI (Eq. (5.14)) for redistribution quantification of continuous prestressed concrete members.

$$\beta = 0.3 - 0.5c/d \tag{5.13}$$

$$\beta = 0.5 - c/d \tag{5.14}$$

The maximum redistribution allowed by CSA and BSI is 20%. The equation suggested by EC2 accounts for the effect of concrete grade and is expressed by

$$\beta = \begin{cases} 0.56 - k_1 c/d \text{ for } f_{ck} \leq 50 \text{ MPa} \\ 0.46 - k_1 c/d \text{ for } f_{ck} > 50 \text{ MPa} \end{cases} \tag{5.15}$$

where

$$k_1 = 1.25(0.6 + 0.0014/\varepsilon_{cu2}) \tag{5.16}$$

Fig. 5.18 illustrates the $\beta$–$c/d$ relationship at ultimate obtained from FEA along with the code curves (BSI, EC2, and CSA). The FEA data are generated by either varying $\omega_1$ ($\omega_2$ is fixed) or $\omega_2$ ($\omega_1$ is fixed) from 0.073 to 0.355, where $\omega_1$ and $\omega_2$ are the combined reinforcement indexes at the midspan and inner support,

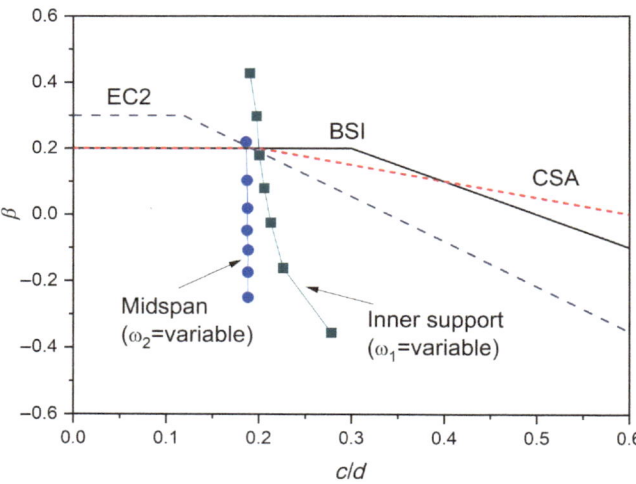

**Figure 5.18** The $\beta$–$c/d$ curves by FEA and different codes.

respectively. According to FEA, as $\omega_2$ varies from 0.073 to 0.355, the value of $c/d$ at the midspan is almost unchanged, accompanied by a significant change of $\beta$ at the midspan from $-25.0\%$ to $21.9\%$. In the case that $\omega_1$ is a variable, the variation of $c/d$ at the inner support is very slight when $\beta$ at the inner support varies remarkably between $-35.5\%$ and $42.8\%$. The aforementioned observations are inconsistent with the design codes, that is, there is a significant change in $c/d$ with the $\beta$ value. Therefore it can be concluded that the stiffness difference between the critical sections cannot be reflected in these codes.

To further confirm this statement, the $\beta-\ln(\omega_1/\omega_2)$ relationship by FEA is compared to the code predictions in Fig. 5.19. The results for the midspan section shown in Fig. 5.19A are produced by varying $\omega_2$ ($\omega_1$ is fixed), while those for the inner support section shown in Fig. 5.19B are produced by varying $\omega_1$ ($\omega_2$ is fixed). The code predictions demonstrate nearly stabilizing $\beta$ with varying $\ln(\omega_1/\omega_2)$. This, however, contradicts with the FEA results, which show a remarkable change in $\beta$ with increasing $\ln(\omega_1/\omega_2)$. Therefore the design codes neglect the structure-related parameter $\omega_1/\omega_2$ and, consequently, they cannot accurately predict the moment redistribution in continuous prestressed concrete members.

To introduce the parameter $\omega_1/\omega_2$ which describes the stiffness difference between the critical sections, CSA Eq. (5.13), BSI Eq. (5.14), and EC2 Eq. (5.15) are modified by introducing coefficients $\lambda_{csa}$, $\lambda_{bsi}$, and $\lambda_{ec2}$, respectively. The modified equations are expressed as follows:

$$\beta = \lambda_{csa}(0.3 - 0.5c/d) \tag{5.17}$$

$$\beta = \lambda_{bsi}(0.5 - c/d) \tag{5.18}$$

$$\beta = \begin{cases} \lambda_{ec2}(0.56 - k_1 c/d) \text{ for } f_{ck} \leq 50 \text{ MPa} \\ \lambda_{ec2}(0.46 - k_1 c/d) \text{ for } f_{ck} > 50 \text{ MPa} \end{cases} \tag{5.19}$$

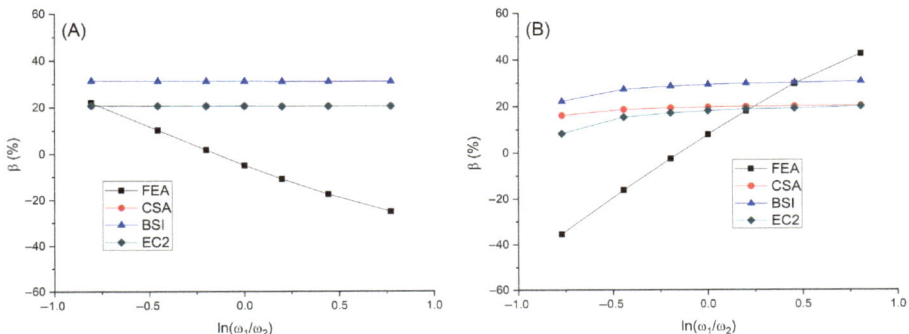

**Figure 5.19** The $\beta-\ln(\omega_1/\omega_2)$ relationships according to FEA and different codes: (A) midspan ($\omega_2 = $ variable); (B) inner support ($\omega_1 = $ variable).

where $\lambda_{csa}$, $\lambda_{bsi}$, and $\lambda_{ec2}$ are coefficients related to $\omega_1/\omega_2$. The code equations, in general, are applied to both sagging and hogging sections. However, the neutral axis evolution against moment redistribution for critical sagging and hogging sections of an externally prestressed concrete member is quite different, as can be seen in Fig. 5.17. It is therefore necessary to obtain the values of $\lambda_{csa}$, $\lambda_{bsi}$, and $\lambda_{ec2}$ separately for the midspan and inner support sections.

The variations of $\lambda_{csa}$ and $\lambda_{bsi}$ against $\ln(\omega_1/\omega_2)$ for the midspan section of the members with varied $\omega_2$ and fixed $\omega_1$ are shown in Fig. 5.20A and B, respectively, while the variations of $\lambda_{ec2}$ for $f_{ck} = 40$ (representing the case of $f_{ck} \leq 50$ MPa) and 60 MPa (representing the case of $f_{ck} > 50$ MPa) are shown in Fig. 5.20C and D, respectively. According to the fit curves, $\lambda_{csa}$, $\lambda_{bsi}$, and $\lambda_{ec2}$ for the midspan section are expressed as follows:

$$\lambda_{csa} = -0.18 - 1.46\ln(\omega_1/\omega_2) \tag{5.20}$$

$$\lambda_{bsi} = -0.12 - 0.96\ln(\omega_1/\omega_2) \tag{5.21}$$

$$\lambda_{ec2} = \begin{cases} -0.13 - 0.98\ln(\omega_1/\omega_2) \text{ for } f_{ck} \leq 50 \text{ MPa} \\ -0.18 - 1.46\ln(\omega_1/\omega_2) \text{ for } f_{ck} > 50 \text{ MPa} \end{cases} \tag{5.22}$$

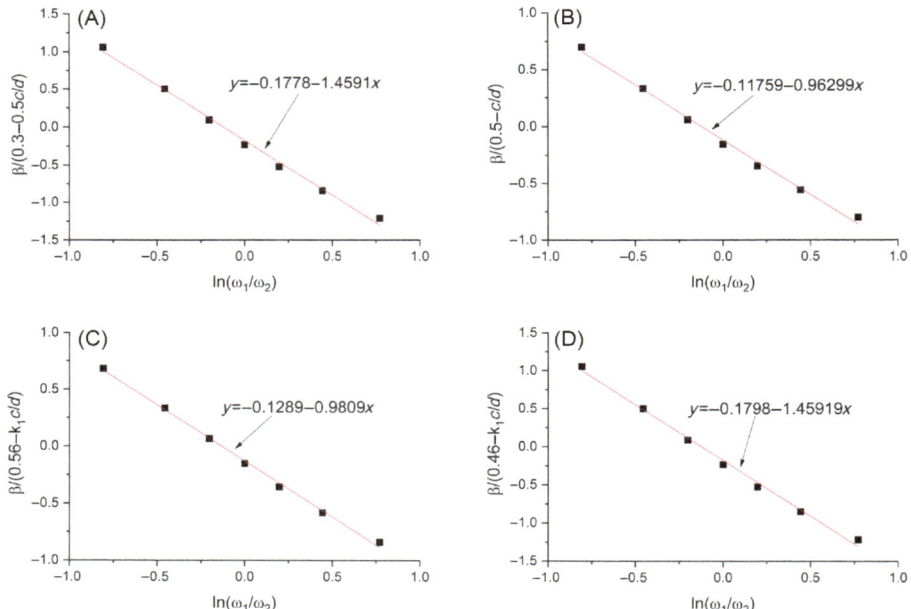

**Figure 5.20** Fit curves for the midspan section: (A) $\lambda_{csa}$; (B) $\lambda_{bsi}$; (C) $\lambda_{ec2}$ for $f_{ck} = 40$ MPa; (D) $\lambda_{ec2}$ for $f_{ck} = 60$ MPa.

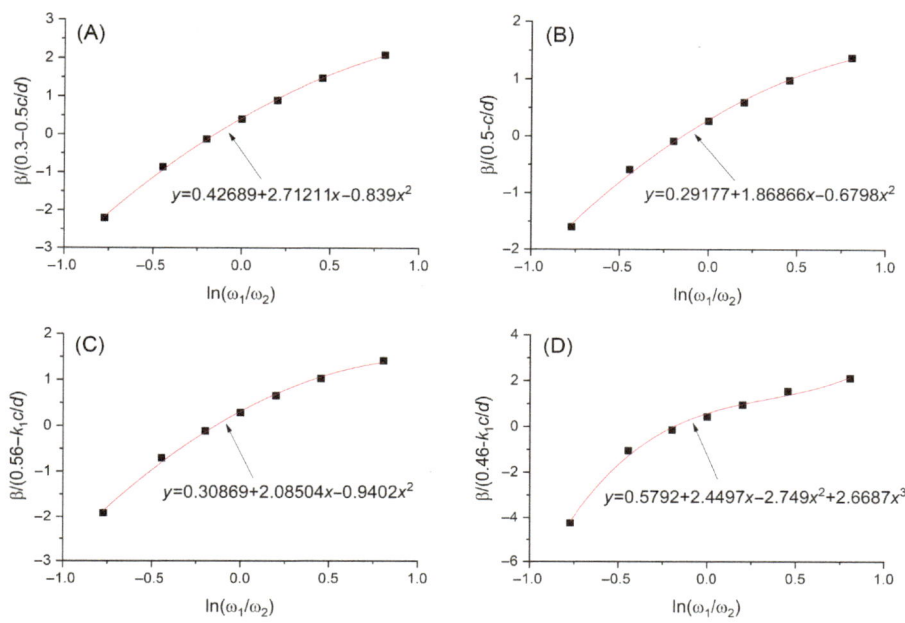

**Figure 5.21** Fit curves for the inner support section: (A) $\lambda_{csa}$; (B) $\lambda_{bsi}$; (C) $\lambda_{ec2}$ for $f_{ck} = 40$ MPa; (D) $\lambda_{ec2}$ for $f_{ck} = 60$ MPa.

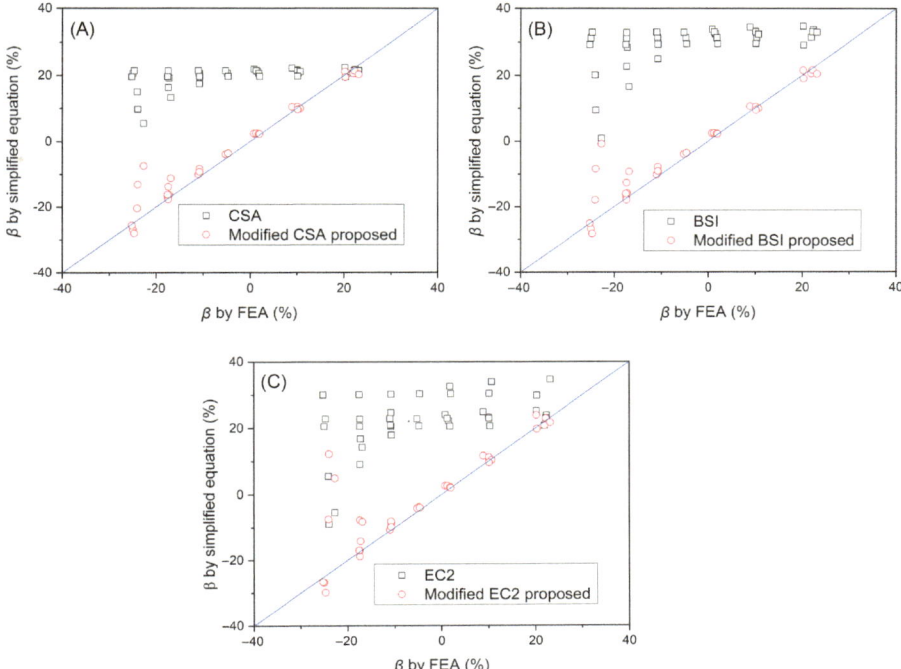

**Figure 5.22** Correlation of simplified equations for the midspan section with $\beta$ by FEA: (A) CSA and modified CSA equations; (B) BSI and modified BSI equations; (C) EC2 and modified EC2 equations.

The $\lambda_{csa}-\ln(\omega_1/\omega_2)$ and $\lambda_{bsi}-\ln(\omega_1/\omega_2)$ relationships for the inner support section of the members with varied $\omega_1$ and fixed $\omega_2$ are shown in Fig. 5.21A and B, respectively, while the $\lambda_{ec2}-\ln(\omega_1/\omega_2)$ relationships for $f_{ck}=40$ and 60 MPa are shown in Fig. 5.21C and D, respectively. According to the fit curves, $\lambda_{csa}$, $\lambda_{bsi}$, and $\lambda_{ec2}$ for the inner support section are expressed as follows:

$$\lambda_{csa} = 0.43 + 2.71\ln(\omega_1/\omega_2) - 0.84\ln^2(\omega_1/\omega_2) \tag{5.23}$$

$$\lambda_{bsi} = 0.29 + 1.87\ln(\omega_1/\omega_2) - 0.68\ln^2(\omega_1/\omega_2) \tag{5.24}$$

$$\lambda_{ec2} = \begin{cases} 0.31 + 2.09\ln(\omega_1/\omega_2) - 0.94\ln^2(\omega_1/\omega_2) \text{ for } f_{ck} \leq 50 \text{ MPa} \\ 0.58 + 2.45\ln(\omega_1/\omega_2) - 2.75\ln^2(\omega_1/\omega_2) + 2.67\ln^3(\omega_1/\omega_2) \text{ for } f_{ck} > 50 \text{ MPa} \end{cases} \tag{5.25}$$

Correlations of simplified equations with the $\beta$ values obtained by FEA for the midspan and inner support are demonstrated in Figs. 5.22 and 5.23, respectively.

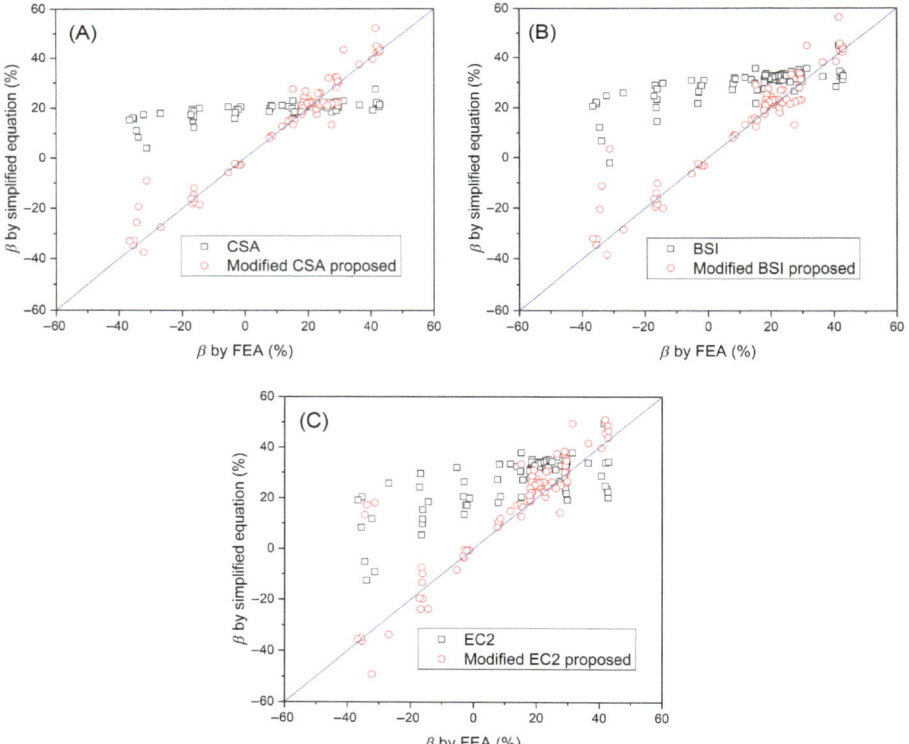

**Figure 5.23** Correlation of simplified equations for the inner support section with $\beta$ by FEA: (A) CSA and modified CSA equations; (B) BSI and modified BSI equations; (C) EC2 and modified EC2 equations.

In Figs. 5.22, a total of 42 external CFRP tendon specimens with varying either $\omega_1$ or $\omega_2$ from 0.073 to 0.355 and $f_{ck}$ from 40 to 90 MPa are used for the correlations. In Fig. 5.23, in addition to the aforementioned 42 specimens, another 49 external CFRP tendon specimens having various variables presented in Section 5.3 are also used for the correlations. It is seen in Figs. 5.22 and 5.23 that the design codes (CSA, BSI, and EC2) exhibit poor fit to the $\beta$ values by FEA. By including the parameter $\omega_1/\omega_2$, the proposed equations (i.e., modified CSA, BSI, and EC2 equations) correlate very well with the FEA predictions for both the midspan and inner support sections.

## 5.7 Conclusions

Based on an extensive parametric analysis conducted on two-span continuous concrete members prestressed with external CFRP tendons, it is concluded that the $A_{s2}/A_{s1}$ ratio is one of the most important factors affecting moment redistribution. The variation of moment redistribution with the amount of non-prestressed steel depends on the combined effects of ductility and stiffness difference between critical sections. The redistribution of moments is significantly reduced when a reinforced concrete member is strengthened by external prestressing. The moment redistribution decreases quickly as the effective prestressing force increases. The eccentricities of external tendons have an important influence on the moment redistribution. A higher span-to-height ratio generally leads to obviously higher moment redistribution. Uniform loading produces much higher moment redistribution than center-point loading. The moment redistribution slightly decreases with the increase of the CFRP tendon modulus of elasticity.

It is shown that varying the stiffness difference leads to a remarkable change in behavior related to moment redistribution, for example, load−reaction relationship, evolution of moments and moment ratio, and development of moment redistribution. On the other hand, varying the stiffness at the midspan does not influence the neutral axis evolution at the inner support, and vice versa. This confirms that the neutral axis depth is only a section-related parameter. Therefore the use of this parameter for the quantification of moment redistribution, which is the case in various codes of practice, is not adequate.

The results show that the current design codes fail to consider the impact of stiffness difference and consequently cannot predict accurately the amount of moment redistribution. Reasonable modifications of code equations are proposed to quantify the moment redistribution in continuous prestressed concrete members with external CFRP tendons. By introducing a parameter reflecting the impact of stiffness difference, the proposed equations show a much better fit to the redistribution values by FEA than that provided by equations in current design codes. It should be noted that for conventional continuous reinforced and prestressed concrete members, the stiffness difference is also a critical factor influencing moment redistribution. Since both moment redistribution and neutral axis evolution highly depend on the structural typology, the modified code equations proposed in this chapter, however, may not be applied to conventional reinforced and prestressed concrete members and require further validation.

# References

ACI Committee 318. (2019). *Building code requirements for structural concrete (ACI 318-19) and commentary (ACI 318R-19)*. Farmington Hills, MI.

ACI committee 440. (2004). *Prestressing concrete structures with FRP tendons*. ACI 440.4R-04, Farmington Hills, MI.

Bennitz, A., Schmidt, J. W., Nilimaa, J., Täljsten, B., Goltermann, P., & Ravn, D. L. (2012). Reinforced concrete T-beams externally prestressed with unbonded carbon fiber-reinforced polymer tendons. *ACI Structural Journal, 109*(4), 521−530. Available from http://www.concrete.org/tempComDocs/-41246/109-s45.pdf.

BSI. (2007). *Structural use of concrete, Part 1: Code of practice for design and construction*. London, UK: BS8110.

Campbell, T. I., & Moucessian, A. (1988). Prediction of the load capacity of two-span continuous prestressed concrete beams. *PCI Journal, 33*(2), 130−151. Available from https://doi.org/10.15554/pcij.03011988.130.151.

CEN. (2004). *Eurocode 2 (EC2): Design of concrete structures. Part 1-1: General rules and rules for buildings*. EN 1992-1-1, Brussels, Belgium.

Cohn, M.Z. (1986). Continuity in prestressed concrete continuity in prestressed concrete partial prestressing. *Partial prestressing, from theory to practice* (Vol. I, pp. 189−256). Survey reports, NATO ASI series. Martinus Nijhoff Publishers, Boston, MA. doi: 10.1007/978-94-009-4438-1_8.

CSA. (2004). Design of concrete structures. *A23.3-04*, Mississauga, Ontario, Canada.

FIB. (2012). *Model Code 2010 (MC10). Bulletins 55 and 56*. Lausanne, Switzerland.

Grace, N. F., & Abdel-Sayed, G. (1998). Behavior of externally draped CFRP tendons in prestressed concrete bridges. *PCI Journal, 43*(5), 88−101. Available from https://doi.org/10.15554/pcij.09011998.88.101.

Kara, I. F., & Ashour, A. F. (2013). Moment redistribution in continuous FRP reinforced concrete beams. *Construction and Building Materials, 49*, 939−948. Available from https://doi.org/10.1016/j.conbuildmat.2013.03.094.

Lou, T., Lopes, S. M. R., & Lopes, A. V. (2014). Factors affecting moment redistribution at ultimate in continuous beams prestressed with external CFRP tendons. *Composites Part B: Engineering, 66*, 136−146. Available from https://doi.org/10.1016/j.compositesb.2014.05.007.

Lou, T., Peng, C., Karavasilis, T. L., Min, D., & Sun, W. (2020). Moment redistribution versus neutral axis depth in continuous PSC beams with external CFRP tendons. *Engineering Structures, 209*, 109927. Available from https://doi.org/10.1016/j.engstruct.2019.109927.

Oehlers, D. J., Liu, I. S. T., Ju, G., & Seracino, R. (2004). Moment redistribution in continuous plated RC flexural members. Part 2: Flexural rigidity approach. *Engineering Structures, 26*(14), 2209−2218. Available from https://doi.org/10.1016/j.engstruct.2004.08.004.

Santos, P., Laranja, G., França, P. M., & Correia, J. R. (2013). Ductility and moment redistribution capacity of multi-span T-section concrete beams reinforced with GFRP bars.

*Construction and Building Materials*, *49*, 949−961. Available from https://doi.org/ 10.1016/j.conbuildmat.2013.01.014.

Tajaddini, A., Ibell, T., Darby, A., Evernden, M., & Silva, P. (2017). Prediction of capacity for moment redistribution in FRP-strengthened continuous RC T-beams. *Journal of Composites for Construction*, *21*(1), 0000719. Available from https://doi.org/10.1061/ (asce)cc.1943-5614.0000719.

Tan, K. H., & Tjandra, R. A. (2007). Strengthening of RC continuous beams by external prestressing. *Journal of Structural Engineering*, *133*(2), 195−204. Available from https:// doi.org/10.1061/(ASCE)0733-9445(2007)133:2(195).

Tichy, M., & Rakosnik, J. (1977). *Plastic analysis of concrete frames (with particular reference to limit states design)*. London, England: Collet (Publishers) Ltd.

Zhou, W., & Zheng, W. Z. (2010). Experimental research on plastic design method and moment redistribution in continuous concrete beams prestressed with unbonded tendons. *Magazine of Concrete Research*, *62*(1), 51−64. Available from https://doi.org/10.1680/ macr.2008.62.1.51.

# Linear transformation and secondary moments

6

## 6.1 Introduction

The tendon line in a continuous prestressed member can be concordant or nonconcordant. Any nonconcordant tendon line can be linearly transformed into a concordant line. Linear transformation is an important feature of continuous prestressed members. Lin and Burns (1981) pointed out that the ultimate load of a continuous prestressed member would not be influenced by a linear transformation, but no theoretical or experimental proofs were provided. Based on experimental observations from three symmetrically and two unsymmetrically loaded externally prestressed members with linearly transformed cable profiles, Aravinthan et al. (2005) concluded that the flexural behavior of continuous members, both at the elastic and ultimate limit states, is independent of the layout of the cables. A similar observation was reported in a numerical study by Lou et al. (2013), which showed that linear transformation does not change the flexural characteristics not only in the linear-elastic stage but also in the inelastic range of loading up to failure.

In a prestressed continuous member with nonconcordant cables, the prestressing induces secondary reactions and moments. However, there has been a great controversy on the prestress secondary moments (reactions) in the postelastic range, and no agreement has yet been reached. A typical viewpoint is that the secondary moments disappear after the formation of plastic hinges because the continuous prestressed member has become statically determinate. This viewpoint was included in an early version of the ACI code (ACI Committee 318, 1971). On the other hand, in the current version of the ACI code (ACI Committee 318, 2019), the secondary moments were taken into account in the calculation of the design moments. Some investigators (Aravinthan et al., 2005) believed that the secondary moments do not change much after the occurrence of cracks. Wyche, Uren, and Reynolds (1992) pointed out that the secondary moments must be considered and that the neglect of secondary moments can be unsafe. In fact, the secondary moments can be beneficial or detrimental, depending on the layout of the cables (Lou et al., 2013). When a cable is below its linearly transformed concordant line, the secondary moment is beneficial to the support sections but detrimental to the span critical section. The phenomenon is the opposite if a cable is above its linearly transformed concordant line.

This chapter introduces linear transformation and secondary moments (reactions) in continuous prestressed concrete members with external CFRP tendons (Lou et al., 2014, 2017). The effect of linear transformation on the general flexural behavior of continuous externally prestressed concrete members is described.

Prestressed Members with External Fiber-Reinforced Polymer (FRP) Tendons. DOI: https://doi.org/10.1016/B978-0-443-23877-2.00006-7

A rational method based on the linear transformation for computing the secondary reactions or moments is proposed. Examples illustrating the development of secondary moments and reactions in continuous prestressed concrete members with external CFRP tendons are given.

## 6.2  Linear transformation

When a tendon line is moved over the interior support(s) without changing its intrinsic shape, as illustrated in Fig. 6.1, this tendon line is termed to be linearly transformed (Lin & Burns, 1981). It was stated that the linear transformation of the tendon line does not change the ultimate load-carrying capacity of continuous members (Lin & Burns, 1981). This statement has been verified by both laboratory and numerical tests.

### 6.2.1  Laboratory test specimens

Aravinthan et al. (2005) tested a series of prestressed concrete members with highly eccentric external cables. Three of the specimens were A-1, B-1, and C-1 with section and structure details shown in Fig. 6.2. The major variable was the layout of the external cable, that is, three different linearly transformed cable profiles were used. The specimens were 10.4 m long, continuous over two equal spans of 5.0 m each. The rectangular cross-section was 400 mm wide and 150 mm deep. Each span was subjected to two concentrated loads with a spacing of 1.25 m. The specimens were pretensioned with four internal bonded steel cables with an area of 51.61 mm$^2$ each and also posttensioned with one external steel cable with an area of 69.68 mm$^2$. The external cable went through the concrete member at certain locations through embedded steel pipes. Vertical struts with various lengths were placed along the member to achieve the desired cable profiles. The ultimate strength and elastic modulus of the cables were 1722 MPa and 196.2 GPa, respectively. The effective prestresses for internal and external cables were 56% and 21% of their ultimate strengths, respectively. The average concrete strength was 54.4 MPa.

Fig. 6.3 shows the experimental results with respect to the load—deflection response and the stress increase in external cable with the deflection for the three specimens, along with the corresponding numerical predictions. According to the experimental results, the responses for the members having different linearly transformed cables are very similar, indicating that the structural behavior is not affected

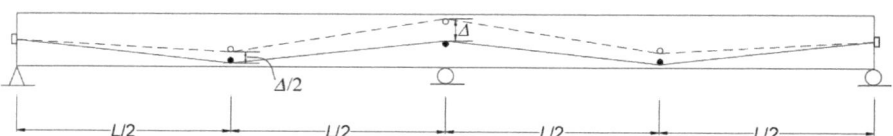

**Figure 6.1** Linear transformation of external cables.

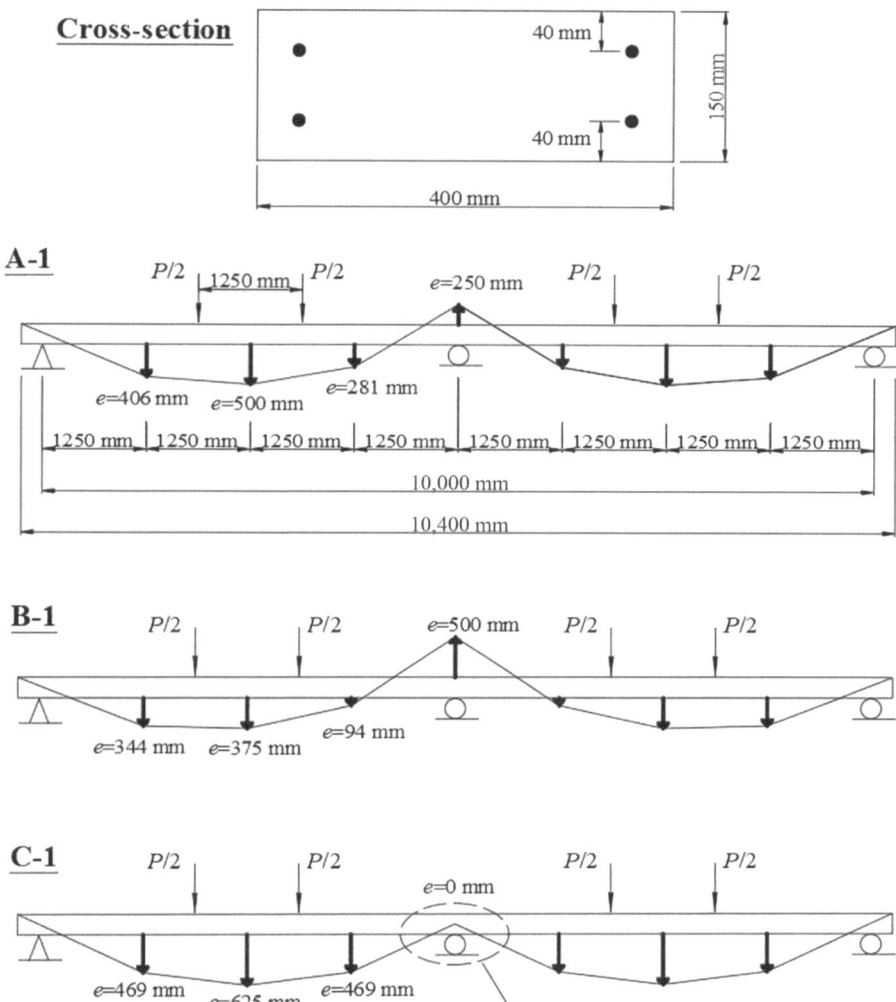

**Figure 6.2** Details of specimens externally prestressed with linearly transformed cables.

by linear transformation. Results of the numerical modeling lead to the same observation.

The above-mentioned statement demonstrates that the cable profile has no influence on the flexural behavior of prestressed continuous members. Linear transformation unaffecting the elastic behavior of prestressed continuous members can be well understood. However, the viewpoint that the postelastic behavior of continuous

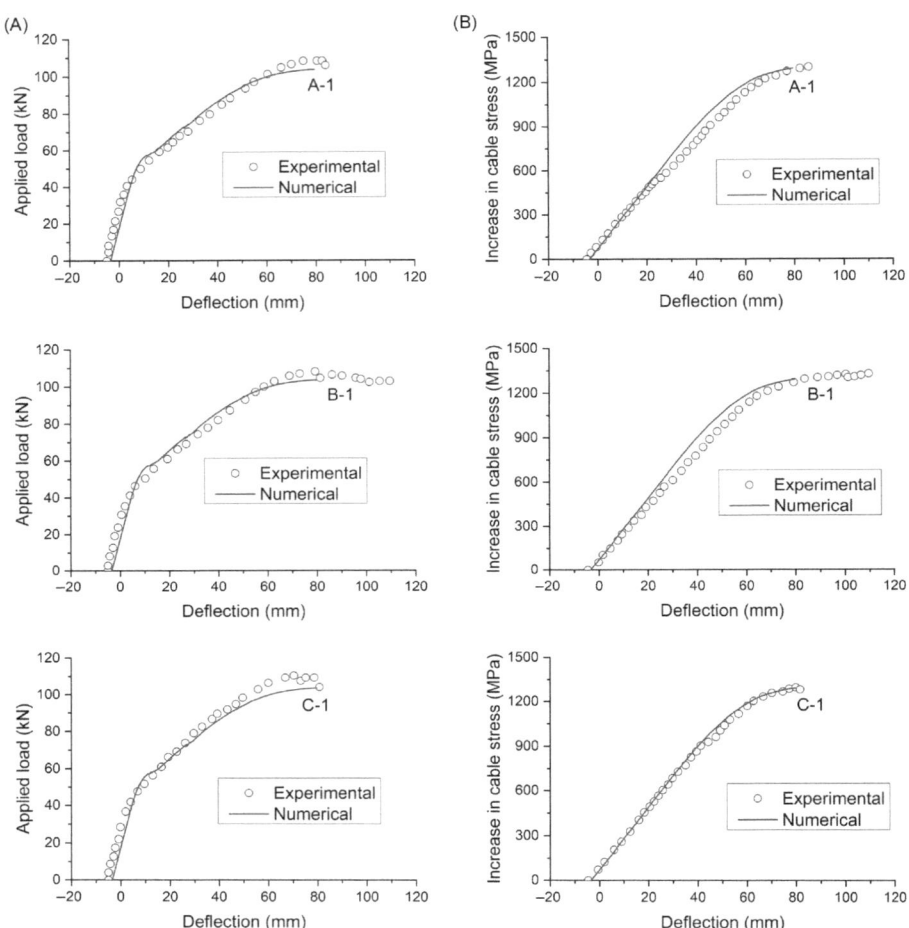

**Figure 6.3** Comparison of numerical predictions and experimental data: (A) load—deflection response; (B) stress increase in external cables with the deflection.

prestressed members is not affected by linear transformation may not be easily acceptable to researchers and engineers. As there is only a little proof focused on steel cables, it is worth providing further results regarding the linear transformation of external FRP cables.

## 6.2.2 Numerical test specimens

A numerical test program is designed here to identify the overall behavior of continuous externally prestressed concrete members with various linearly transformed cable profiles. The structure, section, and loading details of the reference beam are shown in Fig. 6.4A. At the reference cable profile, the tendon eccentricities at the

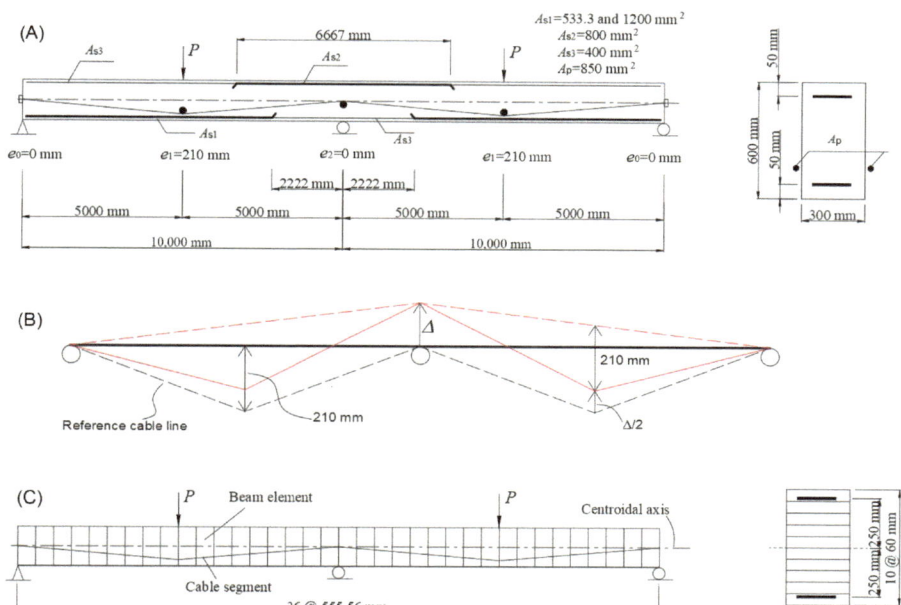

**Figure 6.4** Reference beam and its finite element model: (A) beam details; (B) linear transformation of the cable line; (C) finite element model.

end support $e_0$, midspan $e_1$, and center support $e_2$ are 0, 210, and 0 mm, respectively. Various linearly transformed cable profiles can be produced by increasing the tendon eccentricities by $\Delta$ at the center support and correspondingly by $\Delta/2$ at midspan, as shown in Fig. 6.4B.

The external cables are assumed to be CFRP composites having an ultimate strength $f_f$ of 1840 MPa and an elastic modulus $E_f$ of 147 GPa. The initial prestress $\sigma_{p0}$ before prestress transfer is taken as 1050 MPa. The cable area $A_p$ is taken equal to 850 mm$^2$. The area of nonprestressed compressive steel $A_{s3}$ is taken as 400 mm$^2$; the area of nonprestressed tensile steel over the negative moment region $A_{s2}$ is fixed at 800 mm$^2$; two different nonprestressed tensile steel areas over the positive moment region $A_{s1}$ (i.e., 1200 and 533.3 mm$^2$) are used to generate two levels of the $A_{s2}/A_{s1}$ ratio (i.e., 0.67 and 1.5). The yield strength $f_y$ and elastic modulus $E_s$ of nonprestressed steel are 450 MPa and 200 GPa, respectively. The concrete cylinder's compressive strength $f_{ck}$ is 60 MPa. The finite element model of the member with the reference cable profile is illustrated in Fig. 6.4C.

To identify the effect of linear transformation on the general flexural response of external FRP tendon members, three linearly transformed cable profiles are considered here, namely, a concordant profile and two typical nonconcordant profiles ($\Delta = 0$, 300 mm) as shown in Fig. 6.5. The concordant cable profile is determined by means of the trial-and-error approach as follows: analyze the member assuming

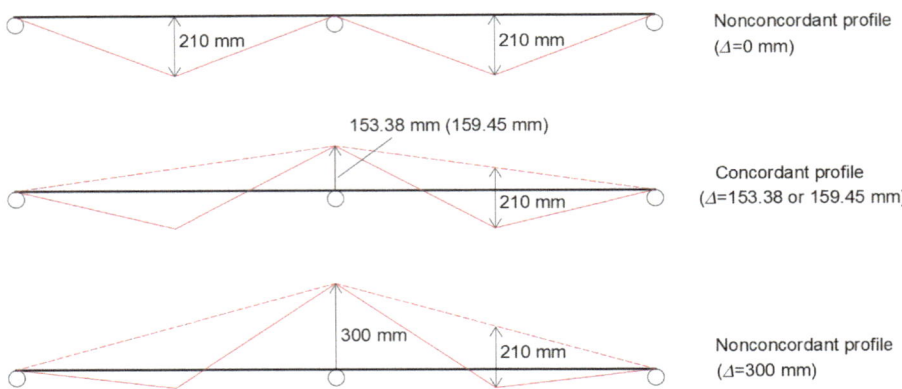

**Figure 6.5** Three typical linearly transformed cable profiles.

zero self-weight and external loads; check the support reactions; adjust the cable profile by increasing (or decreasing) the value of $\Delta$ if the end reaction is positive (or negative); repeat the above process until the support reactions disappear. It is demonstrated that the content of nonprestressed reinforcement has a slight influence on the concordant profile. The concordant profiles for $A_{s2}/A_{s1}$ of 0.67 and 1.5 correspond to $\Delta$ of 153.38 and 159.45 mm, respectively.

Fig. 6.6 illustrates the effect of linear transformation on the typical response characteristics (i.e., load versus stress increase in external FRP cables, deflection, curvature, and neutral axis depth). It can be seen from the graphs that the cable shift by linear transformation does not change the response characteristics throughout the loading process up to the ultimate limit state. Since the members with different linearly transformed cable profiles have the same curvature and neutral axis depth, the strains and stresses in the concrete and nonprestressed steel in these members are also identical at all stages of loading up to failure according to the plane section hypothesis. Therefore linear transformation offers the possibility of various adjustments to the cable layout without changing the general flexural behavior. Undoubtedly, this is very useful when designing such types of members.

It is worth mentioning that the above finding about linear transformation is applicable to more general cases (e.g., unsymmetrical cable layout and loading). To prove this statement, a two-span reference beam with unsymmetrical loading and cable layout is used herein (see Fig. 6.7). One span is subjected to a single concentrated load at midspan while the other span is nonloaded. The cable eccentricities at midspan are 240 mm for the loaded span and 180 mm for the nonloaded span. All the other geometric and material parameters are the same as those of the member of Fig. 6.4 ($A_{s2}/A_{s1} = 0.67$). Two linearly transformed cable profiles are considered, namely, $\Delta = 0$ and 300 mm. It can be seen in Fig. 6.7 that, in the case of the unsymmetrical layout of external cables and pattern of loading, linear

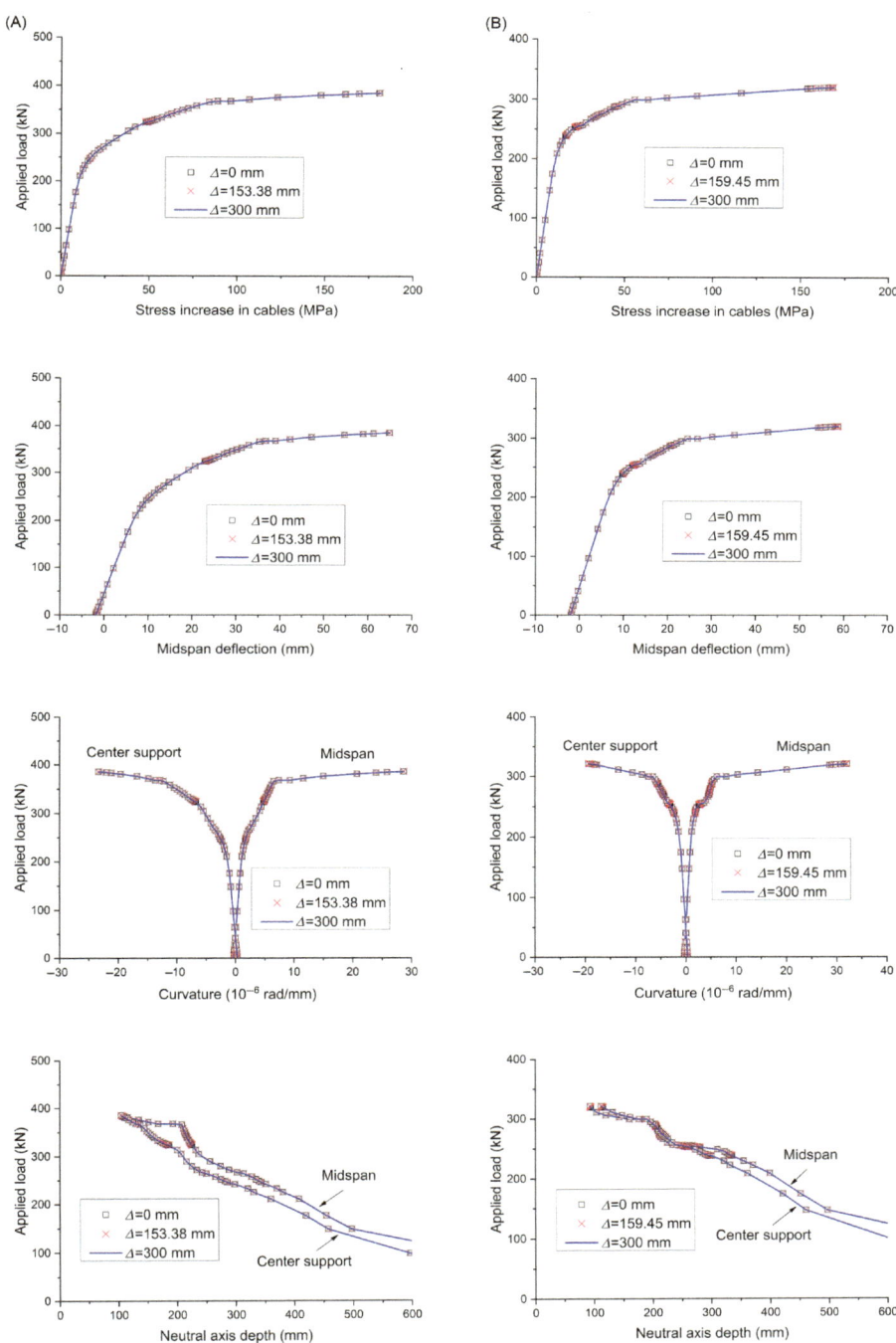

**Figure 6.6** Effect of linear transformation on typical response characteristics: (A) $A_{s2}/A_{s1} = 0.67$; (B) $A_{s2}/A_{s1} = 1.5$.

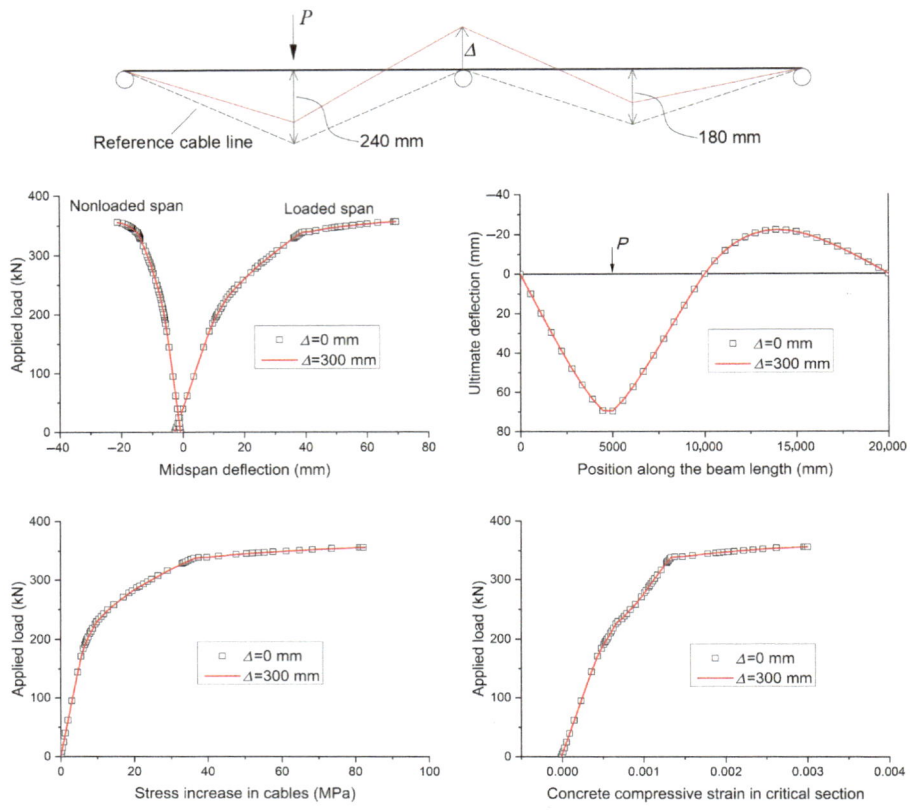

**Figure 6.7** Effect of linear transformation on response at unsymmetrical cable layout and loading.

transformation of the cable line has no influence on the member response over the elastic, inelastic, and ultimate limit states.

## 6.3 Method for computing secondary reactions (moments)

On the other hand, the linear transformation would cause changes of secondary moments, indicating that a nonconcordant tendon profile can be linearly transformed into a concordant profile, which does not produce secondary moments, without changing the basic flexural characteristics of continuous members. This provides an approach to examine the actual secondary moments over the whole loading process as discussed below.

The method for computing the prestress secondary reactions is illustrated in Fig. 6.8. Consider a multispan continuous prestressed concrete member with non-concordant cables. The total reaction at support $i$, $R^i$, at any level of loads $P$ consists of two components, namely, the reaction due to external loads $R^i_{load}$ and the secondary reaction due to prestressing $R^i_{sec}$.

$$R^i = R^i_{load} + R^i_{sec} \tag{6.1}$$

The cables can be linearly transformed into a concordant profile, which produces no secondary reactions. Since linear transformation does not influence the flexural characteristics throughout the loading process, the total reaction at support $i$, $(R^i)_{con}$, for the member with concordant cables, at the load level $P$, is equal to the load induced reaction $R^i_{load}$ for the member with nonconcordant cables.

$$(R^i)_{con} = R^i_{load} \tag{6.2}$$

Combining Eqs. (6.1) and (6.2), the secondary reaction at support $i$ for the member with nonconcordant cables can be expressed as follows:

$$R^i_{sec} = R^i - (R^i)_{con} \tag{6.3}$$

The support reactions $R^i$ and $(R^i)_{con}$ can be computed by a nonlinear computer analysis, and then the secondary reaction $R^i_{sec}$ is determined according to Eq. (6.3).

The secondary reactions of a prestressed concrete continuous member should satisfy the following equation:

$$\sum_i R^i_{sec} = 0 \tag{6.4}$$

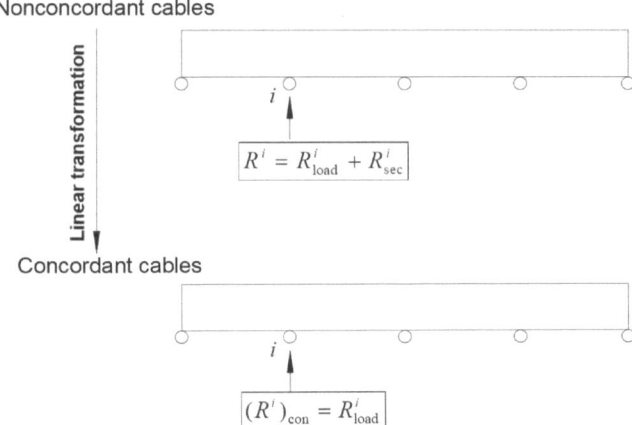

**Figure 6.8** Computation of secondary reactions based on linear transformation.

where the summation is made for all the supports. Also, irrespective of the pattern of loading, the secondary reactions at any symmetrical pair of supports $i$ and $j$, $R^i_{sec}$ and $R^j_{sec}$, should be equal.

$$R^i_{sec} = R^j_{sec} \tag{6.5}$$

Eqs. (6.4) and (6.5) can be used to check the accuracy and correctness of the proposed method for calculating the secondary reactions.

The proposed method is practically important because it provides a rational approach to compute accurately the secondary reactions and moments in continuous prestressed members over the complete loading range up to failure. To demonstrate the accuracy and applicability of the method and to understand the behavior of prestress secondary moments/reactions, this newly developed method is applied in the following sections to examine the secondary moments and reactions in prestressed concrete members with external CFRP tendons.

## 6.4  Example 1—Members with various cable profiles and different nonprestressed steel contents

The results are produced by using the reference beam presented in Fig. 6.4.

At $A_{s2}/A_{s1}$ of 0.67, the development of bending moments and secondary moments for various $\Delta$ values is shown in Fig. 6.9, while the results at $A_{s2}/A_{s1}$ of 1.5 are illustrated in Fig. 6.10. In these figures, $M$ and $M_{con}$ represent the actual moments for a cable profile and its linearly transformed concordant profile, respectively; $M_{sec}$ represents the secondary moment. The elastic moment is also plotted in the graphs (in a straight line). The actual moment is obtained from finite element analysis (FEA) considering both geometric and material nonlinearities. The elastic moment is computed by FEA assuming all the constituent materials are linear elastic but the geometric nonlinearity is considered. The secondary moment is calculated using the newly developed method mentioned above. It is seen in the figures that the actual moments develop linearly with increasing applied load up to the cracking load. After cracking, the moment development exhibits nonlinear behavior as a result of moment redistribution. The nonlinear manner for $A_{s2}/A_{s1}$ of 0.67 is fairly severe while it is not so serious for $A_{s2}/A_{s1}$ of 1.5, indicating that moment redistribution for the former is much more significant than that for the latter. It is also seen that the increase in secondary moments is insignificant in the initial loading stage, but becomes much faster after concrete cracking and nonprestressed steel yielding, attributed to the quicker increase in prestressing force in external cables. At any stage of loading, the secondary moments at the center support are twice the corresponding ones at midspan, as can be seen in Figs. 6.9 and 6.10. This agrees with the common-sense knowledge about secondary moments, thereby validating the method for computing secondary moments.

At $A_{s2}/A_{s1}$ of 0.67, the moment and secondary moment distribution at ultimate along the span for various $\Delta$ values is displayed in Fig. 6.11, while the results at

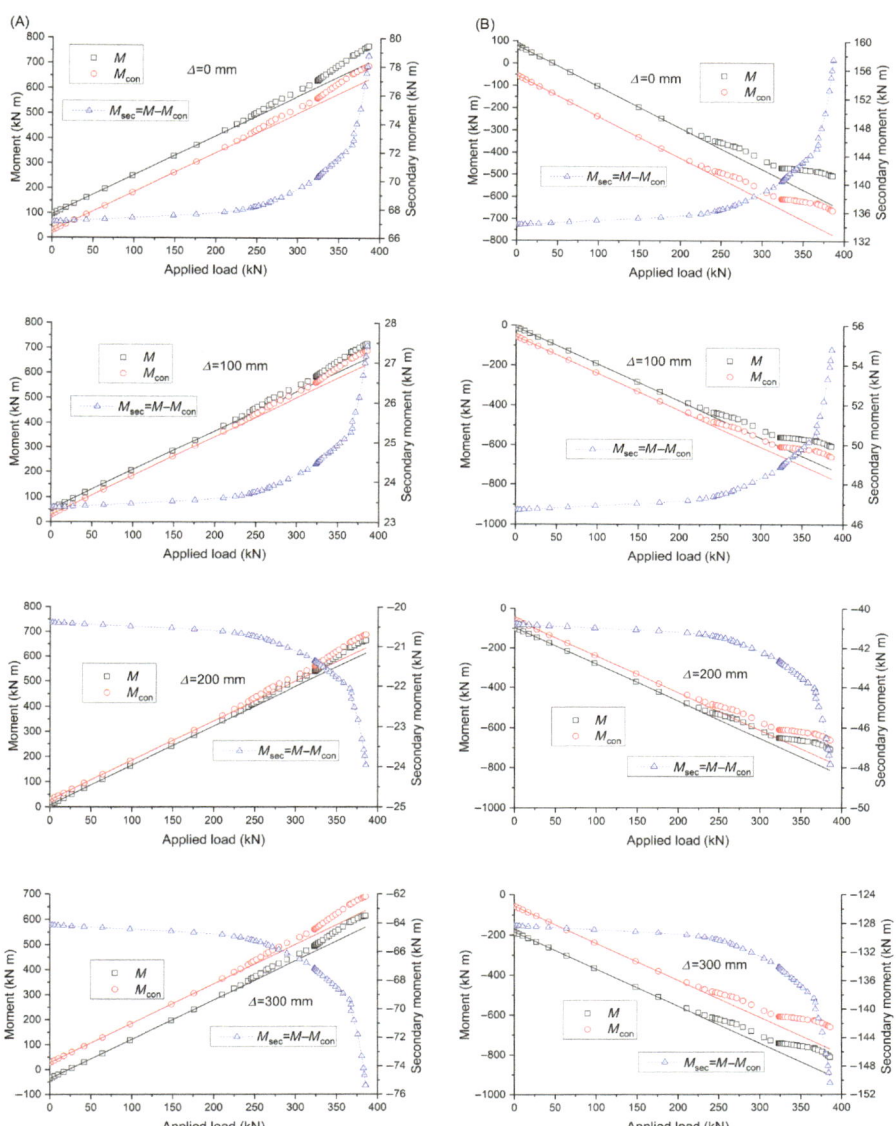

**Figure 6.9** Development of moments and secondary moments for various $\Delta$ values ($A_{s2}/A_{s1} = 0.67$): (A) midspan; (B) center support.

$A_{s2}/A_{s1}$ of 1.5 are presented in Fig. 6.12. In these figures, $M'_u$ and $(M'_u)_{con}$ are the ultimate actual moments excluding the self-weight moment for a cable profile and its linearly transformed concordant profile, respectively; $M'_e$ is the elastic moment excluding the self-weight moment; $M_{sec,u}$ is the secondary moment at ultimate. The graphs of these figures clearly show the difference between the actual moment

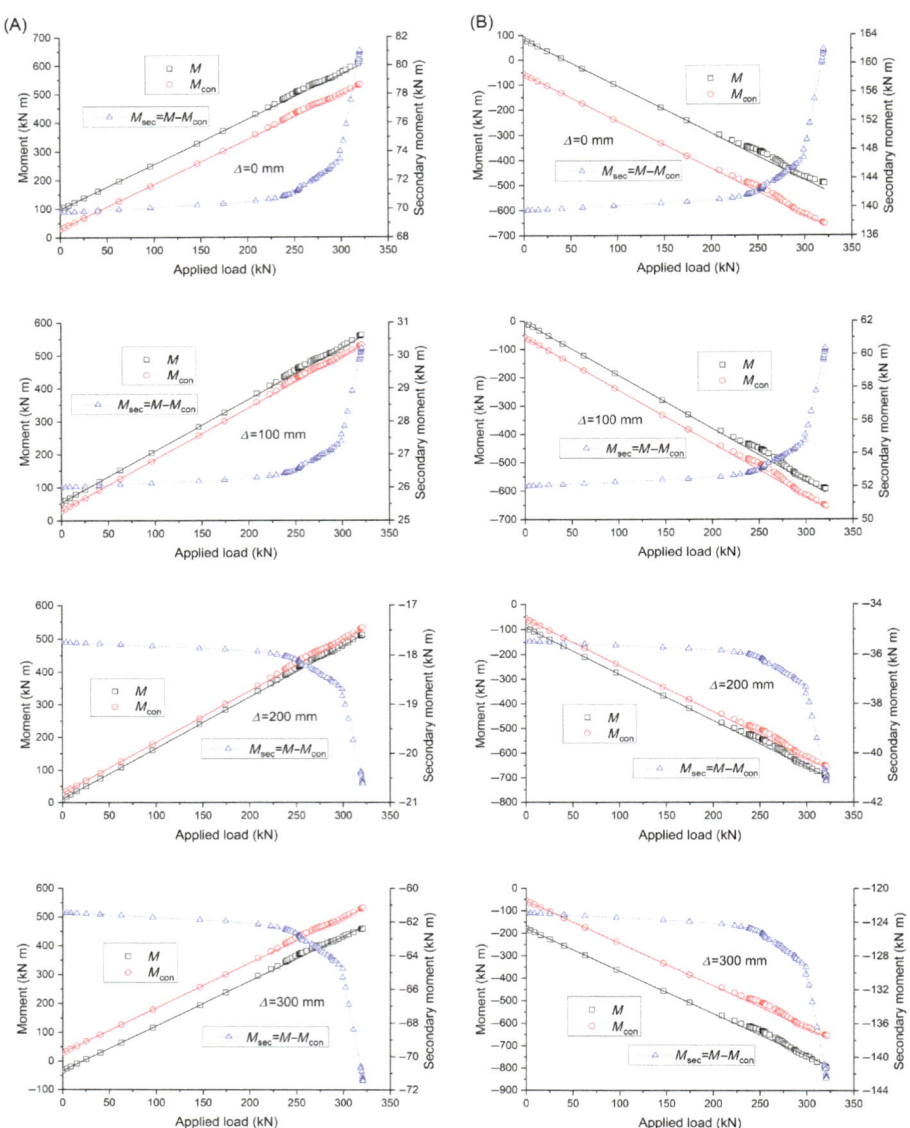

**Figure 6.10** Development of moments and secondary moments for various $\Delta$ values ($A_{s2}/A_{s1} = 1.5$): (A) midspan; (B) center support.

distribution and the distribution obtained by a linear-elastic analysis as well as the relationship between the secondary moments and the applied moments. It is seen that, for $A_{s2}/A_{s1}$ of 0.67, the actual moment at the center support is obviously lower than the corresponding elastic moment while it is opposite at midspan, indicating that these members redistribute moments from the critical negative moment zone to

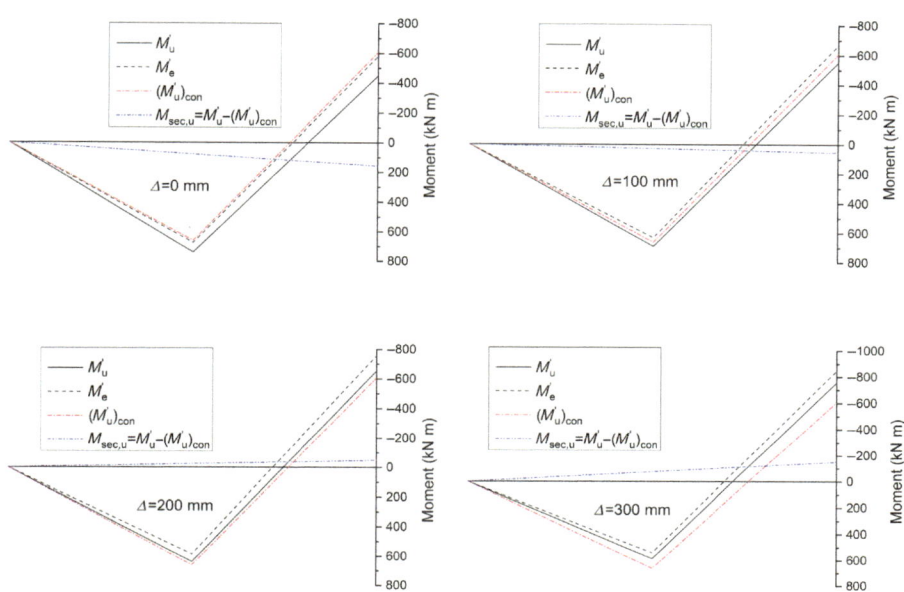

**Figure 6.11** Distribution of moment and secondary moment at ultimate for various $\Delta$ values ($A_{s2}/A_{s1} = 0.67$).

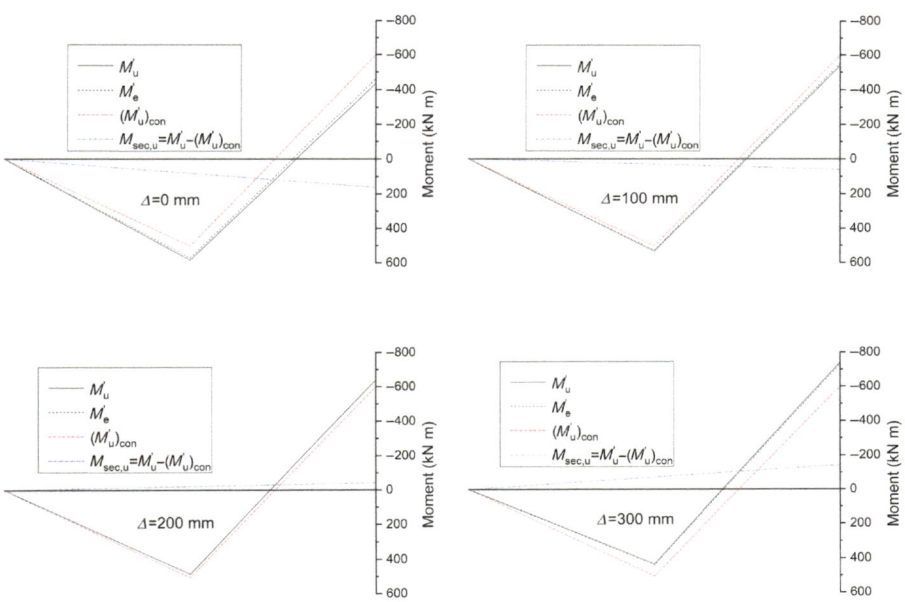

**Figure 6.12** Distribution of moment and secondary moment at ultimate for various $\Delta$ values ($A_{s2}/A_{s1} = 1.5$).

the critical positive moment zone. On the other hand, for $A_{s2}/A_{s1}$ of 1.5, the difference between the actual and elastic moments is slight, implying an insignificant redistribution of moments in these members. In addition, it can be observed that there is a linear distribution of secondary moments over the span as expected. This observation further confirms the reliability of the proposed method for calculating the prestress secondary moments.

Fig. 6.13 shows the variation of secondary moments over the center support at the initial elastic (i.e., zero live loads) and ultimate limit states with the value of $\Delta$. It is seen that the secondary moments at the elastic and ultimate states decrease linearly with the $\Delta$ value, but the rate of decrease in the secondary moment at the ultimate state is faster than that at the elastic state. The $A_{s2}/A_{s1}$ ratio appears to have negligible influence on the decrease rate of secondary moment. At a given value of $\Delta$, an $A_{s2}/A_{s1}$ ratio of 0.67 mobilizes a bit lower secondary moment than an $A_{s2}/A_{s1}$ ratio of 1.5.

## 6.5    Example 2—Members with various prestress levels and different load patterns

A numerical test program is designed to examine the prestress secondary reactions in continuous concrete members prestressed with external CFRP tendons. Two-span

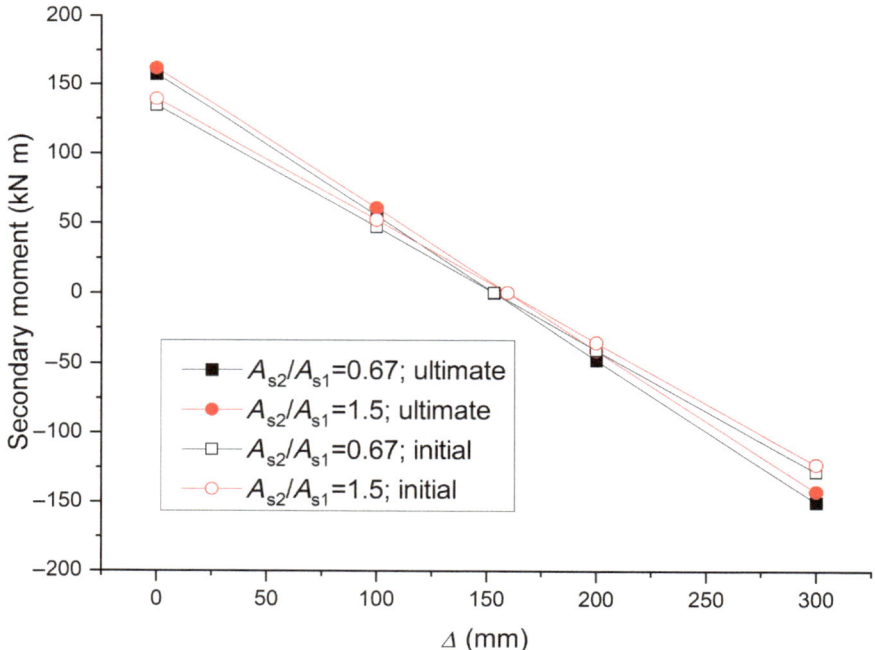

**Figure 6.13** Variation of secondary moment with the $\Delta$ value.

continuous beams shown in Fig. 6.14 are used. Each span is subjected to third-point loads. The loads applied to the right span $P_2$ are either equal to (symmetrical loading) or 50% of (unsymmetrical loading) the loads applied to the left span $P_1 = P$. The external tendons are draped at deviators that are placed at the center support and third points of each span. The tendon eccentricities at the end support $e_0$, outer third point $e_1$, inner third point $e_2$, and center support $e_3$ are 0, 150, 100, and 150 mm, respectively. The external tendons are assumed to be CFRP composites having ultimate strength $f_f$ of 1840 MPa and elastic modulus $E_f$ of 147 GPa. The initial prestress level $\sigma_{p0}/f_f$ varies between 15% and 75%. The tendon area $A_p$ is taken equal to 1000 mm$^2$. The areas of nonprestressed tensile steel over positive moment region $A_{s1}$ and over negative moment region $A_{s2}$ are 1200 and 800 mm$^2$, respectively; and the area of nonprestressed compressive steel $A_{s3}$ is taken as 400 mm$^2$. The yield strength $f_y$ and elastic modulus $E_s$ of nonprestressed steel are 450 MPa and 200 GPa, respectively. The concrete cylinder's compressive strength $f_{ck}$ is 60 MPa.

To compute the secondary reactions using the aforementioned method, the linearly transformed concordant profile of external tendons should be determined first. This concordant cable line is obtained using a trial-and-error method by performing a series of analyses of linearly transformed tendon members subjected to the prestressing force (neglecting the weight of the members and external loads). The cable line is constantly adjusted until the support reactions disappear. The original nonconcordant cables and linearly transformed concordant cables are shown in Fig. 6.15. Linear transformation is made by shifting the original cable line over the center support by $\Delta$, and correspondingly by $(2\Delta)/3$ over the inner third point and $\Delta/3$ over the outer third point, as illustrated in the figure. It is demonstrated that the linearly transformed concordant cable line for the initial prestress level of 75% is slightly different from that for lower initial prestress levels. For the former, the cable shift at the center support is $\Delta = 45.26$ mm, while for the latter $\Delta = 45.39$ mm. This can be explained that the 75% initial prestress level mobilizes an obviously larger axial shortening of the member, thereby causing a slight difference of the cable line when compared to lower initial prestress levels.

The development of support reactions and the evolution of secondary reactions at the end and center supports for symmetrical loading are shown in Fig. 6.16, while the results for unsymmetrical loading are shown in Fig. 6.17. In the figures, $R^1$, $R^2$,

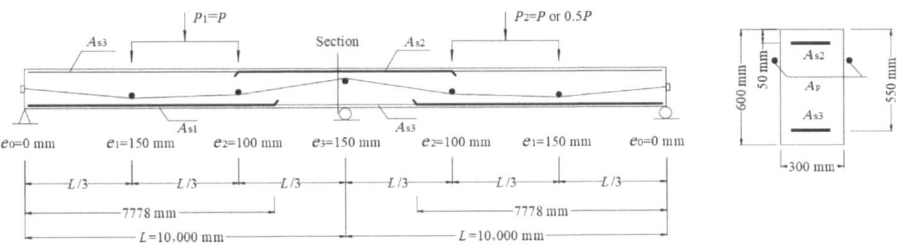

**Figure 6.14** Details of the beams under symmetrical and unsymmetrical loading.

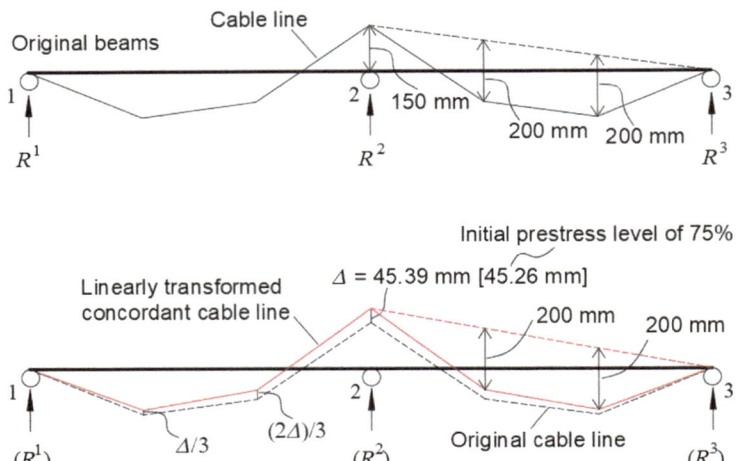

**Figure 6.15** Original nonconcordant cables and linearly transformed concordant cables.

and $R^3$ represent, respectively, the reactions at left-hand, intermediate, and right-hand supports for the members analyzed; $(R^1)_{con}$, $(R^2)_{con}$, and $(R^3)_{con}$ are those for the members with linearly transformed concordant cables; and $R^1_{sec}(=R^1 - (R^1)_{con})$, $R^2_{sec}(=R^2 - (R^2)_{con})$, and $R^3_{sec}(=R^3 - (R^3)_{con})$ are secondary reactions at left, intermediate, and right supports, respectively. It is seen that the support reactions develop linearly with the applied load up to the appearance of flexural cracks. Beyond that, the reaction development exhibits nonlinear behavior due to the redistribution of moments. In addition, this nonlinear behavior is more obvious for a lower prestress level compared to a higher prestress level, and for symmetrical loading compared to unsymmetrical loading, attributed to a more significant redistribution of moments. The response of prestress secondary reactions with the applied load is characterized by three stages with two turning points corresponding to concrete cracking and steel yielding, respectively. This observation is different from some points of view which deemed that the secondary reactions (moments) would remain unchanged, decrease, or disappear after the appearance of cracks or after the formation of plastic hinges (yielding of steel). In contrast, the present example indicates that the secondary reactions increase quicker after cracking or yielding, attributed to a quicker increase in the prestressing force.

    From Figs. 6.16 and 6.17, it can be observed that the secondary reactions at any stage of loading, obtained from the proposed method, satisfy the calibration equations indicated by Eqs. (6.4) and (6.5). For symmetrical loading, the secondary reaction at the center support is twice in magnitude and opposite in direction compared to the secondary reaction at the end support. For unsymmetrical loading, the secondary reaction at the left support is the same as that at the right support, and the summation of the secondary reactions at all three supports is zero. A very slight error at high levels of loading can be attributed to a slight change of the linearly

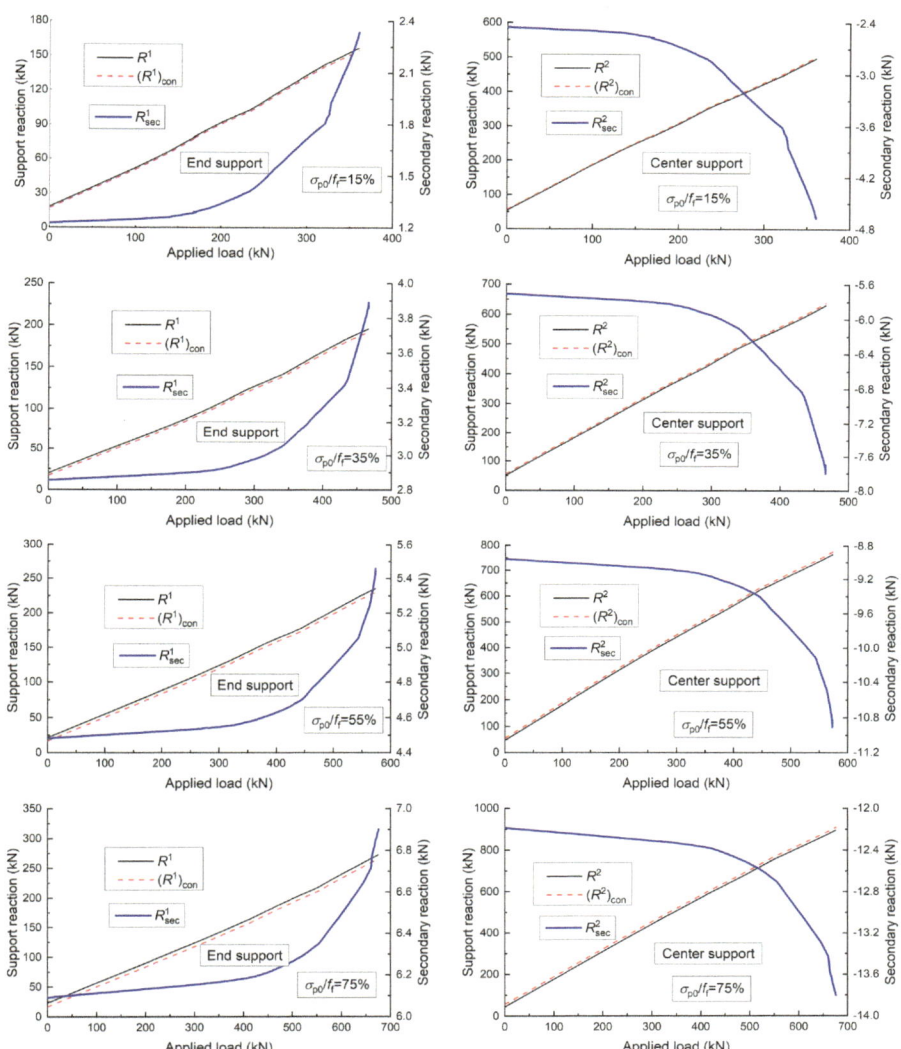

**Figure 6.16** Development of support reactions and secondary reactions for the beams under symmetrical loading.

transformed concordant cable line, caused by additional shortening of the members as a result of external loads.

Figs. 6.18 and 6.19 illustrate the variation of the secondary reaction with the tendon stress for symmetrical loading and unsymmetrical loading, respectively. It is seen that the secondary reaction at a support develops linearly with the tendon stress up to the ultimate limit state. The diagrams for various initial prestress levels are in the same straight line which crosses the zero point. Therefore it is inferred

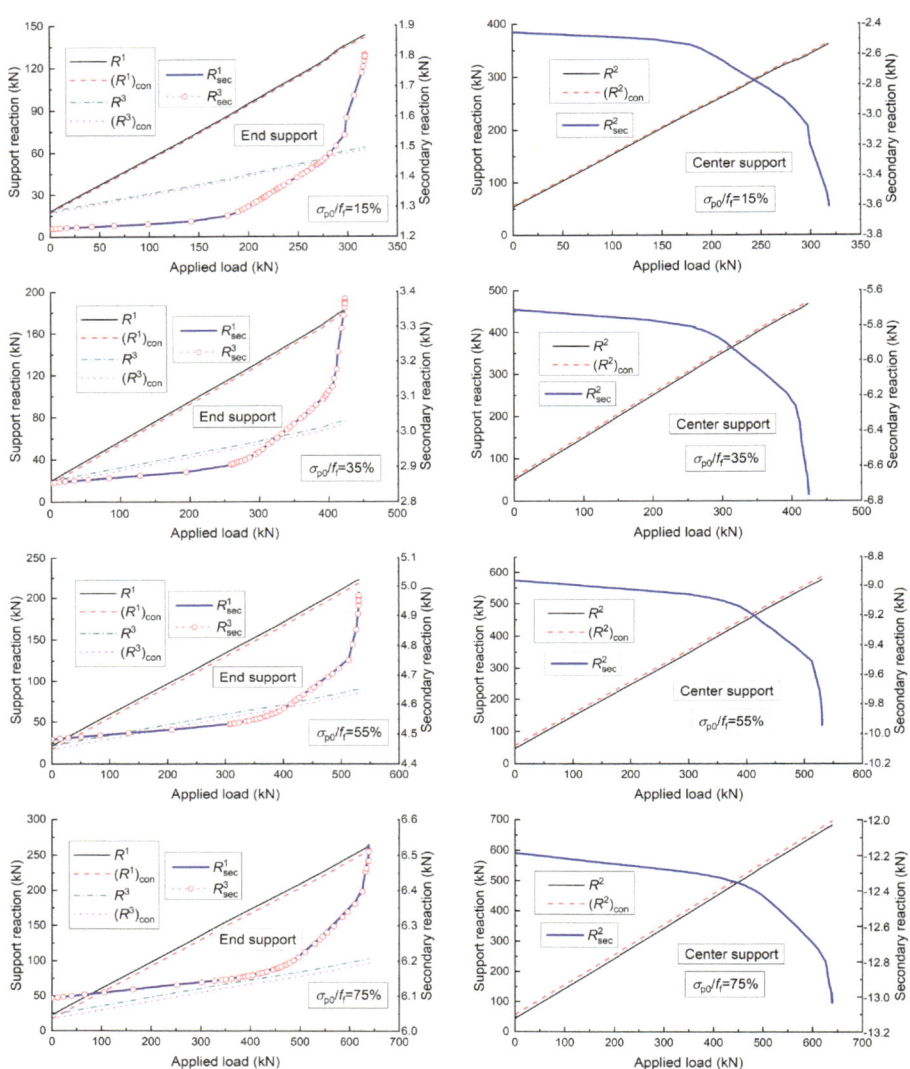

**Figure 6.17** Development of support reactions and secondary reactions for the beams under unsymmetrical loading.

from these results that the secondary reactions or moments at the plastic or ultimate stage can be conveniently calculated by an elastic analysis using the current external prestressing force and neglecting the self-weight of the members. Irrespective of the pattern of loading, the line slopes for the center and end supports are −9.08 and 4.54 N/MPa, respectively. It should be noted that the slope depends on the cable deviation from the linearly transformed concordant line. The larger the deviation, the steeper the slope.

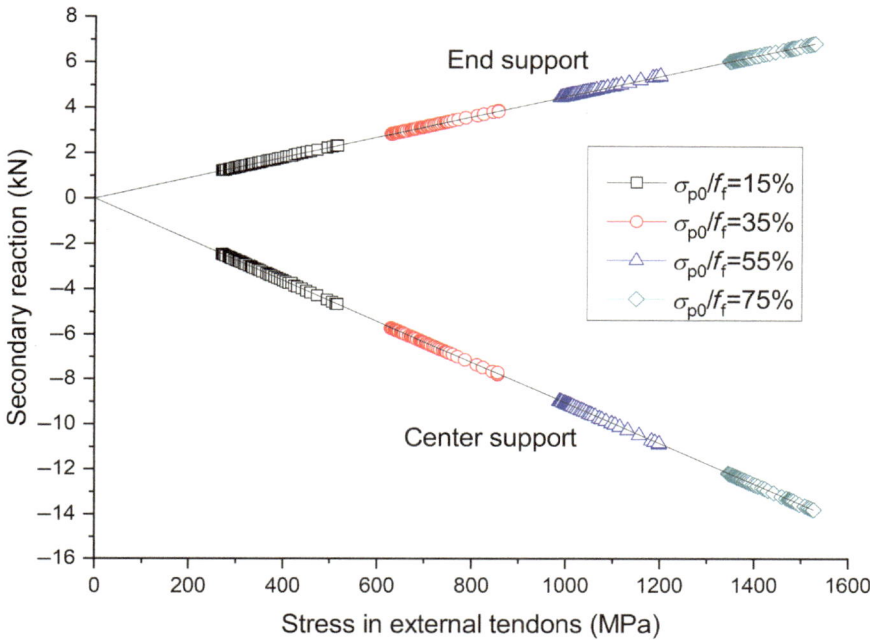

**Figure 6.18** Variation of secondary reactions with the stress in external tendons for the beams under symmetrical loading.

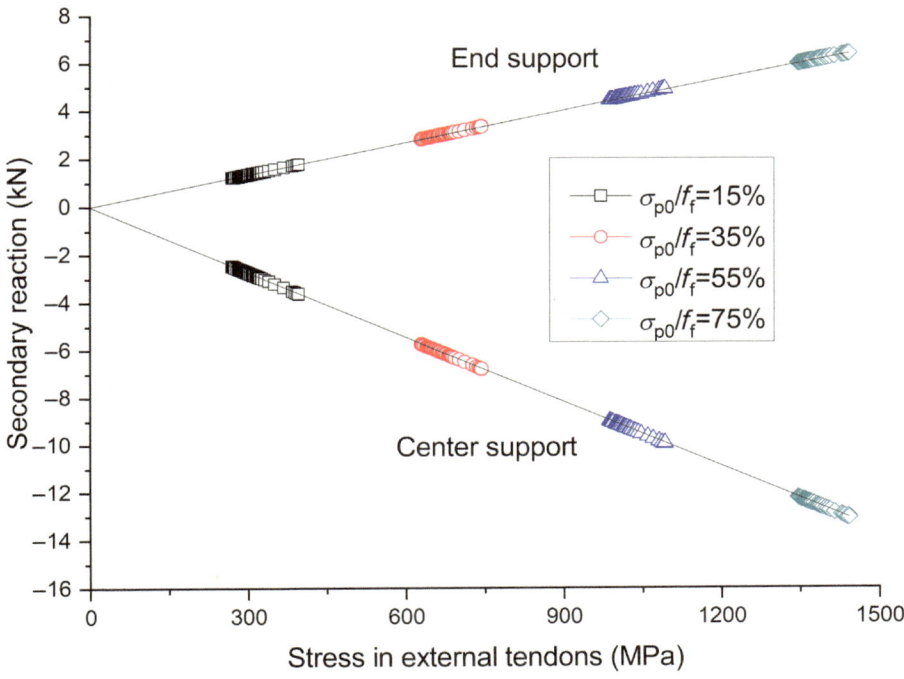

**Figure 6.19** Variation of secondary reactions with the stress in external tendons for the beams under unsymmetrical loading.

At the ultimate limit state, the secondary moments in the members under symmetrical and unsymmetrical loading are shown in Figs. 6.20 and 6.21, respectively, where $X/L$ is the ratio of the distance from the end support to the span. Because the reaction at the end support is positive, the secondary moments produced by the support reaction are also positive over the member. As a consequence, they counteract the negative moment at the center support while accentuating the positive moment at the span critical section. Because the cable line is rather close to the linearly

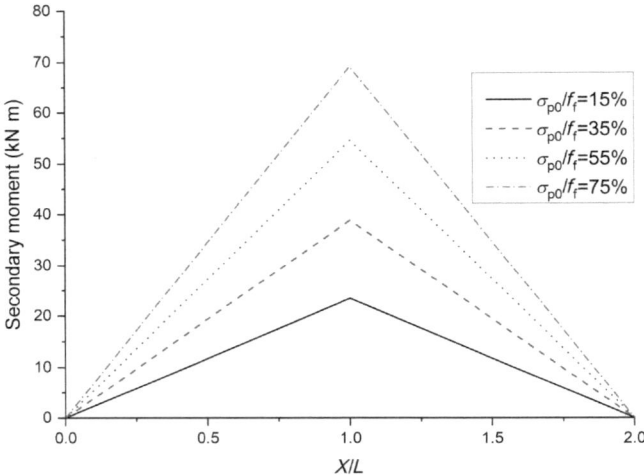

**Figure 6.20** Secondary moments in the beams under symmetrical loading at ultimate.

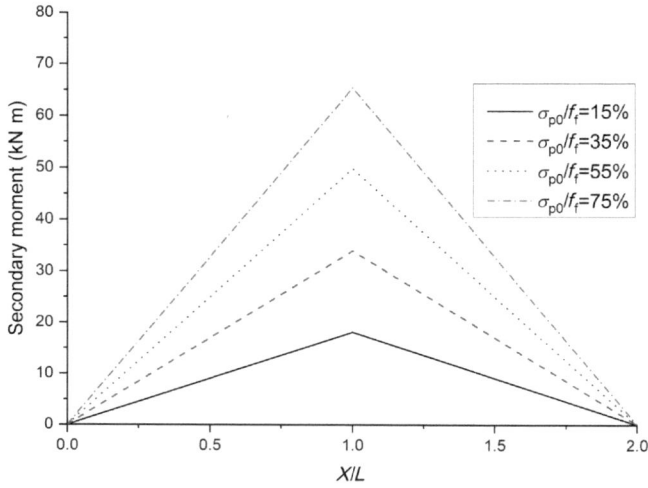

**Figure 6.21** Secondary moments in the beams under unsymmetrical loading at ultimate.

transformed concordant line as illustrated in Fig. 6.15, the prestress secondary moments are not so important. If the cable deviation from the linearly transformed concordant line is larger, the secondary moments would be more important.

## 6.6  Conclusions

Linear transformation does not affect the general flexural behavior (i.e., development of deflection, curvature, neutral axis depth, stresses and strains in concrete, external cables, and reinforcing bars with the applied load) of continuous prestressed members throughout the loading process. By utilizing linear transformation, it is therefore possible to make a flexible layout of prestressing cables while maintaining the basic performance of the structure at all ranges of loading.

Based on the linear transformation concept, a method for predicting the secondary moments in continuous prestressed members with external tendons is proposed. The proposed method is able to predict accurately the secondary reactions or moments in continuous prestressed members at all stages of loading up to the ultimate limit state. The secondary moments for a member with nonconcordant cables are present throughout the loading process. Since the full development of plastic hinges is not likely to happen for prestressed concrete members with external FRP tendons in practice, the inclusion of the secondary moment is necessary when calculating the design moment of this structural typology. It is shown that at a given load level, there is a linear relationship between the secondary moments and the value of $\Delta$, but the slope at the ultimate stage is steeper than that at the elastic stage. In addition, the secondary reactions or moments develop linearly with the stress in external tendons up to the ultimate limit state, and therefore, these reactions or moments can be calculated conveniently based on an elastic analysis using the current prestressing force and neglecting the self-weight of the members.

## References

ACI Committee 318. (1971). Building code requirements for reinforced concrete. ACI 318-71, Farmington Hills, MI.

ACI Committee 318. (2019). *Building code requirements for structural concrete (ACI 318-19) and commentary (ACI 318R-19)*. Farmington Hills, MI.

Aravinthan, T., Witchukreangkrai, E., & Mutsuyoshi, H. (2005). Flexural behavior of two-span continuous prestressed concrete girders with highly eccentric external tendons. *ACI Structural Journal, 102*(3), 402–411.

Lin, T. Y., & Burns, N. H. (1981). *Design of prestressed concrete structures* (3rd ed.). New York: John Wiley & Sons.

Lou, T., Lopes, S. M. R., & Lopes, A. V. (2013). Flexural response of continuous concrete beams prestressed with external tendons. *Journal of Bridge Engineering, 18*(6), 525–537. Available from https://doi.org/10.1061/(ASCE)BE.1943-5592.0000392.

Lou, T., Lopes, S. M. R., & Lopes, A. V. (2014). External CFRP tendon members: Secondary reactions and moment redistribution. *Composites Part B: Engineering*, *57*, 250−261. Available from https://doi.org/10.1016/j.compositesb.2013.10.010.

Lou, T., Lopes, S. M. R., & Lopes, A. V. (2017). Effect of linear transformation on nonlinear behavior of continuous prestressed beams with external FRP cables. *Engineering Structures*, *147*, 410−424. Available from https://doi.org/10.1016/j.engstruct.2017.06.029.

Wyche, P. J., Uren, J. G., & Reynolds, G. C. (1992). Interaction between prestress secondary moments, moment redistribution, and ductility: A treatise on the Australian concrete codes. *ACI Structural Journal*, *89*(1), 57−70.

# Continuous normal- and high-strength concrete members

**7**

## 7.1 Introduction

The high compressive strength is the primary characteristic distinguishing high-strength concrete (HSC) from normal-strength concrete (NSC). Even though there is no definitive boundary between HSC and NSC, concrete with a compressive strength of 55 MPa or above is normally termed as HSC according to ACI (ACI Committee 318, 2019). In addition to high compressive strength, HSC possesses a range of attractive merits such as improved bond between concrete and reinforcement (Buch & Suhail, 2022; Diab et al., 2014), excellent durability (Yuan et al., 2021), and creep resistance (Hong et al., 2023) and reduced size of concrete elements, when compared to NSC. Nowadays, HSC has been widely applied to various civil structures such as tall buildings, long-span bridges, and marine engineering.

On the other hand, HSC is more brittle, showing steeper stress—strain diagram and smaller ultimate strain, when compared to NSC (Sheikh et al., 2013). As ductile characteristics are of crucial importance to structures especially in seismic regions, there arise concerns about the utilization of HSC in practice. Many experimental works (Bernardo & Lopes, 2004; Galano & Vignoli, 2008; Ko et al., 2001; Ma et al., 2019), numerical studies (Ahmed et al., 2023; Arslan & Cihanli, 2010; Lam et al., 2009; Lou et al., 2014), and design proposals (Bouzid & Kassoul, 2016; Kassoul & Bougara, 2010; Shaaban & Mustafa, 2021) were reported regarding the flexural performance and ductility of HSC members. Previous research showed that despite the brittleness of HSC, HSC members with reasonably arranged steel bars exhibited favorable ductile ability.

External prestressing is commonly applied to multispan prestressed bridges, which are often cast with HSC to achieve the desired structural strength and durability. In continuous reinforced and prestressed concrete members, the ability to redistribute bending moments is also a major concern for the use of HSC. Laboratory tests on plastic rotational capacity and moment redistribution of continuous reinforced HSC beams were performed in Coimbra (Carmo & Lopes, 2008). The main variables of the tests were the contents of longitudinal tension steel bars and transverse reinforcement. It was shown that reinforced HSC beams with tensile steel ratios not greater than 2.9% exhibited favorable deformation behavior and adequate capacities to redistribute bending moments.

This chapter presents comprehensive aspects of the flexural behavior of two-span continuous HSC members, and the results are compared to the counterparts made of NSC. First, a comparative analysis of continuous reinforced NSC and HSC members is performed, and formulae relating flexural ductility and moment

Prestressed Members with External Fiber-Reinforced Polymer (FRP) Tendons. DOI: https://doi.org/10.1016/B978-0-443-23877-2.00007-9

redistribution with either neutral axis depth or tensile steel strain are suggested (Shi et al., 2023). Second, a numerical work on continuous bonded prestressed NSC and HSC members is reported, emphasizing the aspects of behavior related to moment redistribution (Lou et al., 2017). Third, the response characteristics of continuous NSC and HSC members prestressed with external carbon fiber-reinforced polymer (CFRP) tendons are presented, and relevant rules in the ACI code are evaluated (Lou et al., 2020).

## 7.2  Reinforced normal- and high-strength concrete members

Fig. 7.1 shows a sketch of two-span reinforced NSC and HSC beams. Three concrete grades are considered, namely $f_{ck}$ = 30, 60, and 90 MPa. A wide range of tensile steel ratios are used, that is, $\rho_{s2}$ = 0.73%$-$4.0%, and $\rho_{s2}/\rho_{s1}$ = 0.8, where $\rho_{s1}$ and $\rho_{s2}$ are tensile steel ratios over positive and negative moment zones, respectively. The compressive steel ratio over both positive and negative moment zones, $\rho_{s3}$, is 0.36%. The elastic modulus and yield strength of steel bars are 200 GPa and 530 MPa, respectively.

### 7.2.1  Load versus deformation

Fig. 7.2A and B shows the load$-$deflection and moment$-$curvature curves of reinforced NSC and HSC members with different steel ratios, respectively. Before final failure caused by concrete crushing at the critical section, the members with a low steel ratio (e.g., $\rho_{s2}$ = 0.73%) typically experience several phases in sequence, that is, first cracking at center support, second cracking at midspan, first yielding at center support and second yielding at midspan. As a result, these members exhibit three distinct stages concerning the moment$-$curvature behavior of a cross-section (see Fig. 7.2B), that is, elastic stage, post-cracking stage, and post-yielding stage. The load$-$deflection behavior appears to be primarily influenced by the second cracking and second yielding (see Fig. 7.2A). In other words, the first cracking and first yielding do not have a notable impact on the load$-$deflection behavior of continuous members. As the steel ratio increases, the rotational capacity of the members

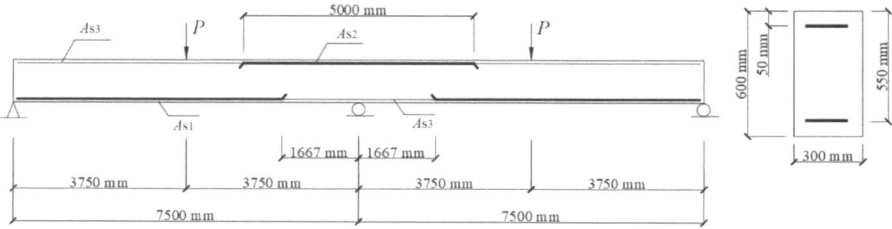

**Figure 7.1** Continuous reinforced normal- and high-strength concrete beams.

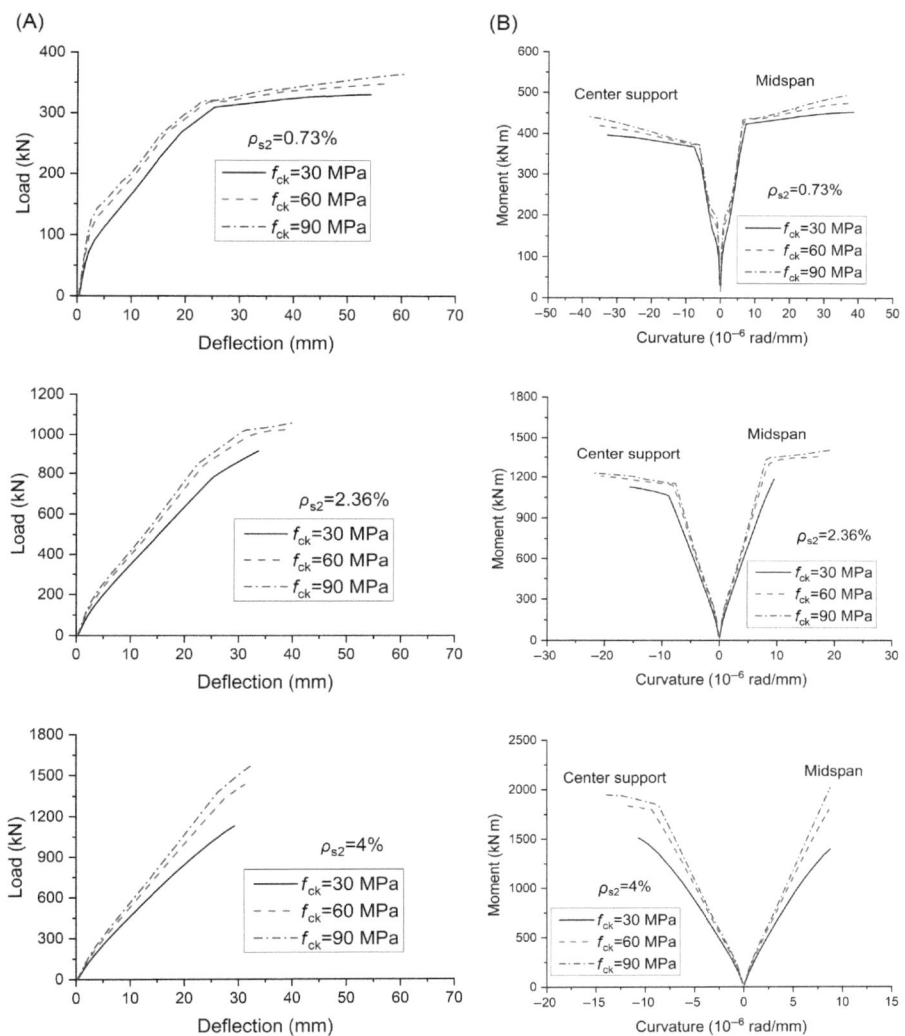

**Figure 7.2** Deformation of reinforced normal- and high-strength concrete beams: (A) load versus midspan deflection; (B) moment versus curvature.

decreases. When the steel ratio reaches or surpasses the balanced ratio, plastic hinges over critical section(s) may not develop. For example, for reinforced NSC members, there is no formation of plastic hinges at midspan when $\rho_{s2} = 2.36\%$, and at both midspan and center support when $\rho_{s2} = 4.0\%$. For reinforced HSC members, plastic hinges do not occur at midspan when $\rho_{s2} = 4.0\%$, as evidenced by the lacking of post-yielding stage of the load–deflection behavior or moment–curvature behavior (midspan). Because of the higher tensile strength and elastic modulus of

HSC, reinforced HSC members exhibit higher cracking load and flexural stiffness than reinforced NSC members.

The variation of ultimate load ($P_u$) and moment ($M_u$) with varying steel ratios for reinforced NSC and HSC members is presented in Table 7.1 and Fig. 7.3, respectively. At $\rho_{s2} = 0.73\%$, the discrepancy of ultimate load or moment between reinforced NSC and HSC members is slight, attributed to the full development of both positive and negative plastic hinges in these members. The increase in ultimate load or moment with increasing steel ratio is more pronounced for reinforced HSC members than for reinforced NSC members, attributed to the fact that reinforced HSC members are able to better develop plastic hinges than reinforced NSC members. As a result, at a high steel ratio (e.g., $\rho_{s2} = 4.0\%$), the ultimate load or moment of reinforced HSC members is substantially higher than that of reinforced NSC members. The variation of ultimate deflection ($\Delta_u$) with varying steel ratios for reinforced NSC and HSC members is presented in Fig. 7.4 and Table 7.1. It is seen that the ultimate deflection decreases with increasing steel ratio. For a given steel ratio, the ultimate deflection of reinforced HSC members is higher than that of reinforced NSC members.

The flexural ductility can be quantified by $\mu_\kappa = \kappa_u/\kappa_y$, where $\mu_\kappa$ is the curvature ductility factor; $\kappa_u$ and $\kappa_y$ are curvature at ultimate and yielding, respectively. The values of $\kappa_y$, $\kappa_u$, and $\mu_\kappa$ (center support) are presented in Table 7.1. Fig. 7.5 shows the variation of curvature ductility factor with varying steel ratios for reinforced NSC and HSC members. The higher the steel ratio, the lower the flexural ductility. Despite the higher brittleness of HSC material, reinforced HSC members exhibit

**Table 7.1** Values of typical response of reinforced normal- and high-strength concrete beams.

| $f_{ck}$ (MPa) | $\rho_{s2}$ (%) | $P_u$ (kN) | $\Delta_u$ (mm) | $\kappa_y$ ($10^{-6}$ rad/mm) | $\kappa_u$ ($10^{-6}$ rad/mm) | $\mu_\kappa$ | $c_u/d$ | $\varepsilon_t$ (%) |
|---|---|---|---|---|---|---|---|---|
| 30 | 0.73 | 330.7 | 54.3 | −6.52 | −32.91 | 5.05 | 0.15 | 1.53 |
|  | 1.55 | 667.0 | 40.1 | −7.71 | −21.55 | 2.80 | 0.25 | 0.89 |
|  | 2.36 | 914.9 | 33.6 | −8.89 | −15.56 | 1.75 | 0.41 | 0.51 |
|  | 3.18 | 1055.5 | 31.2 | −10.36 | −12.12 | 1.17 | 0.52 | 0.32 |
|  | 4.0 | 1131.3 | 29.3 | −12.62 | −10.71 | 0.85 | 0.59 | 0.24 |
| 60 | 0.73 | 348.3 | 57.1 | −6.37 | −35.61 | 5.59 | 0.14 | 1.69 |
|  | 1.55 | 700.4 | 44.9 | −7.19 | −25.60 | 3.56 | 0.18 | 1.15 |
|  | 2.36 | 1027.6 | 39.3 | −7.98 | −20.60 | 2.58 | 0.26 | 0.84 |
|  | 3.18 | 1279.8 | 33.6 | −8.61 | −15.53 | 1.80 | 0.35 | 0.56 |
|  | 4.0 | 1432.3 | 31.2 | −9.34 | −12.28 | 1.32 | 0.44 | 0.38 |
| 90 | 0.73 | 364.4 | 60.4 | −6.28 | −38.28 | 6.10 | 0.13 | 1.83 |
|  | 1.55 | 719.1 | 47.8 | −6.96 | −27.69 | 3.98 | 0.17 | 1.27 |
|  | 2.36 | 1060.7 | 40.4 | −7.67 | −21.75 | 2.84 | 0.22 | 0.94 |
|  | 3.18 | 1374.6 | 36.5 | −8.21 | −18.29 | 2.23 | 0.28 | 0.73 |
|  | 4.0 | 1577.0 | 32.7 | −8.69 | −14.11 | 1.62 | 0.36 | 0.50 |

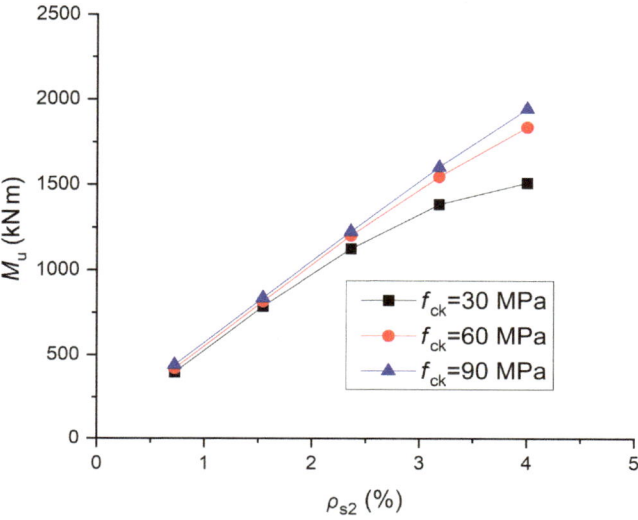

**Figure 7.3** Variation of ultimate moment (center support) with varying steel ratio.

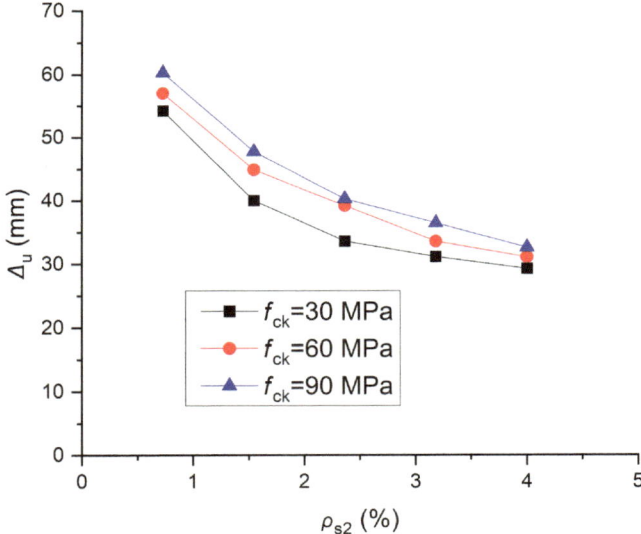

**Figure 7.4** Variation of ultimate deflection at midspan with varying steel ratio.

better ductile behavior than reinforced NSC members. The ductility factors of reinforced HSC members with $f_{ck} = 60$ and 90 MPa are $1.11-1.55$ and $1.21-1.91$ times that of reinforced NSC members ($f_{ck} = 30$ MPa), respectively.

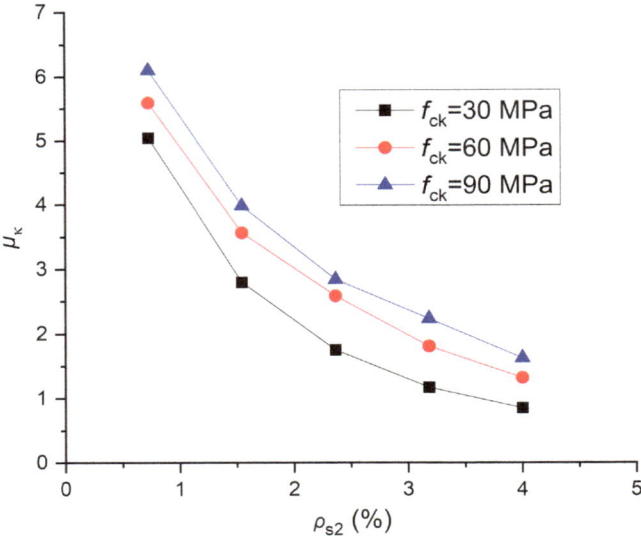

**Figure 7.5** Variation of curvature ductility factor (center support) with varying steel ratio.

## 7.2.2 Neutral axis depth

Fig. 7.6A shows the load versus normalized neutral axis depth relationship for reinforced NSC and HSC members with different steel ratios. The normalized neutral axis depth is defined as the ratio of the neutral axis depth to the effective depth of a cross-section. The neutral axis depth is calculated according to the concrete strain at the extreme compressive fiber and the curvature of a cross-section. Before cracking, the neutral axis does not shift, locating at the centroidal axis of a cross-section (transformed section). For a given steel ratio, the equivalent concrete area contributed by steel bars for reinforced HSC members is smaller than that of reinforced NSC ones. As a result, the initial neutral axis depth in reinforced HSC members is lower than that in reinforced NSC members. The difference is noticeable at a high steel ratio (e.g., $\rho_{s2} = 4.0\%$) but marginal at a low steel ratio (e.g., $\rho_{s2} = 0.73\%$). At a given load level, there is a larger neutral axis depth at midspan compared to that at center support due to a higher steel ratio. Fig. 7.6B shows the curvature versus normalized neutral axis depth relationship for reinforced NSC and HSC members with different steel ratios. The behavior is markedly influenced by concrete cracking and steel yielding. At a given curvature over the post-cracking stage, reinforced HSC members exhibit lower neutral axis depth than reinforced NSC members.

The variation of neutral axis depth at ultimate ($c_u/d$) with varying steel ratios for reinforced NSC and HSC members is shown in Fig. 7.7. The values of $c_u/d$ are also presented in Table 7.1. For a low steel ratio, the $c_u/d$ difference between reinforced NSC and HSC members is slight, for example, at $\rho_{s2} = 0.73\%$, $c_u/d = 0.15$, 0.14,

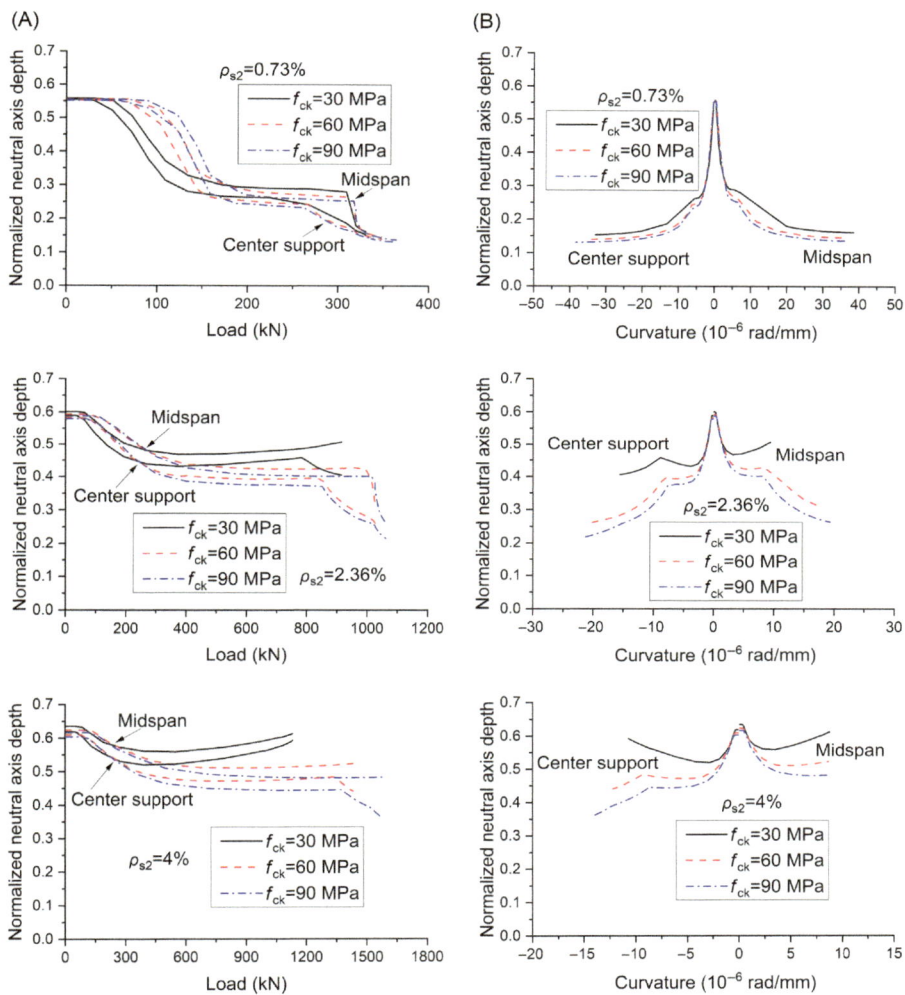

**Figure 7.6** Development of neutral axis depth: (A) load versus neutral axis depth; (B) curvature versus neutral axis depth.

and 0.13 for $f_{ck} = 30$, 60, and 90 MPa, respectively. The increase in $c_u/d$ with increasing steel ratio for reinforced HSC members is less pronounced than that for reinforced NSC members. As a result, reinforced HSC members show smaller $c_u/d$ than reinforced NSC members, especially notable at a high steel ratio. At $\rho_{s2} = 4.0\%$, the $c_u/d$ values of reinforced HSC members with $f_{ck} = 60$ and 90 MPa are 74% and 61% of that of reinforced NSC members ($f_{ck} = 30$ MPa).

Fig. 7.8 shows the $\mu_\kappa - c_u/d$ relationship at the center support of reinforced NSC and HSC members. The graph demonstrates that the ductility decreases as $c_u/d$

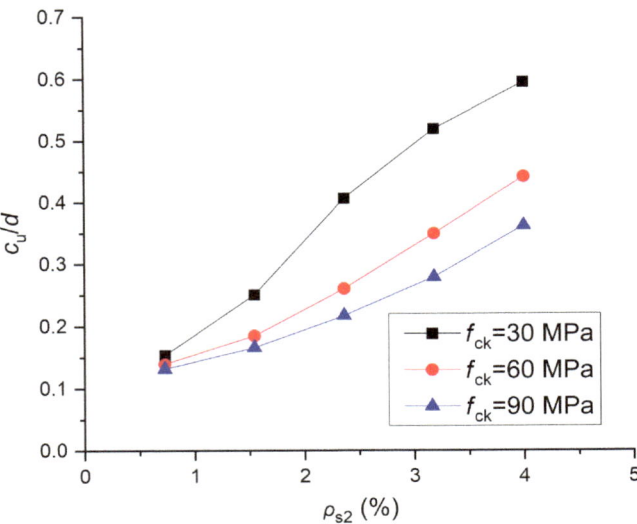

**Figure 7.7** Variation of neutral axis depth at ultimate (center support) with varying steel ratio.

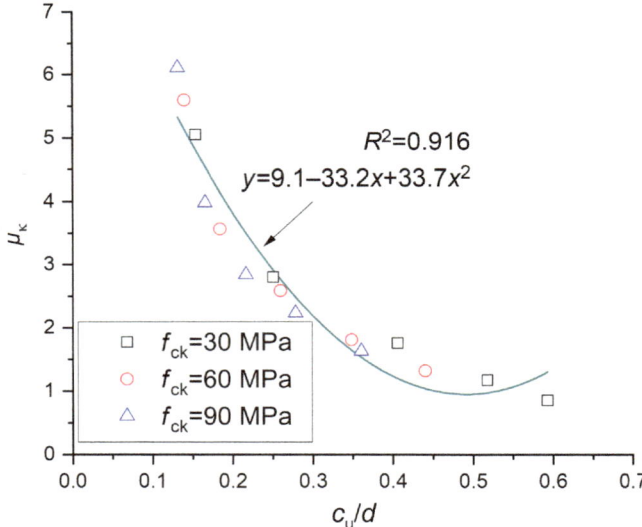

**Figure 7.8** Relationship between curvature ductility factor and neutral axis depth.

increases. According to the fit curve, the relationship between flexural ductility and neutral axis depth may be expressed as follows:

$$\mu_\kappa = 9.1 - 33.2(c_u/d) + 33.7(c_u/d)^2 \tag{7.1}$$

### 7.2.3 Strain in tensile steel bars

Fig. 7.9 shows the strain development of tensile steel bars in reinforced NSC and HSC members with different steel ratios. As demonstrated in Fig. 7.9, yielding at center support leads to a significantly faster increase in the steel strain in this section but has practically no impact on the steel strain development at midspan. On the other hand, yielding at midspan (if any) increases significantly the strain development of steel bars at both midspan and center support. At a given service load level, HSC leads to a lower strain or stress in tensile steel bars than NSC. The variation in strain in tensile steel bars at ultimate ($\varepsilon_t$) with varying steel ratios for reinforced NSC and HSC members is presented in Fig. 7.10 and Table 7.1. The value of $\varepsilon_t$ decreases with increasing steel ratio. At a given steel ratio, HSC leads to a higher $\varepsilon_t$ than NSC. The ultimate strains in tensile steel bars in reinforced HSC members with $f_{ck} = 60$ and $90$ MPa are $1.1-1.73$ and $1.19-2.26$ times that in reinforced NSC members ($f_{ck} = 30$ MPa).

Fig. 7.11 shows the $\mu_\kappa-\varepsilon_t$ relationship at the center support of reinforced NSC and HSC members. It is seen that the ductility factor increases almost linearly with increasing $\varepsilon_t$. Based on the linear fit, the relationship between ductility factor and strain in tensile steel bars may be expressed by

$$\mu_\kappa = -0.011 + 324.8\varepsilon_t \tag{7.2}$$

### 7.2.4 Moment redistribution

Fig. 7.12A and B demonstrates the developments of support reaction and bending moment for reinforced NSC and HSC members with different steel ratios, respectively. In addition to the actual reactions or moments obtained by a nonlinear finite element analysis (FEA), the elastic ones are also plotted. The graphs show that the evolution of actual reactions or moments is identical to the elastic one until the first cracking of concrete. After that, the actual and elastic ones deviate because of the redistribution of moments. This deviation is more pronounced for a lower steel ratio, indicating higher moment redistribution. For certain load levels, the reaction or moment difference between reinforced NSC and HSC members appears to be unimportant, indicating similar moment redistribution. However, the ultimate load of reinforced HSC members may be significantly higher than that of reinforced NSC members as discussed previously, indicating that moment redistribution at ultimate for reinforced NSC and HSC members may be different.

Table 7.2 presents a list of $M_u$, $M_e$, and $\beta_u$ (ultimate moment, elastic moment, and degree of moment redistribution at ultimate) for reinforced NSC and HSC members. Moment redistribution is negative at midspan and positive at center support. Moment redistribution at midspan is around 0.6 times that at center support. The change in moment redistribution with varying steel ratios for reinforced HSC members appears to be less pronounced than that for reinforced NSC members.

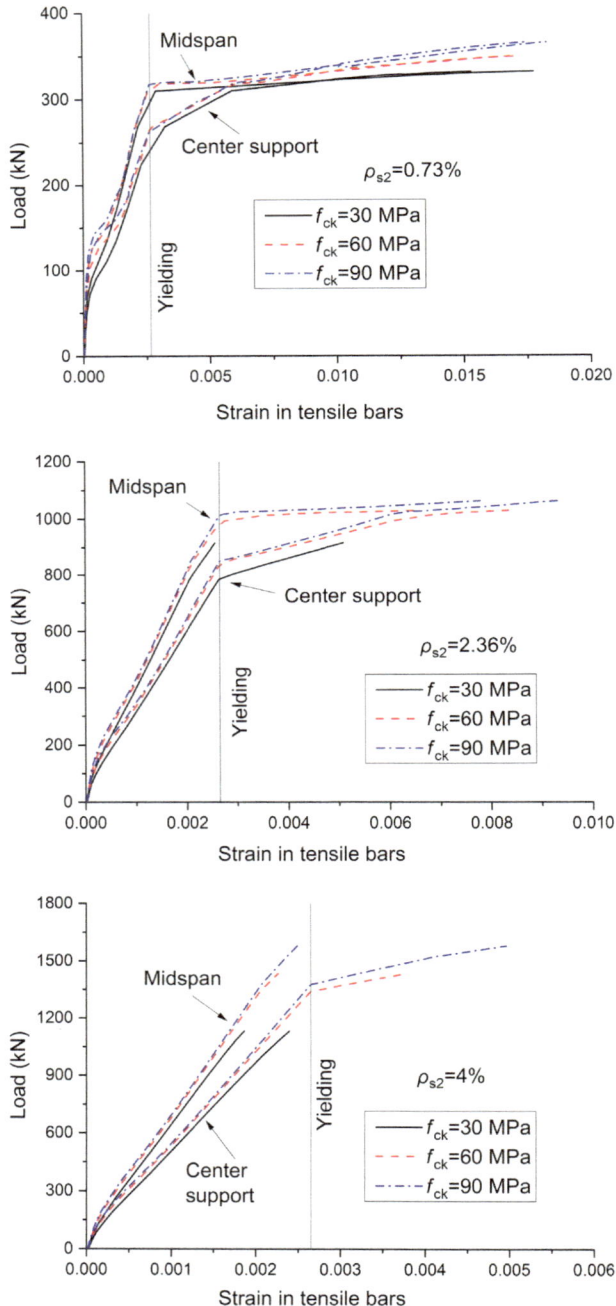

**Figure 7.9** Load versus tensile steel strain.

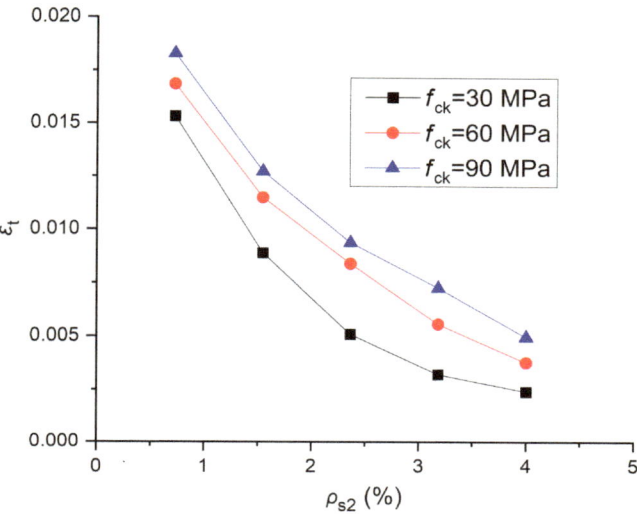

**Figure 7.10** Variation of tensile steel strain at ultimate (center support) with varying steel ratio.

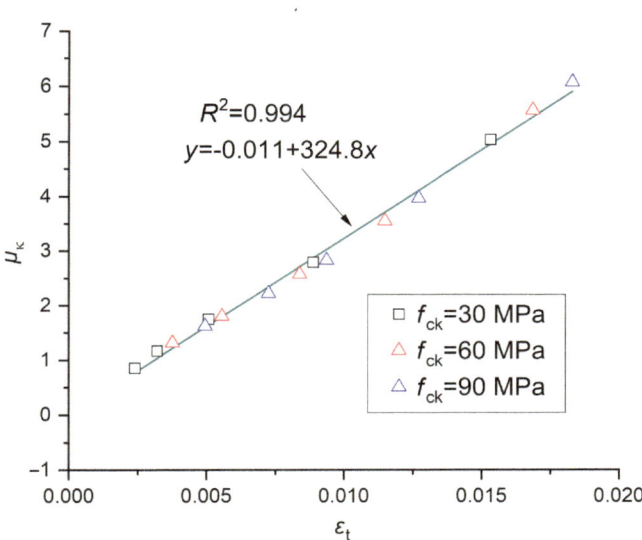

**Figure 7.11** Relationship between curvature ductility factor and tensile steel strain.

Reinforced HSC members generally exhibit higher moment redistribution than reinforced NSC members, particularly at a high steel ratio. The explanation is that reinforced HSC members exhibit better ductile behavior than reinforced NSC members as discussed previously.

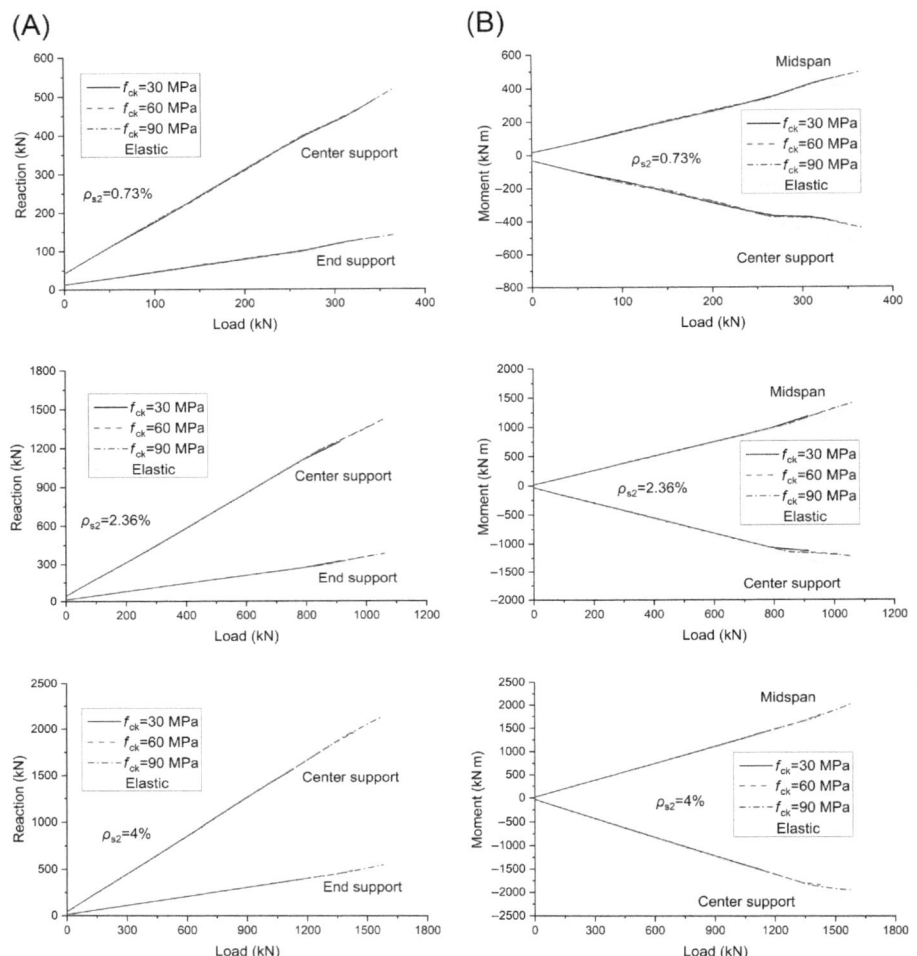

**Figure 7.12** Reaction and moment development: (A) load versus reaction; (B) load versus moment.

Figs. 7.13 and 7.14 show the $\beta_u - c_u/d$ and $\beta_u - \varepsilon_t$ relationships at the center support of reinforced NSC and HSC members, respectively. According to the fit curves, the moment redistribution may be related to the neutral axis depth by

$$\beta_u(\%) = 24.5 - 33.9(c_u/d) \tag{7.3}$$

and to the strain in tensile steel bars by

$$\beta_u(\%) = 2.21 + 3263.5\varepsilon_t - 120{,}054.4\varepsilon_t^2 \tag{7.4}$$

**Table 7.2** Values related to moment redistribution in reinforced normal- and high-strength concrete beams.

| $f_{ck}$ (MPa) | $\rho_{s2}$ (%) | $M_u$ (kN m) | | $M_e$ (kN m) | | $\beta_u$ (%) | |
|---|---|---|---|---|---|---|---|
| | | Midspan | Support | Midspan | Support | Midspan | Support |
| 30 | 0.73 | 453.6 | −396.3 | 406.3 | −491.0 | −11.7 | 19.3 |
| | 1.55 | 888.7 | −786.9 | 805.8 | −952.7 | −10.3 | 17.4 |
| | 2.36 | 1183.9 | −1126.5 | 1102.3 | −1289.7 | −7.4 | 12.7 |
| | 3.18 | 1317.6 | −1386.1 | 1271.6 | −1478.1 | −3.6 | 6.2 |
| | 4.0 | 1396.5 | −1512.9 | 1363.9 | −1578.1 | −2.4 | 4.1 |
| 60 | 0.73 | 475.1 | −419.2 | 426.7 | −516.0 | −11.3 | 18.8 |
| | 1.55 | 936.9 | −815.9 | 844.6 | −1000.5 | −10.9 | 18.5 |
| | 2.36 | 1355.7 | −1205.2 | 1234.7 | −1447.3 | −9.8 | 16.7 |
| | 3.18 | 1657.1 | −1548.4 | 1536.7 | −1789.1 | −7.8 | 13.5 |
| | 4.0 | 1796.9 | −1840.6 | 1720.5 | −1993.5 | −4.4 | 7.7 |
| 90 | 0.73 | 494.6 | −440.8 | 445.6 | −538.7 | −11.0 | 18.2 |
| | 1.55 | 961.0 | −837.9 | 866.2 | −1027.4 | −10.9 | 18.5 |
| | 2.36 | 1406.3 | −1228.5 | 1273.1 | −1494.9 | −10.5 | 17.8 |
| | 3.18 | 1806.1 | −1605.9 | 1648.1 | −1922.0 | −9.6 | 16.4 |
| | 4.0 | 2014.6 | −1947.9 | 1891.2 | −2194.8 | −6.5 | 11.2 |

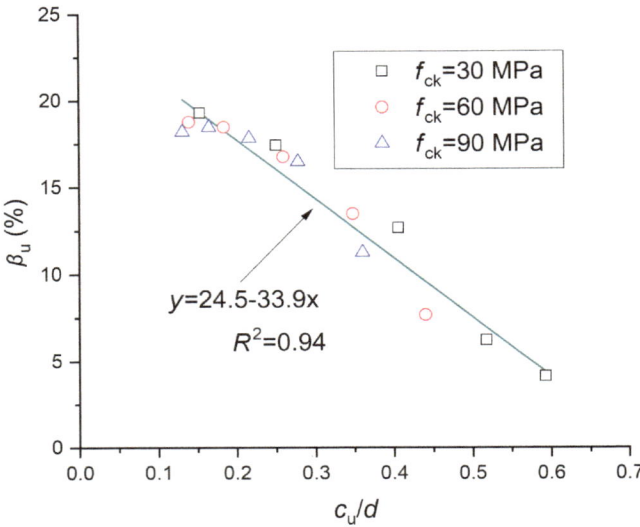

**Figure 7.13** Relationship between moment redistribution and neutral axis depth.

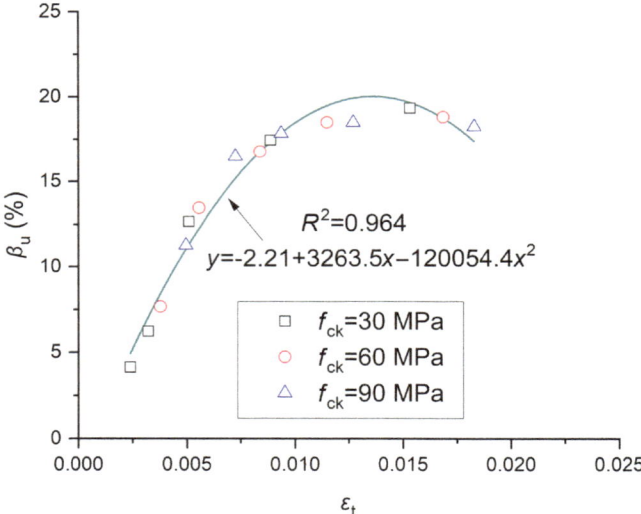

**Figure 7.14** Relationship between moment redistribution and tensile steel strain.

## 7.3 Bonded prestressed normal- and high-strength concrete members

A two-span prestressed concrete reference beam, as shown in Fig. 7.15, is considered. The tendon profile is assumed to be parabolic, with eccentricities of 0, 140,

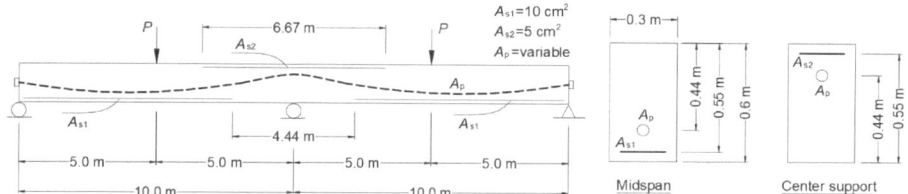

**Figure 7.15** Details of two-span prestressed concrete reference beam.

and 140 mm at the end support, midspan, and center support, respectively. The non-prestressed steel areas are (see Fig. 7.15): $A_{s1} = 10$ cm$^2$ and $A_{s2} = 5$ cm$^2$, while the tendon area $A_p$ is a variable that leads to the prestressing reinforcement ratio $\rho_p$ ranging widely from 0.15% to 1.29%. $\rho_p = A_p/(bd_p)$, where $b$ = cross-sectional width and $d_p$ = distance from the centroid of the tendon to the extreme compressive fiber at the critical section (center support or midspan). The material properties are as follows: $f_y = 450$ MPa; $E_s = 200$ GPa; $f_{pu} = 1860$ MPa; $f_{py} = 1674$ MPa; $E_p = 195$ GPa; and $f_{ck} = 40$, 60, and 90 MPa, covering both NSC and HSC. The initial prestress is taken as 1050 MPa. The finite element model consists of 36 elements with the same length.

## 7.3.1  Failure mode and crack pattern

Crushing failure is observed for all members. At failure, all the nonprestressed tensile steels over the critical sections have yielded, while the prestressing steel tendon may or may not reach its yield stress, depending on the prestressing reinforcement ratio and concrete strength as shown in Fig. 7.16. It is seen in the figure that at a prestressing reinforcement ratio lower than 0.44%, the tendon stresses at ultimate mobilized by different levels of concrete strength are almost identical. However, the discrepancy between the tendon stresses for different concrete grades becomes more and more pronounced as the prestressing reinforcement ratio increases. At moderate to high prestressing reinforcement ratios, a higher concrete strength mobilizes considerably higher ultimate tendon stress, indicating better exploitation of the tensile strength of prestressing tendons.

The crack pattern may be deduced from the strain distribution at ultimate in continuous prestressed members as illustrated in Fig. 7.17, where $X/L$ is the ratio of the distance from the end support to the span length. At a low prestressing reinforcement ratio, there are large tensile strains in concrete over the center support and midspan sections against slight ones over other regions, especially the region near the center support. This indicates there is crack concentration in the members, which is heavier over the center support than over the midspan, attributed to a lower amount of nonprestressed tensile steel provided over the negative moment region. As the prestressing reinforcement ratio increases, the maximum tensile strains at the critical sections are obviously reduced and, meanwhile, over the non-critical regions there appear more tensile strains greater than the cracking strain, indicating that the crack pattern is improved, namely, smaller crack width but more

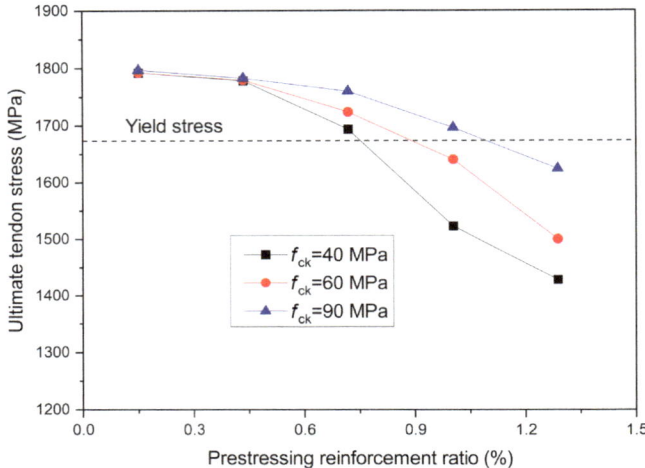

**Figure 7.16** Variation of ultimate tendon stress at midspan with prestressing reinforcement ratio for different levels of concrete strength.

extent of the crack zone. At a low prestressing reinforcement ratio of 0.15%, the concrete strength of 40 MPa registers a bit larger crack width over the midspan while the concrete strength of 90 MPa mobilizes an obviously bigger crack width over the center support, compared to other concrete grades investigated. On the other hand, at prestressing reinforcement ratios of 0.72% and 1.29%, a higher concrete strength leads to an obviously larger crack width over both the center support and midspan sections.

### 7.3.2 Load—deflection behavior

Fig. 7.18 shows the load—deflection curves of continuous prestressed members with different concrete grades and various prestressing reinforcement ratios. At a low prestressing reinforcement ratio of 0.15%, the load—deflection responses exhibit an obvious ductile stage due to the yielding of nonprestressed steel. This stage, however, is not so clear at high prestressing reinforcement ratios. This can be explained by the fact that at a high partial prestressing ratio, the structural behavior in the inelastic ranges is controlled by prestressing steel tendons that do not have a yield plateau as in the case of ordinary reinforcing steel. It is also seen that a higher concrete strength leads to higher flexural stiffness as well as larger cracking and ultimate loads, particularly obvious at a high prestressing reinforcement ratio.

Fig. 7.19 illustrates the variation of ultimate deflection with the prestressing reinforcement ratio for different levels of concrete strength. The ultimate deflection quickly decreases as the prestressing reinforcement ratio increases; the decrease in ultimate deflection of prestressed NSC members is quicker than that of prestressed HSC ones. Although the ultimate compressive strain of HSC is smaller than that of

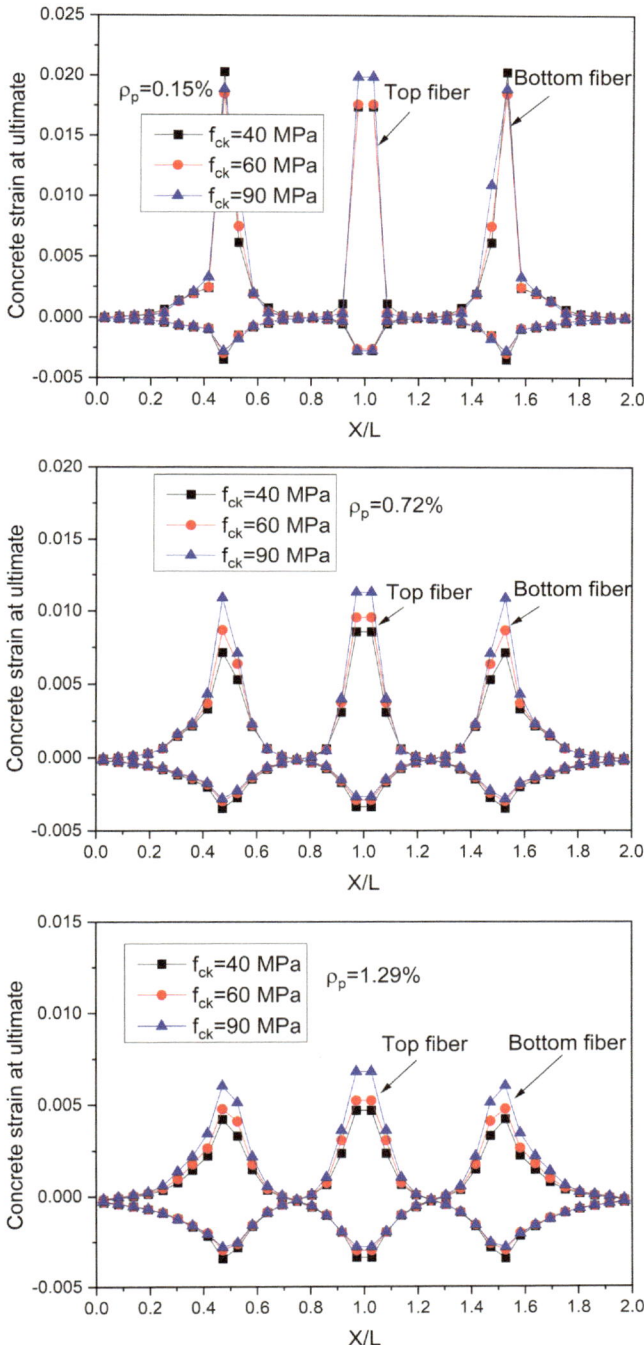

**Figure 7.17** Distribution of concrete strain at ultimate for different levels of concrete strength and various prestressing reinforcement ratios.

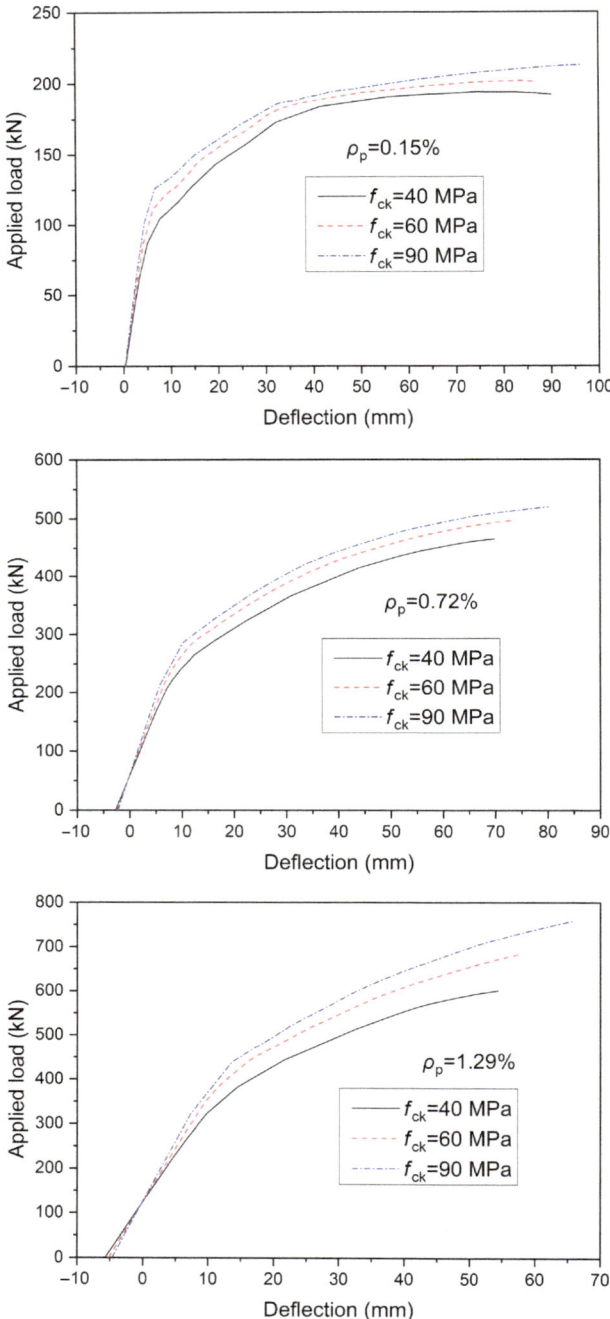

**Figure 7.18** Load−deflection behavior at midspan for different levels of concrete strength and various prestressing reinforcement ratios.

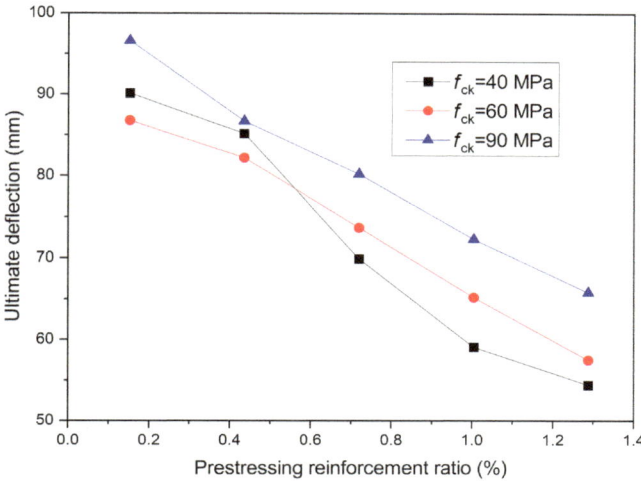

**Figure 7.19** Variation of ultimate midspan deflection with prestressing reinforcement ratio for different levels of concrete strength.

NSC, a prestressed HSC member generally exhibits a higher deformation capacity than a prestressed NSC one, especially at a high prestressing reinforcement ratio.

### 7.3.3 Variation of neutral axis depth

The neutral axis depth is recognized as a critical parameter related to ductility and moment redistribution in prestressed concrete members. The shift of the neutral axis within the center support or midspan section against the curvature development for continuous prestressed members with different concrete grades and various prestressing reinforcement ratios is shown in Fig. 7.20. It is observed that the neutral axis moves rapidly toward the extreme compressive fiber until the crack evolution stabilizes. After that, the movement of the neutral axis slows down. The responses of curvature versus neutral axis depth for different concrete grades are almost identical at first, but they differ after stabilization of crack development. At a given curvature, a higher concrete strength results in a lower neutral axis depth, especially at a high prestressing reinforcement ratio.

Fig. 7.21 illustrates the variation of neutral axis depth at ultimate with the prestressing reinforcement ratio for different levels of concrete strength. As the prestressing reinforcement ratio increases, the increase in neutral axis depth for a lower concrete strength tends to be faster than that for a higher concrete strength. The difference between the neutral axis depths for different levels of concrete strength appears to be insignificant at a low prestressing reinforcement ratio but is increasingly important with the increase of the prestressing reinforcement ratio.

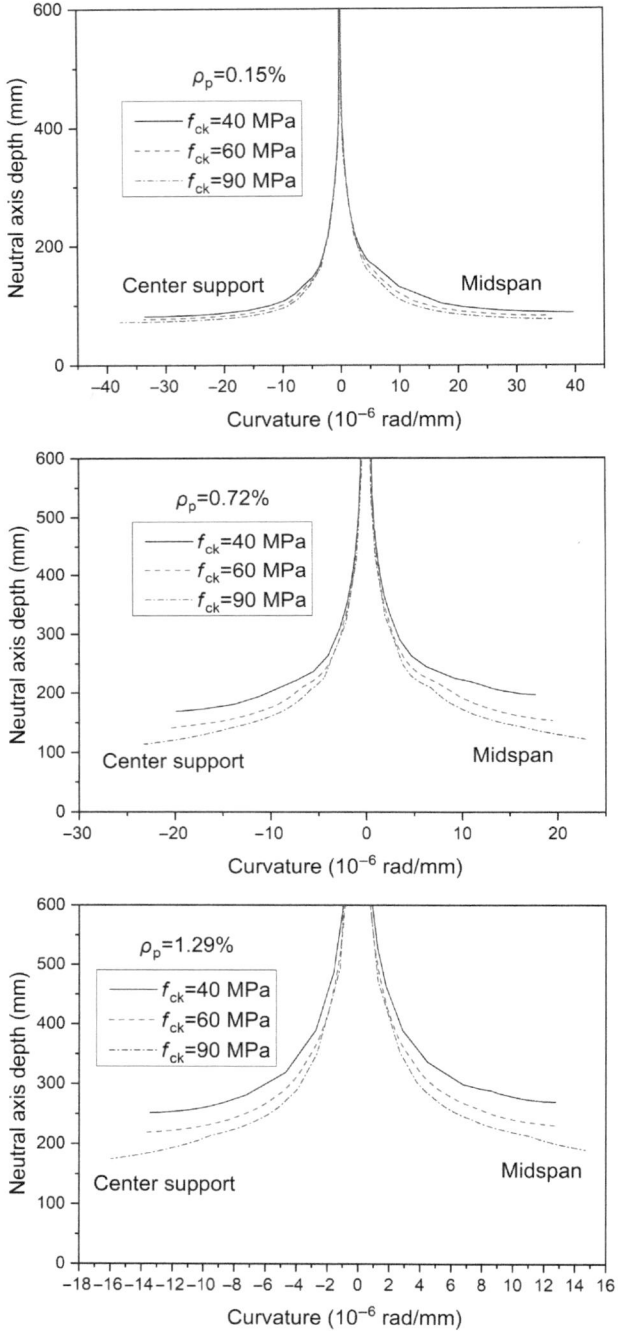

**Figure 7.20** Response of neutral axis depth versus curvature for different levels of concrete strength and various prestressing reinforcement ratios.

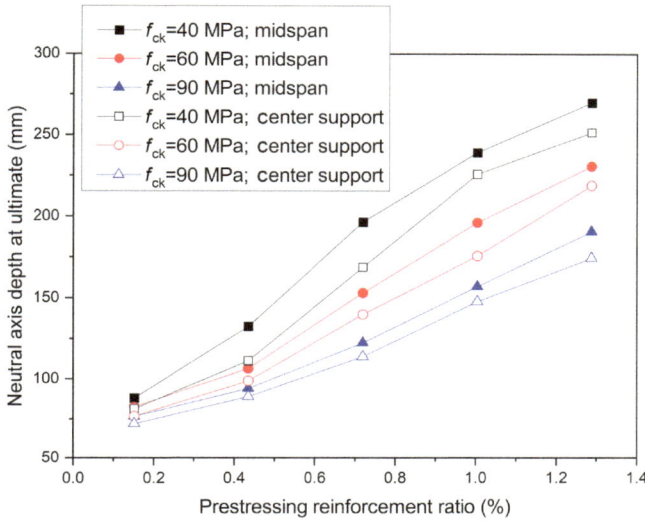

**Figure 7.21** Variation of neutral axis depth at ultimate with prestressing reinforcement ratio for different levels of concrete strength.

### 7.3.4 Strain in nonprestressed steel

In partially prestressed concrete members, the strain in nonprestressed tensile reinforcement is another important parameter, apart from the neutral axis depth, related to ductility and moment redistribution. Fig. 7.22 shows the development of nonprestressed steel strain for continuous prestressed members with different concrete grades and various prestressing reinforcement ratios. It is seen that the increase in nonprestressed steel strain is small in the elastic stage of loading. After cracking and yielding, the strain in nonprestressed steel increases much faster than before. Before second yielding (i.e., yielding at the midspan), the increase in nonprestressed steel strain at the center support is faster than that at the midspan. Thereafter, an opposite phenomenon is observed. NSC tends to mobilize faster increase in nonprestressed steel strain than HSC.

Fig. 7.23 shows the variation of nonprestressed steel strain at ultimate with the prestressing reinforcement ratio for different levels of concrete strength. It is seen that the ultimate strain in nonprestressed steel decreases quickly with the increase of prestressing reinforcement ratio. At a given prestressing reinforcement ratio, the ultimate strain in nonprestressed steel at a section for a lower concrete strength appears to be smaller than that for a higher concrete strength, except at a low prestressing reinforcement ratio.

### 7.3.5 Development of bending moments

The bending moment in a continuous prestressed member with nonconcordant tendon profile consists of the load-induced moment and prestress secondary moment.

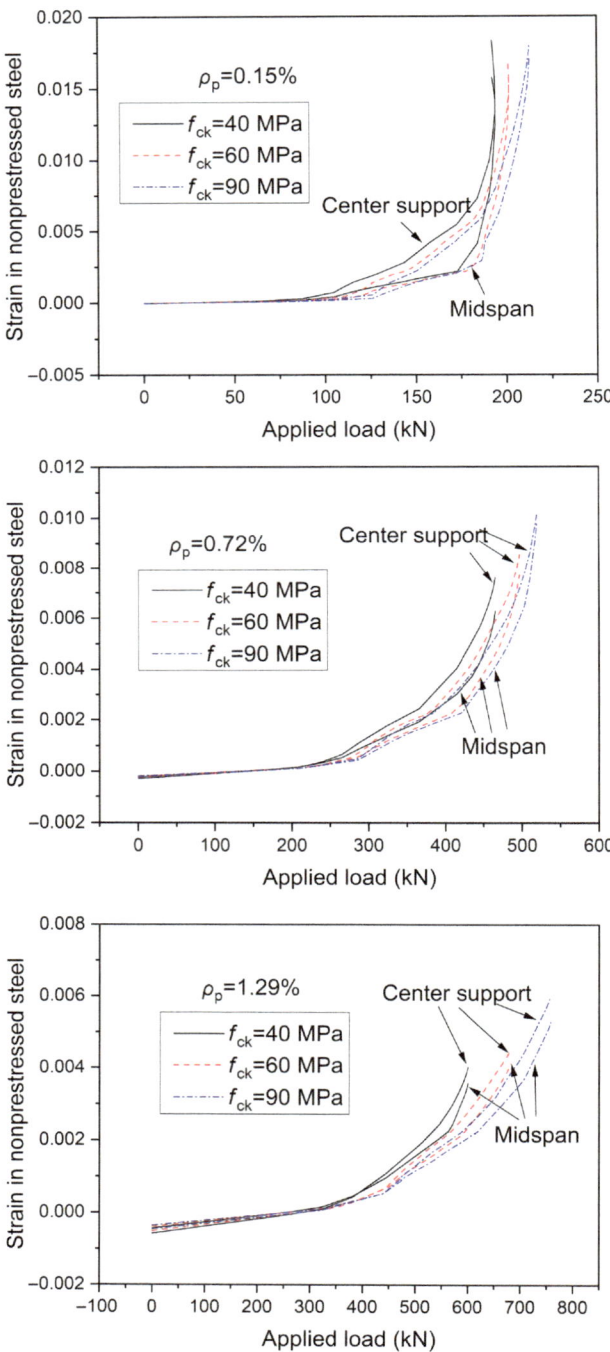

**Figure 7.22** Development of nonprestressed steel strain for different levels of concrete strength and various prestressing reinforcement ratios.

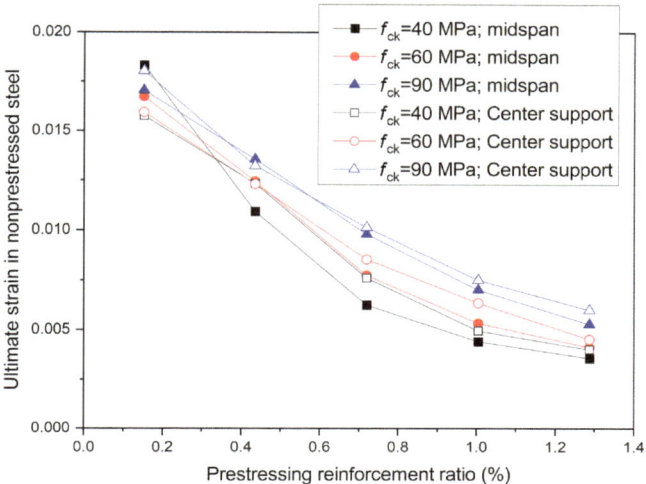

**Figure 7.23** Variation of ultimate strain in nonprestressed steel with prestressing reinforcement ratio for different levels of concrete strength.

For the tendon profile illustrated in Fig. 7.15, the prestressing produces positive secondary moments over the member. On the other hand, the (dead and live) loads produce positive moments at the midspan but negative ones at the center support. Therefore the presence of prestress secondary moments enlarges the positive moment at the midspan but reduces the negative moment at the center support.

Fig. 7.24 shows the development of bending moments in continuous prestressed members with different concrete grades and various prestressing reinforcement ratios. The continuous members exhibit linear load−moment behavior up to cracking, followed by nonlinear behavior due to the redistribution of moments. This nonlinear manner is rather pronounced at a low prestressing reinforcement ratio while it is obviously mitigated as the prestressing reinforcement ratio increases. The level of concrete strength appears to have no important influence on the moment development.

At the ultimate limit state, the distribution of actual and elastic moments (self-weight moment is not included) for different concrete grades and various prestressing reinforcement ratios is illustrated in Fig. 7.25. It can be seen that the actual moment is smaller than the elastic one over the center support while it is opposite over the midspan, indicating bending moments are redistributed from the center support toward the midspan. The maximum deviation takes place at the center support, and the deviation is reduced as the prestressing reinforcement ratio increases.

## 7.3.6 Degree of moment redistribution

Fig. 7.26 shows the development of the degree of moment redistribution for continuous prestressed members with different concrete grades and various prestressing

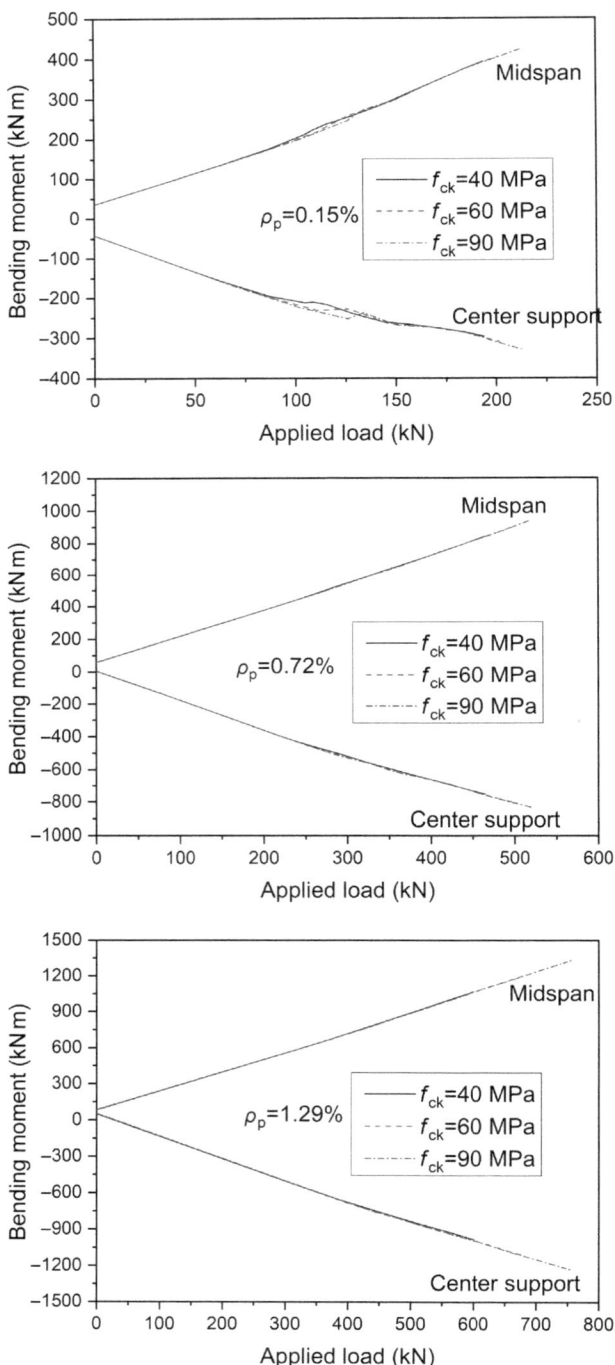

**Figure 7.24** Development of bending moment for different levels of concrete strength and various prestressing reinforcement ratios.

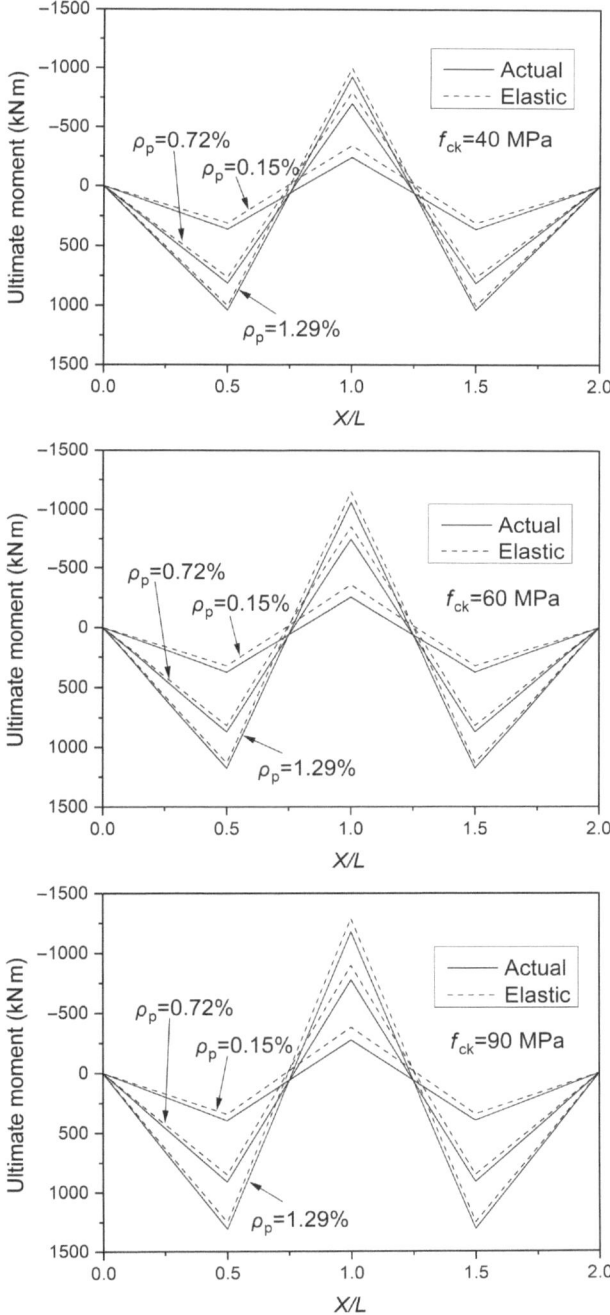

**Figure 7.25** Distribution of bending moment at ultimate for different levels of concrete strength and various prestressing reinforcement ratios.

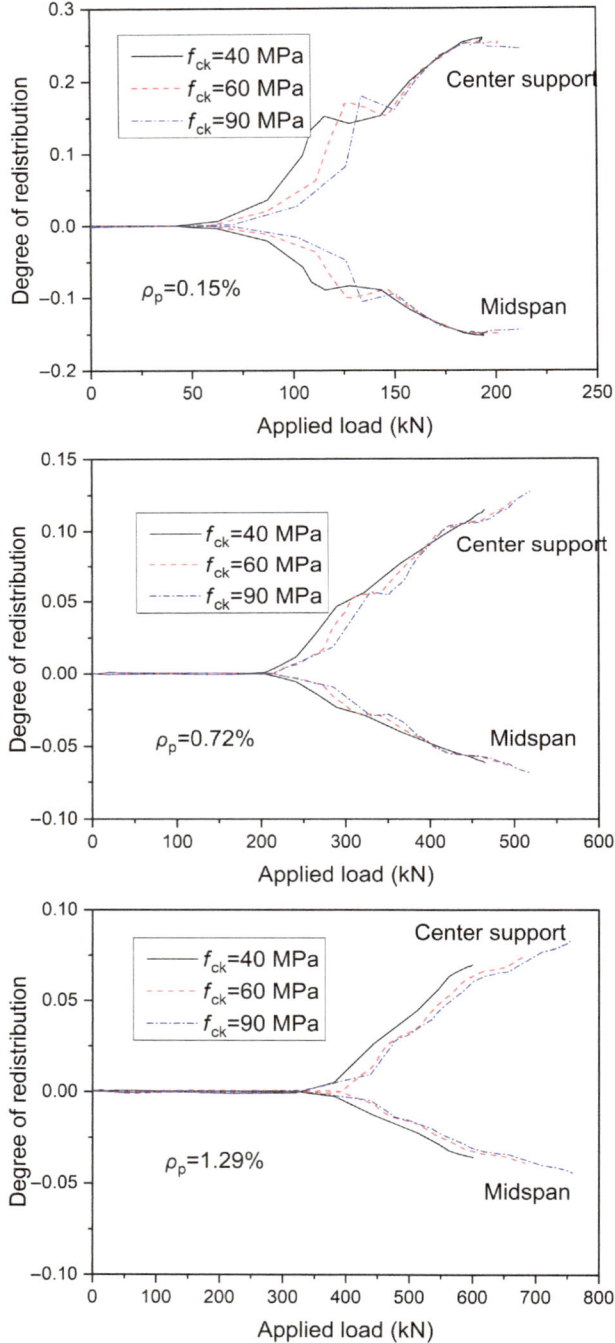

**Figure 7.26** Development of the degree of moment redistribution for different levels of concrete strength and various prestressing reinforcement ratios.

reinforcement ratios. After cracking, the evolution of moment redistribution at a low prestressing reinforcement ratio is typically influenced by some phases, namely, stabilizing crack evolution and first and second yielding of nonprestressed steel. The influence of these stages is alleviated as the prestressing reinforcement ratio increases.

Fig. 7.27 shows the variation of the degree of moment redistribution at ultimate ($\beta_u$) with the prestressing reinforcement ratio for different levels of concrete strength. It is seen that moment redistribution decreases with increasing prestressing reinforcement ratio, attributed to the reduction in flexural ductility and also to the change in relative stiffness between the center support and midspan sections. A higher concrete strength mobilizes a slower decrease in moment redistribution. At a low prestressing reinforcement ratio, the redistribution discrepancy between different concrete grades is negligible. At a high prestressing reinforcement ratio, HSC results in a bit higher redistribution of moments than NSC.

Fig. 7.28 illustrates the $\beta_u-\varepsilon_t$ curve by the ACI code as well as the FEA data at the center support. It can be seen that the ACI code can satisfactorily reflect the actual trend regarding the variation of $\beta_u$ with varying $\varepsilon_t$. Also, the FEA data are located above the ACI curve, indicating a conservative prediction by the ACI code. According to the fit curve as depicted in the figure, the following equation with the parameter $\varepsilon_t$ may be used to calculate the actual redistribution of moments in prestressed concrete continuous members:

$$\beta_u(\%) = 0.53 + 1382.39\varepsilon_t \tag{7.5}$$

Fig. 7.29 displays the $\beta_u-c_u/d$ curves of the CSA code (CSA, 2004) and EC2 (CEN, 2004) as well as FEA data at the center support. It is generally observed that both the CSA code and EC2 are able to reflect the tendency with respect to the variation of $\beta_u$ with varying $c_u/d$. However, it may be unsafe at a low value of $c_u/d$ for the CSA code and for EC2 when NSC is used. According to the fit curve as depicted in the figure, the following equation with the parameter $c_u/d$ may be used

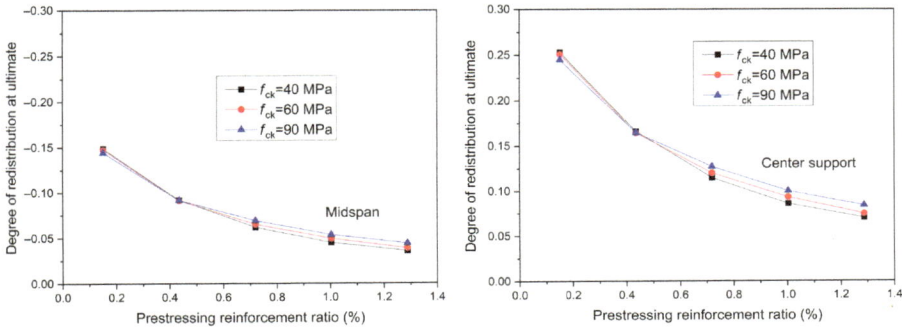

**Figure 7.27** Variation of the degree of moment redistribution at ultimate with prestressing reinforcement ratio for different levels of concrete strength.

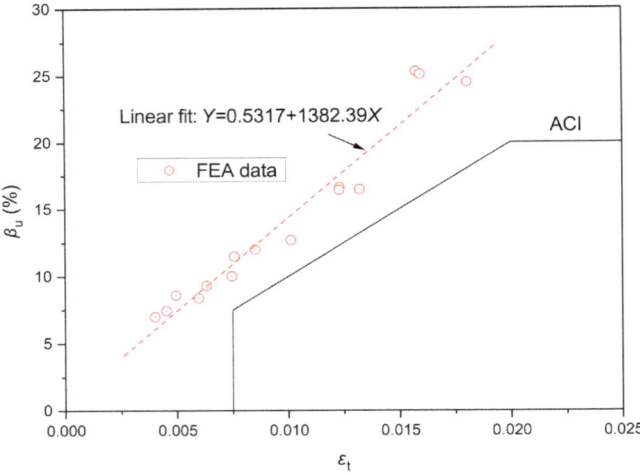

**Figure 7.28** ACI curve regarding $\beta_u - \varepsilon_t$ relationship and FEA data. *FEA*, Finite element analysis.

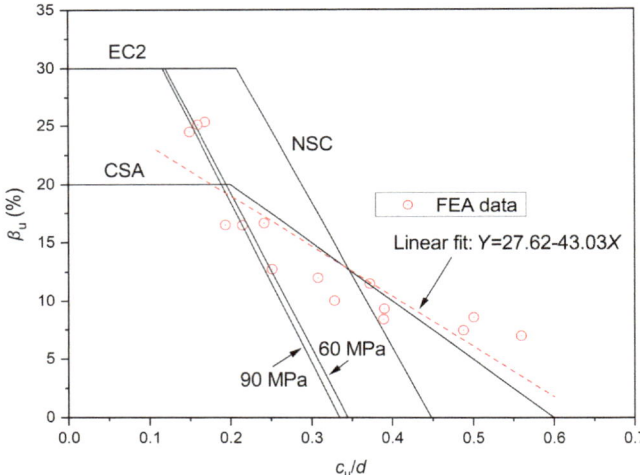

**Figure 7.29** CSA and EC2 curves regarding $\beta_u - c_u/d$ relationship and FEA data. *FEA*, Finite element analysis.

to calculate the actual redistribution of moments in prestressed concrete continuous members:

$$\beta_u(\%) = 27.62 - 43.03c_u/d \tag{7.6}$$

A comparison between the codes and numerical predictions regarding the variation of $\beta_u$ at the center support with the prestressing reinforcement ratio is illustrated in Fig. 7.30 and presented in Table 7.3. The ACI code well reflects the concrete strength

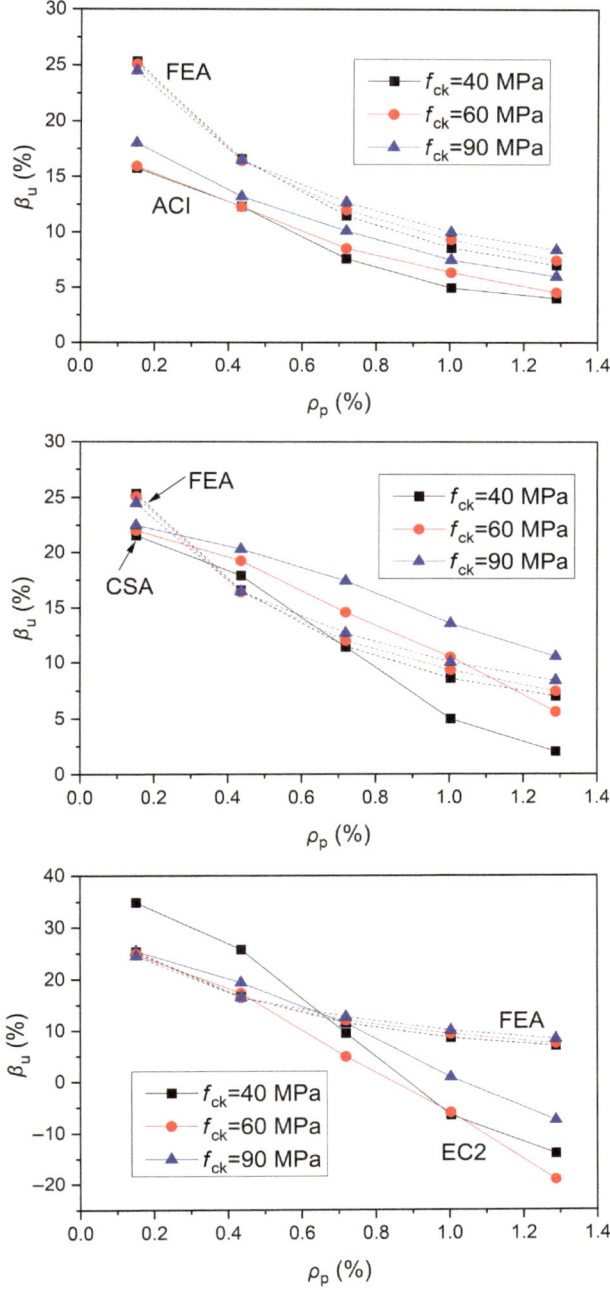

**Figure 7.30** Comparison between $\beta_u - \rho_p$ relationships by FEA and design codes for different levels of concrete strength. *FEA*, Finite element analysis.

**Table 7.3** Results in relation to moment redistribution at ultimate over the center support section for different concrete grades and various prestressing reinforcement ratios.

| $f_{ck}$ (MPa) | $\rho_p$ (%) | $c_u/d$ (%) | $\varepsilon_t$ (%) | $\beta_u$ (%) | | | |
|---|---|---|---|---|---|---|---|
| | | | | ACI | CSA | EC2 | FEA |
| 40 | 0.15 | 16.90 | 1.58 | 15.76 | 21.55 | 34.88 | 25.33 |
| | 0.44 | 24.19 | 1.23 | 12.32 | 17.91 | 25.76 | 16.60 |
| | 0.72 | 37.20 | 0.76 | 7.59 | 11.40 | 9.50 | 11.49 |
| | 1.00 | 50.02 | 0.49 | 4.95 | 4.99 | −6.53 | 8.58 |
| | 1.29 | 55.96 | 0.40 | 4.01 | 2.02 | −13.95 | 7.00 |
| 60 | 0.15 | 15.96 | 1.59 | 15.94 | 22.02 | 24.73 | 25.11 |
| | 0.44 | 21.51 | 1.23 | 12.31 | 19.25 | 17.32 | 16.43 |
| | 0.72 | 30.82 | 0.85 | 8.53 | 14.59 | 4.91 | 12.00 |
| | 1.00 | 39.01 | 0.63 | 6.35 | 10.50 | −6.01 | 9.32 |
| | 1.29 | 48.75 | 0.45 | 4.53 | 5.62 | −19.01 | 7.45 |
| 90 | 0.15 | 14.95 | 1.80 | 18.04 | 22.52 | 25.44 | 24.48 |
| | 0.44 | 19.38 | 1.32 | 13.22 | 20.31 | 19.36 | 16.47 |
| | 0.72 | 25.13 | 1.01 | 10.13 | 17.44 | 11.45 | 12.71 |
| | 1.00 | 32.83 | 0.75 | 7.48 | 13.59 | 0.87 | 10.03 |
| | 1.29 | 38.91 | 0.60 | 5.99 | 10.55 | −7.50 | 8.39 |

*FEA*, Finite element analysis.

effect and it is safe over the entire range of prestressing reinforcement ratio. However, the influence of concrete grade is not adequately included in the CSA code and EC2. It is also generally observed that the CSA code is safe for continuous prestressed NSC members but it is nonconservative for continuous prestressed HSC members. EC2 is conservative for continuous prestressed HSC members but it may be unsafe for continuous prestressed NSC members at a low prestressing reinforcement ratio.

## 7.4 Prestressed normal- and high-strength concrete members with external carbon fiber-reinforced polymer tendons

Fig. 7.31 illustrates a sketch of continuous beams with external CFRP tendons. The main variables are the concrete strength $f_{ck}$ and the tendon area $A_p$ (or tendon ratio $\rho_p$): $f_{ck} = 40$, 60, and 90 MPa; and $A_p = 200-1700$ mm$^2$ or equivalently $\rho_p = 0.15\%-1.29\%$. CFRP tendons have a tensile strength of 1840 MPa and an elastic modulus of 147 GPa. The initial prestress is 1104 MPa. The tensile reinforcing steel areas over sagging and hogging regions, $A_{s1}$ and $A_{s2}$, are 1600 and 800 mm$^2$, respectively. The compressive reinforcing steel area, $A_{s3}$, is 360 mm$^2$. All the steels have the same yield strength of 450 MPa and elastic modulus of 200 GPa.

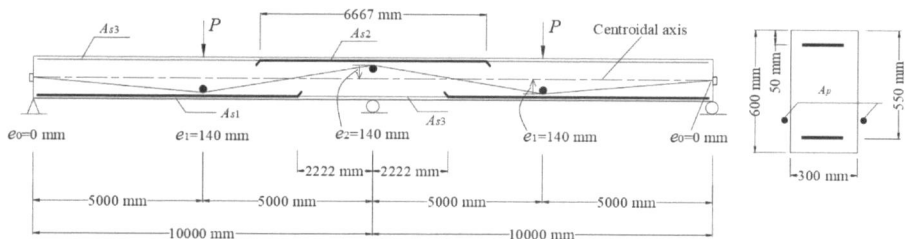

**Figure 7.31** Structure and cross-section of NSC and HSC members with external CFRP tendons. *CFRP*, Carbon fiber-reinforced polymer; *NSC*, normal-strength concrete; *HSC*, high-strength concrete.

## 7.4.1 Failure and cracking modes

All the members fail due to concrete crushing at one of the critical sections. Sufficient development of plastic hinges is observed in the members but the NSC member with a tendon area of 1700 mm$^2$. At failure, the external CFRP tendons are well below their rupture capacity. Crushing failure may take place at the midspan or inner support, depending on the concrete grade and tendon area. For $f_{ck} = 40$ MPa, crushing failure takes place at the midspan at a low tendon area while at the inner support at a high tendon area. For $f_{ck} = 60$ MPa, the midspan section is collapsed when $A_p$ is not greater than 1325 mm$^2$ and failure takes place at the inner support when $A_p = 1700$ mm$^2$. For $f_{ck} = 90$ MPa, failure always happens at the midspan, and the exploitation of the critical hogging region is improved as $A_p$ increases.

The concrete strain distribution at failure is shown in Fig. 7.32. At a low tendon area ($A_p = 200$ mm$^2$), the concrete tensile strains at the midspan and inner support are very large while in the other zones the strains are small. This indicates that there are big crack widths appearing in the critical sections against small crack widths in the noncritical regions. At a low tendon area, the crack width at the midspan appears to be larger than that at the inner support, while a high tendon area results in an opposite observation. At a low tendon area ($A_p = 200$ mm$^2$), NSC (i.e., $f_{ck} = 40$ MPa) leads to a slightly larger crack width at the critical section than HSC (i.e., $f_{ck} = 60$ and 90 MPa). As the tendon area increases, the maximum crack width at the critical section reduces. The reduction of the crack width is more effective for NSC than for HSC. As a result, when the tendon area increases to 595 mm$^2$ or above, NSC turns to produce a smaller crack width than HSC. The difference between the crack widths for NSC and HSC becomes increasingly apparent with the increasing area of external tendons. All the members with different concrete strengths and tendon areas exhibit almost the same crack zone.

## 7.4.2 Moment—curvature and load—deflection behavior

Fig. 7.33A and B shows the moment—curvature and load—deflection curves for different concrete strengths and tendon areas, respectively. Before the application of live loads (i.e., for the members under self-weight only), there is an initial moment

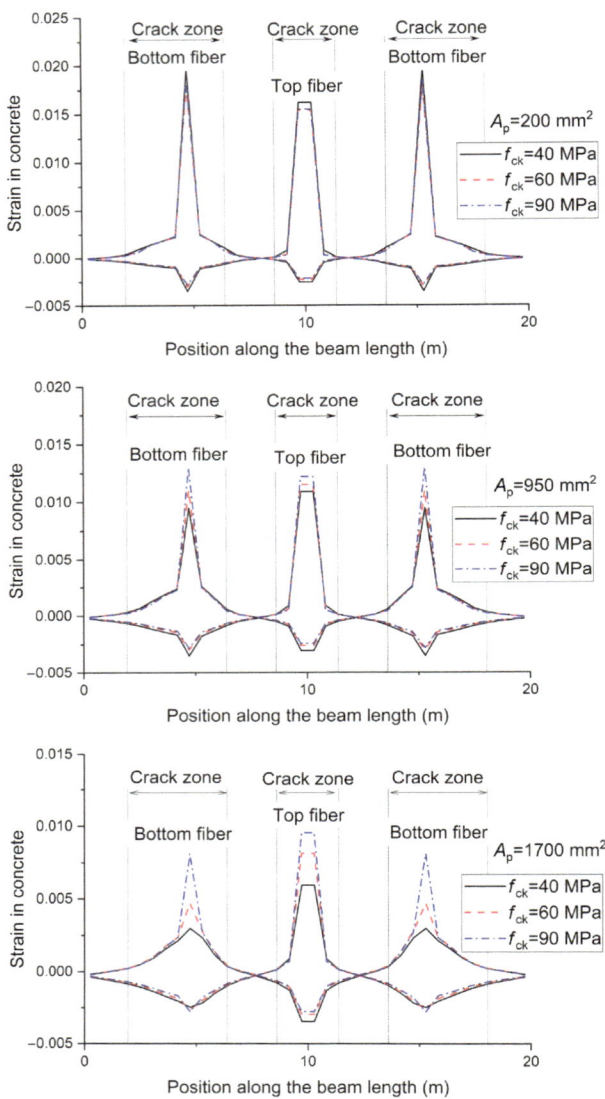

**Figure 7.32** Concrete strain distribution for different concrete strengths and tendon areas.

consisting of a self-weight-induced moment and a prestress secondary moment. Since external tendons are placed a bit below the concordant profile, the secondary moment produced by external prestressing is positive and the magnitude is rather small. Therefore the initial moment is a bit lower over the inner support and a bit higher over the midspan than the self-weight-induced moment. Because $A_{s1}$ is greater than $A_{s2}$, the sagging yield or ultimate moment is greater than the hogging

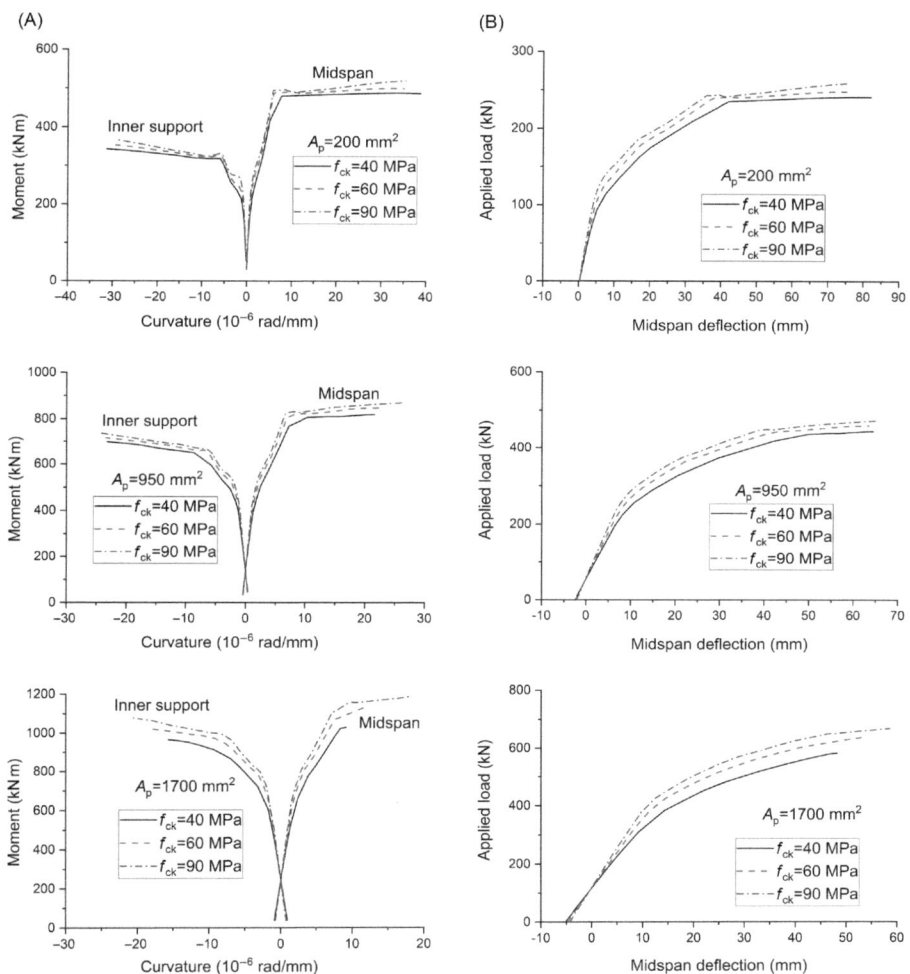

**Figure 7.33** Deformation development for different concrete strengths and tendon areas: (A) moment–curvature behavior; (B) load–deflection behavior.

one. The difference is notable at a low tendon area while it diminishes with increasing tendon area as a result of decreasing contribution of reinforcing steel. As expected, HSC produces higher flexural stiffness, that is, a smaller curvature or deflection at a given moment or load, when compared to NSC. All the members exhibit favorable ductile behavior, except for the NSC member with a tendon area of 1700 mm$^2$. The latter member does not exhibit an apparent yielding plateau as the tensile reinforcing steel at the midspan just gets to its yield strength at failure.

Fig. 7.34 displays variations of $P_u$ (ultimate load) and $\delta_u$ (ultimate deflection) with varying $\rho_p$ for different concrete strengths. Fig. 7.34A demonstrates that HSC

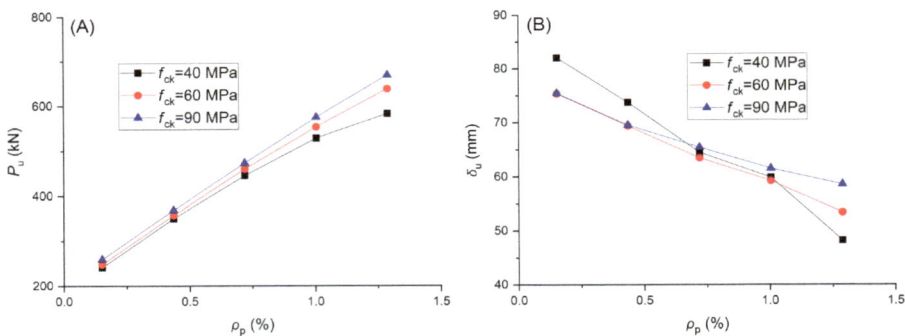

**Figure 7.34** Variations of $P_u$ and $\delta_u$ with varying $\rho_p$ for different concrete strengths: (A) $P_u$; (B) $\delta_u$.

produces a greater ultimate load than NSC. The load difference is slight at low $\rho_p$ levels and increasingly notable with increasing $\rho_p$. At low $\rho_p$ levels, the ultimate deflection for NSC ($f_{ck} = 40$ MPa) is obviously higher than that for HSC ($f_{ck} = 60$ or 90 MPa), as can be observed in Fig. 7.34B. The ultimate deflection decreases as $\rho_p$ increases. The decrease rate for NSC is greater than that for HSC. As a consequence, at a high $\rho_p$ level of 1.29%, the ultimate deflection for NSC turns out to be obviously lower than that for HSC.

### 7.4.3 Increase in tendon stress

The stress increase in external tendons for different concrete strengths and tendon areas is shown in Fig. 7.35. At given load levels, HSC produces a smaller stress increase than NSC. On the other hand, at a given deflection level, HSC results in a higher stress increase than NSC, which is particularly notable at a high $\rho_p$ level.

Fig. 7.36A shows that there is almost a linear relationship between $\Delta\sigma_p$ (ultimate tendon stress increase) and $\rho_p$. As $\rho_p$ increases, the value of $\Delta\sigma_p$ for NSC decreases rapidly, much quicker than that for HSC. HSC generally leads to a higher $\Delta\sigma_p$ than NSC, except at a low $\rho_p$ level.

ACI 318-19 recommends an equation for calculating the ultimate stress increase in unbonded tendons ($L/d_p \leq 35$):

$$\Delta\sigma_p = 70 + \frac{f_{ck}}{100\rho_p} \tag{7.7}$$

The value of $\Delta\sigma_p$ should not be greater than 414 MPa.

Fig. 7.36B shows that for $\rho_p$ greater than 0.44%, the ACI 318-19 predictions correspond well with the numerical results. Both the effects of the tendon area and concrete strength are satisfactorily reflected in ACI 318-19. However, when $\rho_p$ increases up to 0.44%, the ACI code significantly overestimates the rate of decrease of the tendon stress and also fails to reflect the concrete strength effect. Moreover,

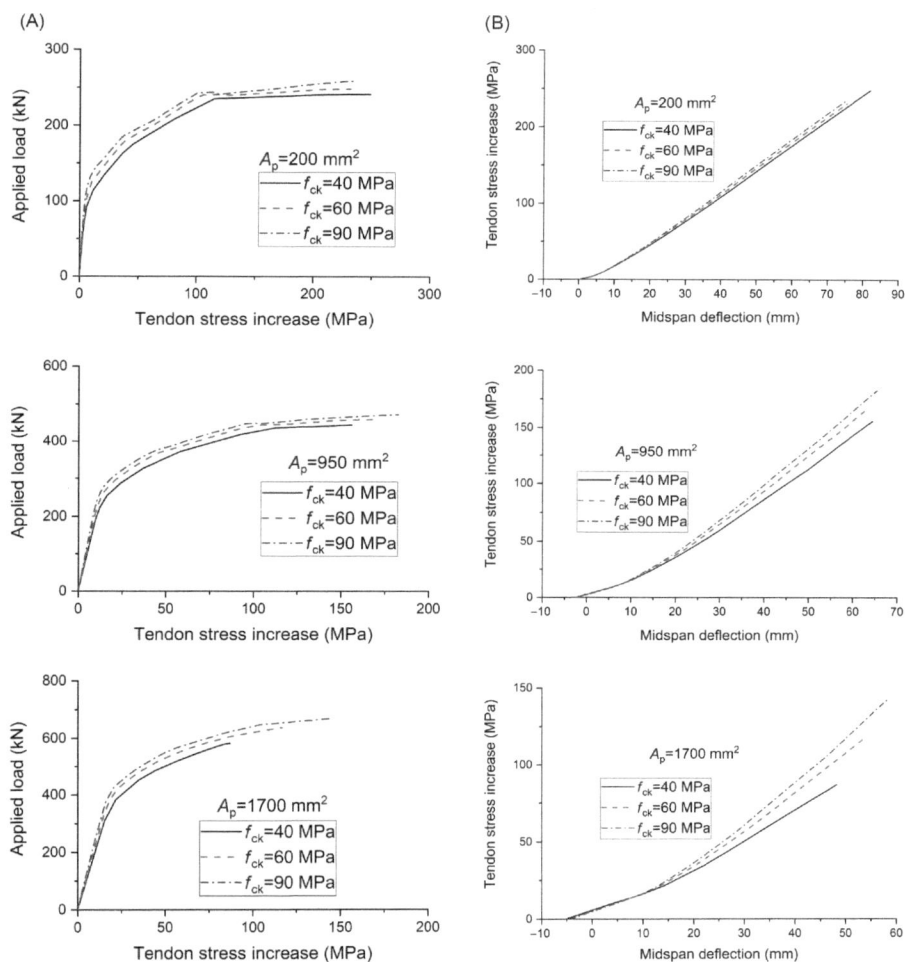

**Figure 7.35** Stress increase in external tendons for different concrete strengths and tendon areas: (A) with the applied load; (B) with the deflection.

at $\rho_p = 0.15\%$, ACI 318-19 leads to a substantial overestimate of the tendon stress, especially when HSC is used.

## 7.4.4 Neutral axis depth

The initial neutral axis depth is highly dependent on the tendon area. At a low tendon area ($A_p = 200$ mm$^2$), the prestressing effect cannot resist the self-weight effect. In this case, the initial neutral axis depth at the critical section is positive, with a large value. At a medium tendon area ($A_p = 950$ mm$^2$), the initial neutral axis depth is negative at the critical section. At a high tendon area ($A_p = 1700$ mm$^2$), the initial

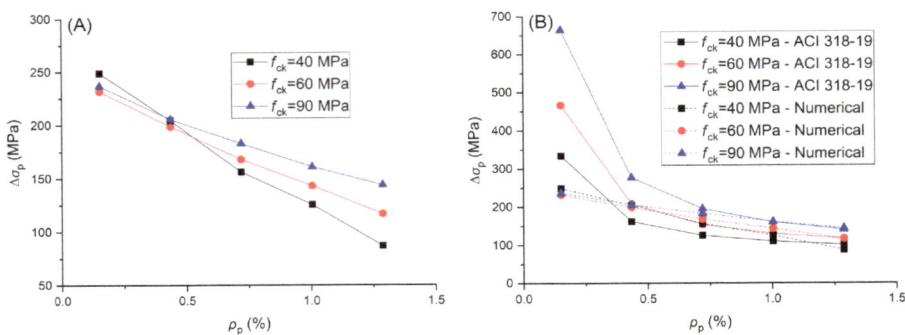

**Figure 7.36** Variation of $\Delta\sigma_p$ with varying $\rho_p$ for different concrete strengths: (A) numerical predictions; (B) comparison with ACI 318-19.

hogging and sagging curvatures are considerable. In this case, the initial neutral axis depth at the critical section is also positive but its magnitude is small. Fig. 7.37A and B shows the neutral axis depth versus the applied load and curvature relationships for different concrete strengths and tendon areas, respectively. It is seen that for specific loading levels, the inner support section exhibits a smaller neutral axis depth than the midspan section. For specific curvature levels, HSC leads to lower neutral axis depth than NSC, especially at a higher tendon area.

Fig. 7.38 shows the relation between $c_u$ (neutral axis depth at ultimate) and $\rho_p$ for different concrete strengths. The $c_u$ value increases with increasing $\rho_p$, and the rate of increase for NSC is much faster than that for HSC. In addition, at given $\rho_p$ levels, HSC produces lower $c_u$ than NSC. The difference is slight for a low $\rho_p$ level and increasingly pronounced with increasing $\rho_p$.

### 7.4.5 Stress and strain in reinforcing steel

The load versus stresses in reinforcing steel curves for different concrete strengths and tendon areas are illustrated in Fig. 7.39. It is seen that concrete cracking affects remarkably the stress evolution in tensile reinforcing steel, but its influence on compressive reinforcing steel is not apparent. The stress increase in the tensile reinforcing steel is slow at first while it turns to be rapid after cracking. The tensile reinforcing steel at the inner support reaches its yield strength first, followed by yielding at the midspan. For HSC members, all the compressive reinforcing steels do not get to their yield strength at ultimate. For NSC members, on the other hand, the compressive reinforcing steel may reach its yield strength at a high tendon area. In general, a high tendon area is more likely to cause the yielding of compressive steels at the critical sections. At given load levels, NSC leads to a higher tensile or compressive steel strain (stress also if not having yielded) than HSC.

Fig. 7.40 demonstrates the relation between $\varepsilon_{su}$ (strain in reinforcing steel at ultimate) and $\rho_p$ for different concrete strengths. As $\rho_p$ increases, the tensile steel strain decreases while the compressive steel strain increases. The midspan section exhibits larger strain in tensile steel at low $\rho_p$ levels while a smaller one at high $\rho_p$ levels

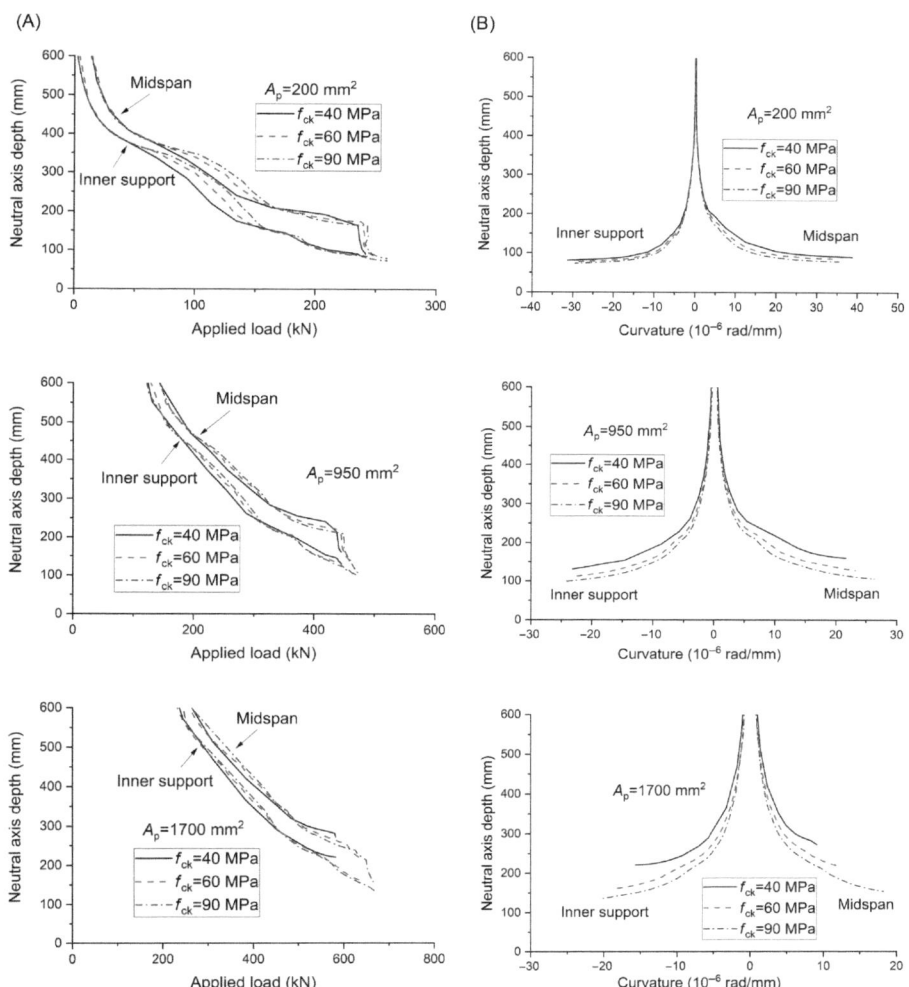

**Figure 7.37** Development of neutral axis depth for different concrete strengths and tendon areas: (A) with the applied load; (B) with the curvature.

than the inner support section. In addition, the midspan section generally produces higher strain in compressive steel than the inner support section, except for NSC at a high $\rho_p$ level of 1.29%. NSC generally leads to a lower strain in tensile steel and a higher strain in compressive steel than HSC.

### 7.4.6  Moment redistribution

Fig. 7.41A and B shows the development of moments (actual $M$ and elastic $M_e$) and moment redistribution ($\beta = 1 - M/M_e$), respectively. The actual load−moment

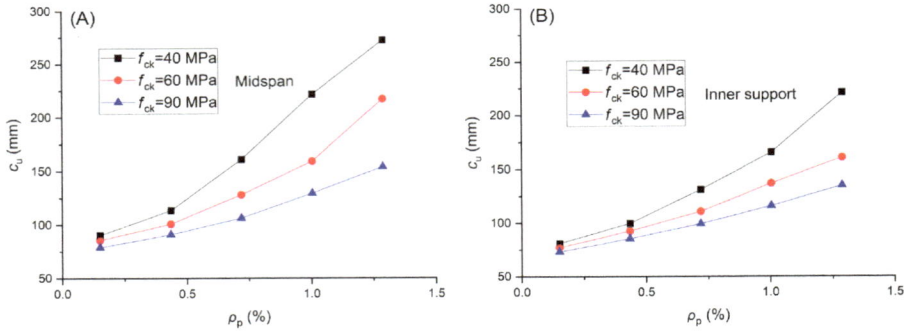

**Figure 7.38** Variation of $c_u$ with varying $\rho_p$ for different concrete strengths: (A) midspan; (B) inner support.

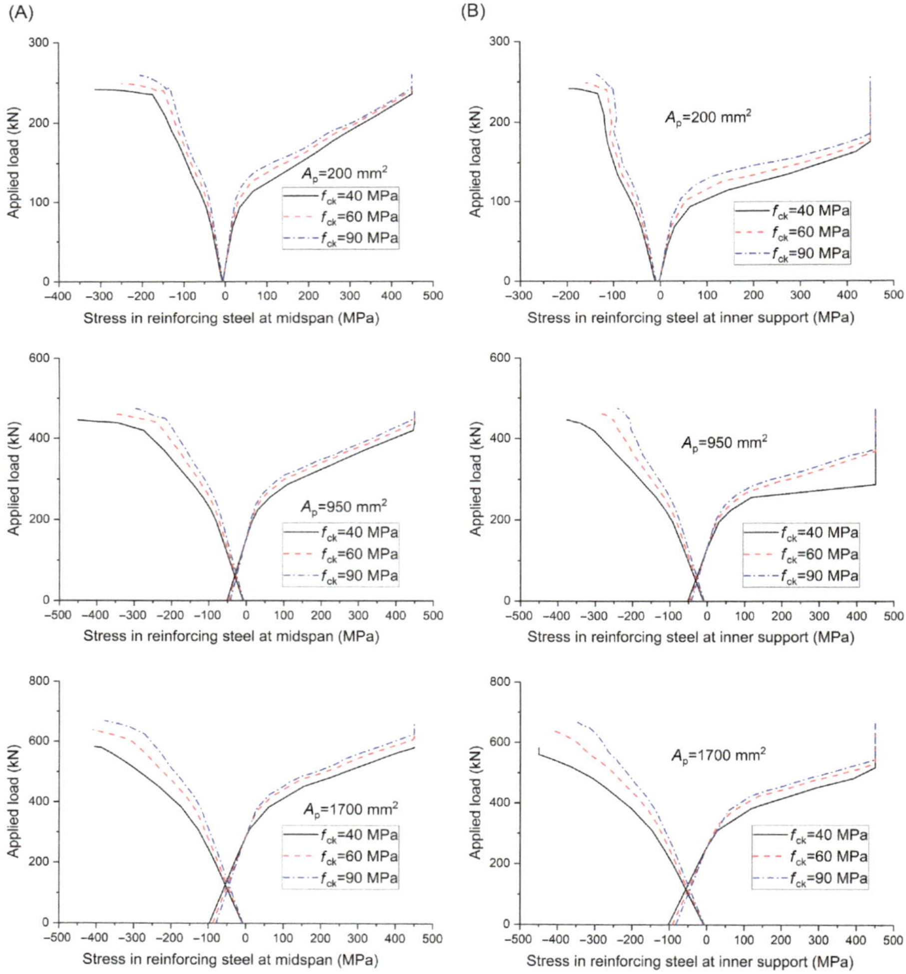

**Figure 7.39** Development of the stress in reinforcing steel for different concrete strengths and tendon areas: (A) midspan; (B) inner support.

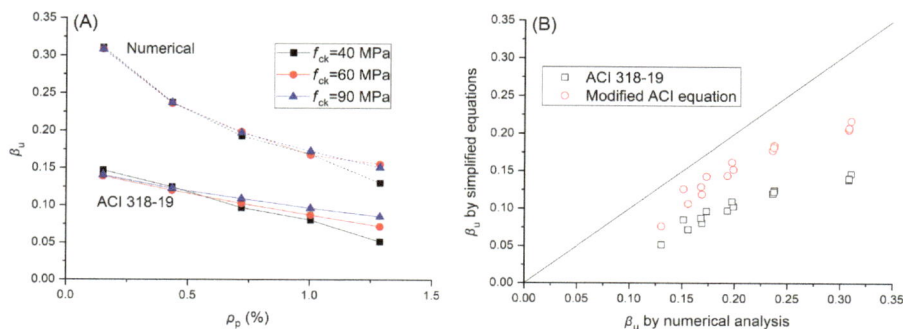

**Figure 7.43** Comparison of $\beta_u$ values: (A) ACI 318-19 and numerical predictions; (B) simplified equations against numerical analysis.

where $\lambda_{aci}$ is a parameter related to $A_{s2}/A_{s1}$, that is,
  when $A_{s2}/A_{s1} \leq 1$,

$$\lambda_{aci} = 0.65 - 1.2\ln(A_{s2}/A_{s1}) \tag{7.9a}$$

and when $A_{s2}/A_{s1} > 1$,

$$\lambda_{aci} = 0.65 + 0.67\ln(A_{s2}/A_{s1}) - 2.76\ln^2(A_{s2}/A_{s1}) \tag{7.9b}$$

A comparison of the simplified equations (i.e., ACI and modified ACI equations) against the numerical predictions for $\beta_u$ of the prestressed NSC and HSC members with external CFRP tendons is illustrated in Fig. 7.43B. It is seen that by introducing the structure-related factor $A_{s2}/A_{s1}$, the modified ACI equation, that is, Eq. (7.8), demonstrates a better correlation than ACI 318-19.

# 7.5 Conclusions

Reinforced HSC members exhibit higher ultimate load and deflection than reinforced NSC members. The load difference between NSC and HSC members is increasingly notable with the increasing steel ratio. Reinforced HSC members have higher flexural ductility despite the higher brittleness of HSC material compared to reinforced NSC members. For $\rho_{s2} = 0.73\% - 4.0\%$, reinforced HSC members ($f_{ck} = 90$ MPa) show $21\% - 91\%$ higher curvature ductility factor than reinforced NSC members ($f_{ck} = 30$ MPa). Reinforced HSC members exhibit lower neutral axis depth at ultimate than reinforced NSC members, especially notable at a high steel ratio. The ultimate strain in tensile steel bars in reinforced HSC members is obviously higher than that of reinforced NSC ones. At the same load level, the difference between support reactions or bending moments in reinforced HSC and NSC members is marginal, indicating similar moment redistribution in these members.

At ultimate, reinforced HSC members have higher moment redistribution than reinforced NSC members, except for $\rho_{s2} = 0.73\%$. Both flexural ductility and moment redistribution are closely related to the neutral axis depth or strain in tensile steel bars at ultimate. Based on the data of the analysis, practical formulae reflecting their relationships are suggested.

With respect to bonded prestressed NSC and HSC members, it is demonstrated that increasing the prestressing reinforcement ratio improves effectively the crack pattern, that is, the crack width is reduced whereas the crack zone is extended. HSC can better exploit the tendon tensile strength at moderate to high prestressing reinforcement ratios and generally mobilizes higher member deformation capacity than NSC. A higher concrete strength mobilizes a lower neutral axis depth at ultimate. The discrepancy between neutral axis depths by different concrete grades is small at a low prestressing reinforcement ratio but becomes increasingly important with an increasing prestressing reinforcement ratio. Prestressed HSC members generally exhibit higher nonprestressed steel strain than prestressed NSC ones. Increasing the prestressing reinforcement ratio results in a quick decrease in moment redistribution, attributed to the reduction in flexural ductility and to the change in relative stiffness between critical sections. A higher strength generally leads to a bit higher moment redistribution in continuous prestressed members. The ACI code, where the parameter $\varepsilon_t$ is adopted, can well reflect the effect of concrete strength on the moment redistribution in continuous prestressed concrete members but the CSA code and EC2, where the parameter $c_u/d$ is adopted, cannot. Therefore the strain $\varepsilon_t$ seems to be better than the $c_u/d$ ratio when used for predicting moment redistribution. When predicting moment redistribution, the ACI code is conservative; the CSA code is safe for prestressed NSC members but unsafe for prestressed HSC members; EC2 is conservative for prestressed HSC members but nonconservative for prestressed NSC members at a low prestressing reinforcement ratio.

The results of analysis of continuous prestressed NSC and HSC members with external CFRP tendons show that NSC members with a high tendon area may fail to exhibit favorable ductile behavior, while the counterpart members made of HSC are able to develop a sufficient rotation of plastic hinges. HSC produces larger crack widths at the critical sections than NSC except at a low $\rho_p$ level. HSC leads to higher ultimate load-carrying capacity than NSC, particularly obvious at high $\rho_p$ levels. HSC produces a smaller ultimate deflection at low $\rho_p$ levels but a larger one at high $\rho_p$ levels than NSC. HSC generally produces higher ultimate stress in external tendons than NSC. In calculating the ultimate tendon stress, ACI 318-19 satisfactorily reflects the impact of concrete grade and tendon area for $\rho_p$ greater than 0.44%. However, ACI 318-19 is nonconservative at a low $\rho_p$ level, especially when HSC is used. HSC produces a lower $c_u$ value than NSC, and the difference is increasingly significant with increasing $\rho_p$. The $c_u$ value for the inner support is lower than that for the midspan. HSC leads to higher ultimate strain in tensile reinforcing steel than NSC, except at low $\rho_p$ levels. The ultimate strain in compressive reinforcing steel for HSC is lower than that for NSC. The inner support section exhibits smaller strain in tensile reinforcing steel than the midspan section at low $\rho_p$ levels, while the observation is opposite at high $\rho_p$ levels. The concrete strength has negligible impact on moment redistribution in these members provided that the members have sufficient plastic rotational capacities. The predictive model of Eq. (7.8) is more accurate than ACI 318-19 when predicting the moment redistribution in this structural typology.

# References

ACI Committee 318. (2019). *Building code requirements for structural concrete (ACI 318-19) and commentary (ACI 318R-19)*. Farmington Hills, MI.

Ahmed, M., Sheikh, M. N., Hadi, M. N. S., & Liang, Q. Q. (2023). Numerical simulation of axially loaded square high-strength concrete short columns with steel equal-angles as longitudinal reinforcement. *Engineering Structures*, *276*, 115391. Available from https://doi.org/10.1016/j.engstruct.2022.115391.

Arslan, G., & Cihanli, E. (2010). Curvature ductility prediction of reinforced high-strength concrete beam sections. *Journal of Civil Engineering and Management*, *16*(4), 462−470. Available from https://doi.org/10.3846/jcem.2010.52.

Bernardo, L. F. A., & Lopes, S. M. R. (2004). Neutral axis depth versus flexural ductility in high-strength concrete beams. *Journal of Structural Engineering*, *130*(3), 452−459. Available from https://doi.org/10.1061/(asce)0733-9445(2004)130:3(452).

Bouzid, H., & Kassoul, A. (2016). Curvature ductility of high strength concrete beams according to Eurocode 2. *Structural Engineering and Mechanics*, *58*(1), 1−19. Available from https://doi.org/10.12989/sem.2016.58.1.001.

Buch, S. H., & Suhail, R. (2022). Bond-slip behaviour of deformed rebars in normal and high strength concrete at elevated temperature-a review. *Australian Journal of Structural Engineering*, *23*(3), 189−204. Available from https://doi.org/10.1080/13287982.2022.2055816.

Carmo, R. N. F., & Lopes, S. M. R. (2008). Available plastic rotation in continuous high-strength concrete beams. *Canadian Journal of Civil Engineering*, *35*(10), 1152−1162.

CEN. (2004). *Eurocode 2 (EC2): Design of concrete structures. Part 1-1: General rules and rules for buildings*. EN 1992-1-1, Brussels, Belgium.

CSA. (2004). Design of concrete structures. A23.3-04, Mississauga, Ontario, Canada.

Diab, A. M., Hussein, M. A., Elyamany, H. E., & Al Ashy, H. M. (2014). Properties of pull-out bond strength and concept to assess ultimate bond stress of NSC and HSC. *Magazine of Concrete Research*, *66*(17), 877−895. Available from https://doi.org/10.1680/macr.14.00009.

Galano, L., & Vignoli, A. (2008). Strength and ductility of HSC and SCC slender columns subjected to short-term eccentric load. *ACI Structural Journal*, *105*(3), 259−269.

Hong, S. H., Choi, J. S., Yuan, T. F., & Yoon, Y. S. (2023). A review on concrete creep characteristics and its evaluation on high-strength lightweight concrete. *Journal of Materials Research and Technology*, *22*, 230−251. Available from https://doi.org/10.1016/j.jmrt.2022.11.125.

Kassoul, A., & Bougara, A. (2010). Maximum ratio of longitudinal tensile reinforcement in high strength doubly reinforced concrete beams designed according to Eurocode 8. *Engineering Structures*, *32*(10), 3206−3213. Available from https://doi.org/10.1016/j.engstruct.2010.06.009.

Ko, M. Y., Kim, S. W., & Kim, J. K. (2001). Experimental study on the plastic rotation capacity of reinforced high strength concrete beams. *Materials and Structures, 34*(5), 302−311. Available from https://doi.org/10.1007/bf02482210.

Lam, J. Y. K., Ho, J. C. M., & Kwan, A. K. H. (2009). Flexural ductility of high-strength concrete columns with minimal confinement. *Materials and Structures, 42*(7), 909−921. Available from https://doi.org/10.1617/s11527-008-9431-5.

Lou, T., Liu, M., Lopes, S. M. R., & Lopes, A. V. (2017). Moment redistribution in two-span prestressed NSC and HSC beams. *Materials and Structures, 50*(6), 1−10. Available from https://doi.org/10.1617/s11527-017-1116-5.

Lou, T., Lopes, S. M. R., & Lopes, A. V. (2014). Evaluation of moment redistribution in normal-strength and high-strength reinforced concrete beams. *Journal of Structural Engineering, 140*(10), 04014072. Available from https://doi.org/10.1061/(asce)st.1943-541x.0000994.

Lou, T., Min, D., Sun, W., & Chen, B. (2020). Numerical assessment of continuous prestressed NSC and HSC members with external CFRP tendons. *Composite Structures, 234*, 111671. Available from https://doi.org/10.1016/j.compstruct.2019.111671.

Ma, C. K., Sulaiman, M. F., Apandi, N., Awang, A. Z., Omar, W., & Jaw, S. W. (2019). Ductility and stiffness of slender confined reinforced high-strength concrete columns under monotonic axial load. Measurement, 146, 838-845.

Shaaban, I. G., & Mustafa, T. S. (2021). Towards efficient structural and serviceability design of high-strength concrete T-beams. *Proceedings of the Institution of Civil Engineers: Structures and Buildings, 174*(10), 836−848. Available from https://doi.org/10.1680/jstbu.19.00081.

Sheikh, S. A., Laine, D., & Cui, C. (2013). Behavior of normal- and high-strength confined concrete. *ACI Structural Journal, 110*(6), 989−999. Available from https://doi.org/10.14359/51686154.

Shi, S., Pang, M., & Lou, T. (2023). Numerical assessment on continuous reinforced normal-strength concrete and high-strength concrete beams. *Buildings, 13*(5), 1157. Available from https://doi.org/10.3390/buildings13051157.

Yuan, T. F., Hong, S. H., Choi, J. S., & Yoon, Y. S. (2021). Evaluation on the microstructure and durability of high-strength concrete containing electric arc furnace oxidizing slag. *Materials, 14*(5), 1304. Available from https://doi.org/10.3390/ma14051304.

# Using FRP rebars instead of steel rebars in simply supported concrete members with external tendons

**8**

## 8.1 Introduction

Steel corrosion would lead to the deterioration of reinforced or prestressed concrete members (Ahmad, 2003). An effective solution to this problem is to replace steel reinforcement with fiber-reinforced polymer (FRP). In addition to their noncorrosive property, FRP composites are high-strength, lightweight, and nonmagnetic. These composite materials are increasingly used for strengthening structural elements (Mostofinejad et al., 2019; Wang et al., 2019; Zhou et al., 2018). Different types of FRP composites are available, such as aramid fiber-reinforced polymer (AFRP), basalt fiber-reinforced polymer (BFRP), carbon fiber-reinforced polymer (CFRP), and glass fiber-reinforced polymer (GFRP). Unlike steel reinforcement with ductile characteristics, FRP composites are linear-elastic materials without yielding (ACI Committee 440, 2004, 2006). In addition, the FRP modulus of elasticity is usually lower than that of steel reinforcement (ACI Committee 440, 2004, 2006). Hence, some concerns about the use of FRP reinforcement instead of steel reinforcement may arise, for example, ductility and deflection issues due to the lack of yielding and low modulus of elasticity for FRP composites.

External prestressing has been widely used in the retrofit or construction of various concrete structures. Extensive research has been conducted concerning prestressed concrete members with external FRP tendons. Grace and Abdel-Sayed (1998) carried out tests on four specimens to examine the behavior of prestressed concrete beams with combined externally unbonded and internally bonded CFRP tendons. Their test results showed that the ductility can be improved by strengthening using draped external CFRP tendons. Ghallab and Beeby (2005) presented the test results of 16 prestressed concrete members with external AFRP tendons. They studied several factors influencing the ultimate stress in external FRP or steel tendons and proposed a modification to the BS8110 equation. Aziz et al. (2005) developed an analytical method to predict the load−deflection behavior of prestressed concrete members with external CFRP tendons. Wang et al. (2015) tested four specimens including three prestressed concrete members with external BFRP tendons and one control reinforced concrete member. Their study showed that using BFRP composites as external tendons for strengthening reinforced concrete beams can effectively improve structural performance. Zhu et al. (2017) experimentally

Prestressed Members with External Fiber-Reinforced Polymer (FRP) Tendons. DOI: https://doi.org/10.1016/B978-0-443-23877-2.00008-0

investigated the influence of bending angle and radius on the performance of external FRP tendons. They concluded that the load capacities of FRP tendons are reduced as the bending angle increases or the bending radius decreases. Both numerical and experimental investigations (Bennitz et al., 2012; Lou et al., 2012) showed that behaviors for prestressed concrete members with external CFRP and steel tendons are similar.

In prestressed concrete members with external tendons, some bonded rebars need to be provided to avoid behaving as a tied arch and to limit the crack width and spacing (ACI Committee 318, 2019). Previous studies demonstrated that externally prestressed concrete members without bonded rebars exhibit significant crack concentration, which can be effectively relieved by providing bonded rebars (Lou et al., 2012). In existing works on prestressed concrete members with external FRP tendons, steel rebars were usually provided. The performance of conventional reinforced concrete members with FRP rebars instead of steel ones has been extensively studied (Alam & Hussein, 2017; Barris et al., 2017; Dundar et al., 2015; Ju et al., 2018). However, the findings obtained from the study of reinforced concrete members with FRP/steel rebars may not be valid for prestressed concrete members with external FRP tendons as the latter is a different structural system from the former due to the presence of external prestressing. The influence of using FRP rebars instead of steel ones on the behavior of prestressed concrete members with external FRP tendons needs to be addressed.

This chapter presents a numerical and analytical work performed to evaluate the feasibility of providing FRP rebars instead of steel ones in single-span prestressed concrete members with external CFRP tendons (Pang et al., 2020). A comprehensive numerical assessment of simply supported externally prestressed concrete members with various types and areas of bonded rebars is introduced. An analytical model able to predict the tendon stress at ultimate and flexural strength in simply supported externally prestressed concrete members with FRP or steel rebars is also presented.

## 8.2    Numerical assessment

A simply supported prestressed concrete beam with external tendons draped at two deviators, as shown in Fig. 8.1, is used. The depths of external tendons at the end supports and deviators are 300 and 500 mm, respectively. The prestressing tendons are 1000 mm$^2$ in area, made of CFRP composites having a tensile strength of 1840 MPa and a modulus of elasticity of 147 GPa. The initial prestress before prestress transfer is 1104 MPa, namely 60% of the tensile strength. The area of tensile rebars, $A_r$, varies from 360 to 3560 mm$^2$. Thus the value of $\rho_r$ (ratio of tensile rebars) ranges from 0.22% to 2.16%, where $\rho_r = A_r/(bd_r)$ in which $b$ is the cross-sectional width; and $d_r$ is the depth of tensile rebars. The area of compressive rebars is 360 mm$^2$. Three types of bonded rebars are considered, namely, steel, CFRP, and GFRP. The mechanical properties and stress—strain characteristics of the rebars are presented in Table 8.1 and Fig. 8.2, respectively. The concrete strength $f_{ck}$ is 60 MPa.

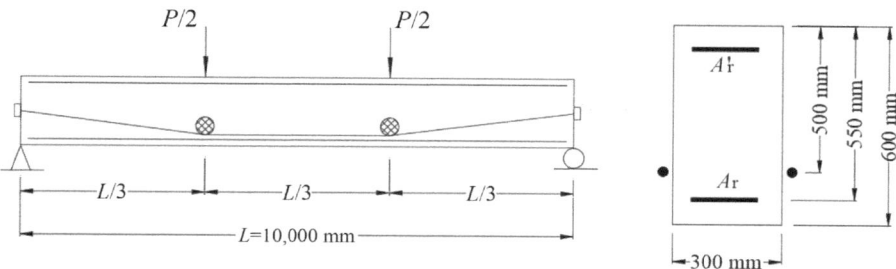

**Figure 8.1** Details of simply supported beams with fiber-reinforced polymer/steel rebars.

**Table 8.1** Mechanical properties of rebars.

| Rebars | Ultimate strength (MPa) | Yield strength (MPa) | Elastic modulus (GPa) | Rupture strain (%) |
|--------|-------------------------|----------------------|-----------------------|---------------------|
| Steel  | 450  | 450 | 200 | >3.5 |
| CFRP   | 1840 | –   | 147 | 1.25 |
| GFRP   | 750  | –   | 40  | 1.88 |

## 8.2.1  Failure and cracking modes

Fig. 8.3 shows the ultimate strain distribution in the top and bottom fibers of the members with different $\rho_r$ levels. The failure and cracking modes can be seen in the graphs of this figure. Failure of all the members occurs due to concrete crushing at midspan when the specified ultimate compressive strain of 0.003 is reached. For a member with no rebars ($\rho_r = 0.0\%$), the noncritical sections are still far below their ultimate strain capacity at failure. When a minimum content of tensile rebars is provided ($\rho_r = 0.22\%$), the exploitation of noncritical sections is improved. Such improvement is pronounced for FRP (especially CFRP) rebars and relatively not so notable for steel rebars. However, at a high $\rho_r$ level of 2.16%, the exploitation of noncritical sections for steel rebars is nearly comparable to that for FRP rebars. At failure, both FRP tendons and rebars do not reach their rupture strength while the tensile steel rebars have yielded.

Cracks of concrete occur once the tensile strain reaches its cracking strain. Over the cracking zone, the crack width may be represented by the ultimate tensile strain. In the case of $\rho_r = 0\%$, there appears a huge tensile strain at the midspan against marginal ones over the other zones. This indicates an unfavorable crack mode, that is, the member has only one large crack at the midspan, and the concrete over other zones is nearly uncracked. By providing a minimum content of tensile rebars ($\rho_r = 0.22\%$), the crack mode is substantially improved, that is, the crack width at the midspan is reduced and more cracks occur at the noncritical

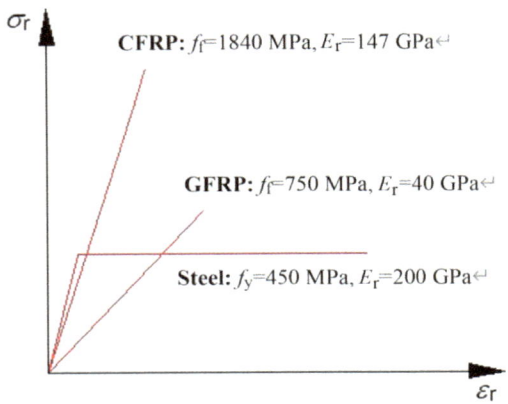

**Figure 8.2** Stress−strain characteristics of rebars ($\sigma_r$ = rebar stress; $\varepsilon_r$ = rebar strain; $f_f$ = FRP rupture strength; $E_r$ = rebar elastic modulus; and $f_y$ = steel rebar yield strength). *FRP*, Fiber-reinforced polymer

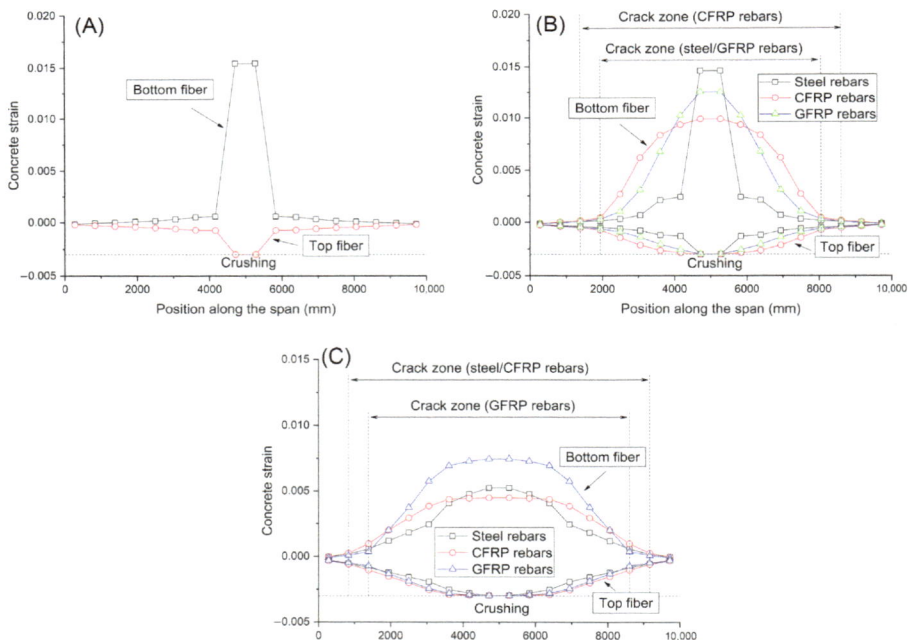

**Figure 8.3** Concrete strain distribution at ultimate: (A) $\rho_r = 0.0\%$; (B) $\rho_r = 0.22\%$; (C) $\rho_r = 2.16\%$.

zones. The use of CFRP rebars is shown to be more effective than the use of GFRP or steel rebars in improving the crack mode. At a high $\rho_r$ level of 2.16%, the crack modes for the members with CFRP and steel rebars are similar, while GFRP rebars lead to smaller crack zones and larger crack width over the flexural span than steel rebars.

### 8.2.2 Tendon stress development

Fig. 8.4A shows the stress increase in external tendons versus midspan deflection curves for the members with different types of rebars ($\rho_r = 1.19\%$). There is a roughly linear relationship between the tendon stress and the deflection. The slopes for the members with steel, CFRP, and GFRP rebars are 2.12, 2.2, and 2.46 MPa/mm, respectively. Fig. 8.4B shows the stress increase in external tendons with the applied load. The members with FRP rebars exhibit bilinear behavior with a turning point due to concrete cracking, while the members with steel rebars exhibit trilinear

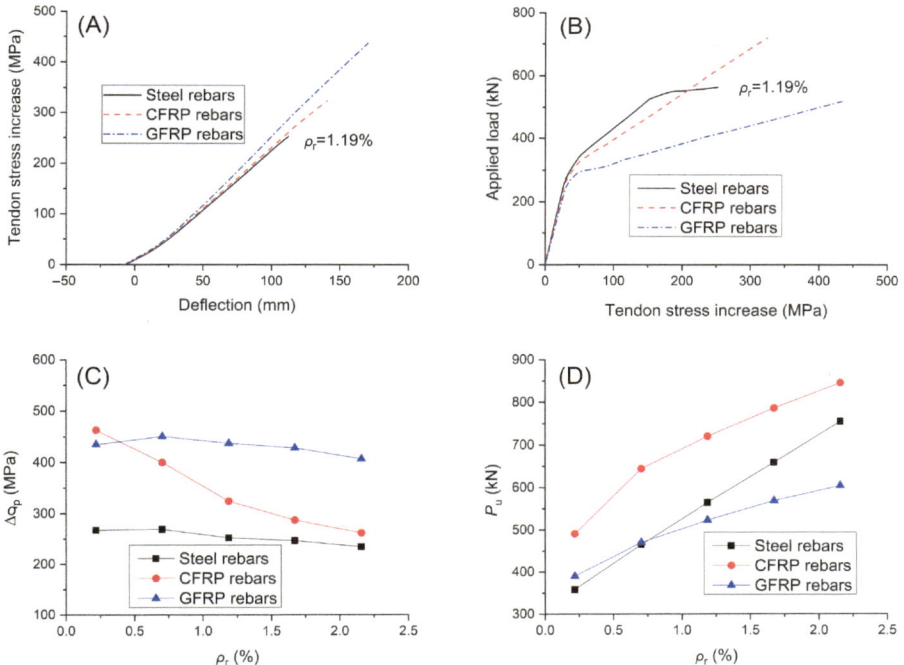

**Figure 8.4** Tendon stress development for the beams with different types of rebars: (A) midspan deflection versus tendon stress increment; (B) load versus tendon stress increase; (C) variation of ultimate tendon stress increment with varying $\rho_r$; (D) variation of ultimate load with varying $\rho_r$.

behavior with turning points due to concrete cracking and steel yielding, respectively. Since the elastic behavior is dominated by concrete, the type of rebars appears to have practically no influence on the tendon stress evolution in the elastic range of loading. Beyond cracking, GFRP rebars develop substantially lower rebar stress because of the smaller elastic modulus and, therefore, a higher stress increase in external tendons is required at a given load level to satisfy the force equilibrium, when compared to CFRP or steel rebars.

The variations of the ultimate stress increment in external tendons ($\Delta\sigma_p$) and the ultimate load ($P_u$) with the $\rho_r$ level are illustrated in Fig. 8.4C and D, respectively. As $\rho_r$ increases, the value of $\Delta\sigma_p$ quickly decreases for the members with CFRP rebars while decreasing slightly for the members with GFRP or steel rebars. As expected, GFRP rebars result in substantially higher (around 70% higher) $\Delta\sigma_p$ than steel rebars. At $\rho_r = 0.22\%$, the $\Delta\sigma_p$ ratio between the members with CFRP and steel rebars is as high as 1.73, attributed to the fact that CFRP rebars lead to much better exploitation of noncritical sections than steel rebars. This ratio is reduced to 1.12 at $\rho_r = 2.16\%$. This can be explained by the comparable exploitation of noncritical sections of the members with CFRP and steel rebars as mentioned in the previous section. The ultimate load is dependent upon the ultimate stresses in external tendons and tensile rebars. Due to higher reinforcement stresses at ultimate, CFRP rebars lead to substantially higher (37.1% higher) ultimate load than steel rebars at $\rho_r = 0.22\%$. The difference tends to decrease with increasing $\rho_r$ due to the reduced difference between reinforcement stresses in the members with CFRP and steel rebars. At a low $\rho_r$ level of 0.22%, GFRP rebars lead to a slightly higher ultimate load than steel rebars. As $\rho_r$ increases, the increase in ultimate load for the members with GFRP rebars is slower than that for the members with steel rebars. As a result, when $\rho_r$ increases to a level greater than 0.77%, the ultimate load for the members with GFRP rebars turns out to be lower than that for the members with steel rebars.

## 8.2.3  Deformation behavior

The moment−curvature and load−deflection curves for the members with different types of rebars ($\rho_r = 1.19\%$) are shown in Fig. 8.5A and B, respectively. In the precracking stage, the effect of rebars on the response characteristics is negligible due to slight stress increments in rebars. The responses for the members with FRP and steel rebars differ after cracking because the rebar contribution becomes increasingly important. The reduction in flexural stiffness due to cracking is highly dependent on the rebar modulus of elasticity, that is, the higher the rebar modulus of elasticity, the less the reduction in member stiffness. Therefore at a given load level, GFRP rebars lead to higher postcracking deformation than CFRP or steel rebars. The postcracking deformation for the members with FRP rebars develops linearly until failure. For the member with steel rebars, a significantly further reduction in flexural stiffness occurs on steel yielding.

Fig. 8.5C and D illustrates the variation of ultimate midspan curvature ($\kappa_u$) and deflection ($\Delta_u$) with varying $\rho_r$, respectively. It is seen in Fig. 8.5C that the value

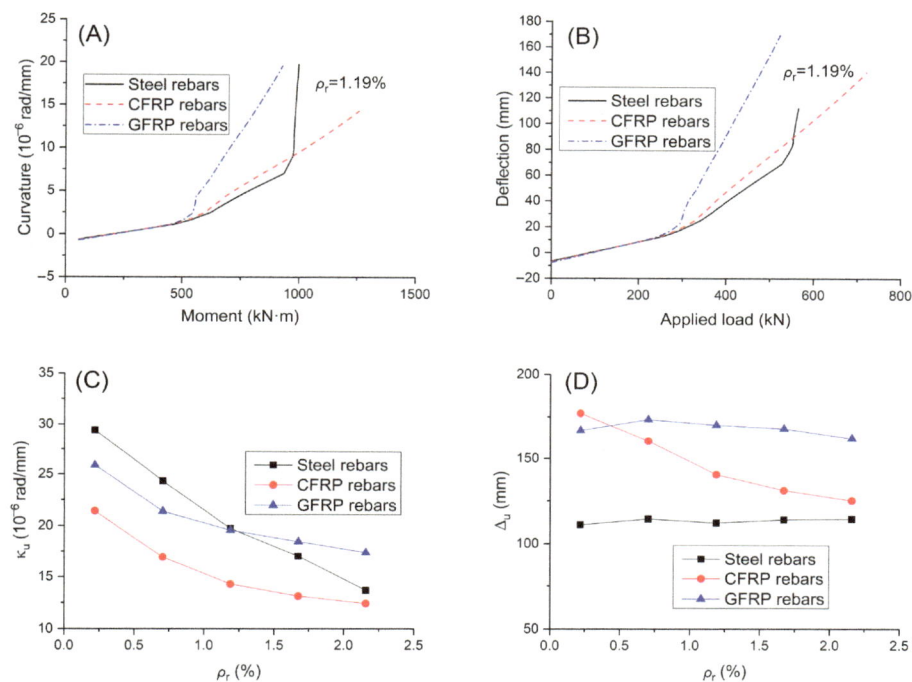

**Figure 8.5** Deformation behavior for the beams with different types of rebars: (A) moment versus curvature at midspan; (B) load versus midspan deflection; (C) variation of ultimate curvature with varying $\rho_r$; (D) variation of ultimate deflection with varying $\rho_r$.

of $\kappa_u$ decreases as $\rho_r$ increases. GFRP rebars mobilize a smaller $\kappa_u$ at low $\rho_r$ levels whereas a larger $\kappa_u$ at high $\rho_r$ levels in comparison with steel rebars. CFRP rebars mobilize lower $\kappa_u$ than steel rebars; the difference is substantial at low $\rho_r$ levels but reduced with increasing $\rho_r$. It is noted that the above observation is similar to the effect of rebars on the crack width (represented by the ultimate concrete tensile strain) as discussed previously. This is attributed to that the ultimate curvature is directly proportional to the ultimate concrete tensile strain in terms of the plane section hypothesis. As shown in Fig. 8.5D, GFRP rebars register substantially higher $\Delta_u$ than steel rebars because of a significantly lower modulus of elasticity. CFRP rebars register considerably higher $\Delta_u$ than steel rebars at low $\rho_r$ levels, while the difference is reduced with increasing $\rho_r$. This could also be attributed to the rebar effect on the exploitation of noncritical sections, as explained previously.

## 8.2.4 Neutral axis depth and rebar strain

Since the neutral axis depth and rebar strain are key parameters describing flexural ductility, it is important to well understand their behavior. The movement of the neutral axis, after it rises to the bottom fiber of the midspan section, with the

moment for the members with different types of rebars ($\rho_r = 1.19\%$) is illustrated in
Fig. 8.6A. For the members with FRP rebars, the neutral axis shifts rapidly with
increasing moments and then the shift gradually slows down. Similar behavior is
observed for the member with steel rebars until yielding. Thereafter a fast movement of the neutral axis is resumed due to steel yielding. Fig. 8.6B shows the strain
development in FRP and steel rebars with the bending moment ($\rho_r = 1.19\%$). The
behavior of the member with steel rebars exhibits three stages with transitions
caused by cracking and yielding, respectively. Two-stage behavior is observed for
the members with FRP rebars because of the lack of yielding. Due to a lower modulus of elasticity, GFRP rebars exhibit a significantly faster increase in strain after
cracking but a slower stress development, when compared to steel or CFRP rebars.

Fig. 8.6 C and D shows the variation of neutral axis depth ($c_u$) and tensile bar
strain at ultimate ($\varepsilon_r$) with varying $\rho_r$, respectively. Comparing these graphs to the
graph of Fig. 8.5C, it is seen that the effect of rebars on $\kappa_u$ is opposite to that on $c_u$
while coincident to that on $\varepsilon_r$. This observation can be explained by their theoretical
relationships. According to the plane section hypothesis, $c_u$ and $\varepsilon_r$ are related to $\kappa_u$
by: $c_u = \varepsilon_u/\kappa_u$; $\varepsilon_r = \kappa_u(d_r - c_u) = \kappa_u d_r - \varepsilon_u$, where $d_r$ is the depth of tensile rebars,
equal to 550 mm; $\varepsilon_u$ is the ultimate concrete compressive strain, equal to 0.003.

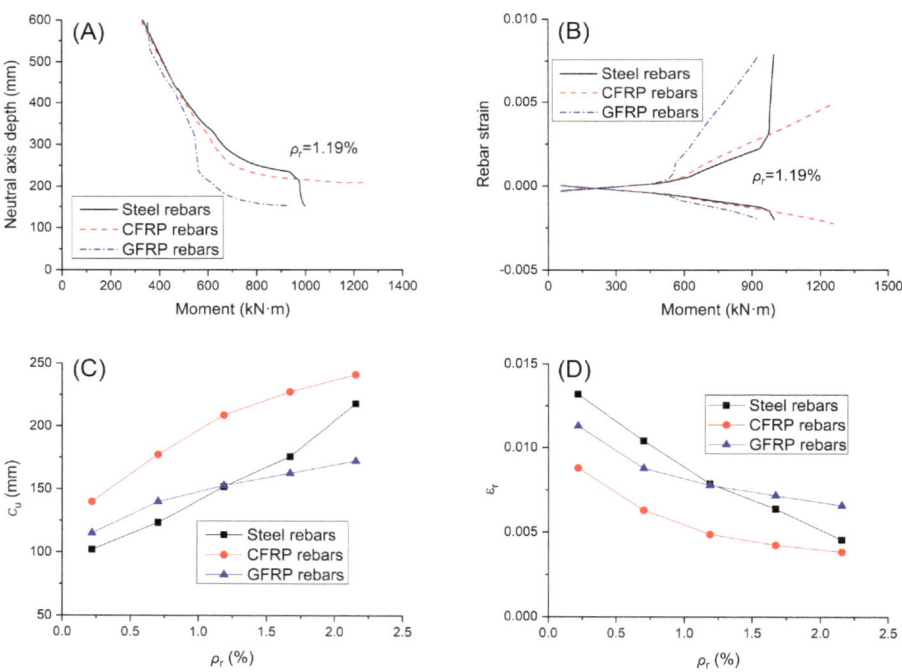

**Figure 8.6** Neutral axis depth and bar strain for different types of rebars: (A) moment versus
neutral axis depth at midspan; (B) moment versus rebar strain at midspan; (C) variation of
ultimate neutral axis depth with varying $\rho_r$; (D) variation of ultimate tensile rebar strain with
varying $\rho_r$.

Because of a smaller value of $\kappa_u$, CFRP rebars mobilize a higher $c_u$ value and a lower $\varepsilon_r$ value than steel rebars. Likewise, the values of $c_u$ and $\varepsilon_r$ mobilized by GFRP rebars could be higher or lower than those by steel rebars, depending on the $\rho_r$ level.

# 8.3 Analytical modeling

## 8.3.1 Existing models using combined reinforcement index for prediction of ultimate stress in unbonded tendons

Due to strain incompatibility, the stress in external or unbonded tendons is member-dependent. The quantification of the tendon stress is a key task in design practice. The combined reinforcement index is considered one of the best parameters used for calculating the tendon stress, as this parameter involves a number of important factors including the tendon area and depth, effective prestress, rebar area, and concrete strength. The combined reinforcement index for the members with steel rebars is defined by

$$\omega_0 = \frac{A_p \sigma_{pe} + A_r f_y}{b d_p f_{ck}} \tag{8.1a}$$

where $d_p$ is the tendon depth. At ultimate, FRP rebars often reach a stress level far below their rupture strength. Therefore the combined reinforcement index for the members with FRP rebars could be expressed by

$$\omega_0 = \frac{A_p \sigma_{pe} + A_r \sigma_r}{b d_p f_{ck}} \tag{8.1b}$$

where $\sigma_r$ is the stress in tensile rebars at ultimate.

Fig. 8.7A shows the numerical results regarding the relationship between $\Delta\sigma_p$ and $\omega_0$ for the members with FRP or steel rebars. It is interesting to note that the variations of $\Delta\sigma_p$ with varying $\omega_0$ for the members with CFRP and GFRP rebars appear to be consistent. Since CFRP represents the highest modulus of elasticity and GFRP the lowest one among FRP groups, it can be concluded that the members with different types of FRP rebars follow approximately the same $\Delta\sigma_p - \omega_0$ response. At a given $\omega_0$ value, the members with FRP rebars lead to substantially higher tendon stress than the members with steel rebars. The tendon stress tends to decrease with increasing $\omega_0$, while the decrease for the members with FRP rebars is much quicker than that for the members with steel rebars.

By performing laboratory tests of 22 prestressed concrete members with unbonded tendons, Du and Tao (1985) found that there was an approximately linear relationship between the tendon stress and combined reinforcement index. They recommended the following expression for calculating $\Delta\sigma_p$:

$$\Delta\sigma_p = 786 - 1920\omega_0 \tag{8.2}$$

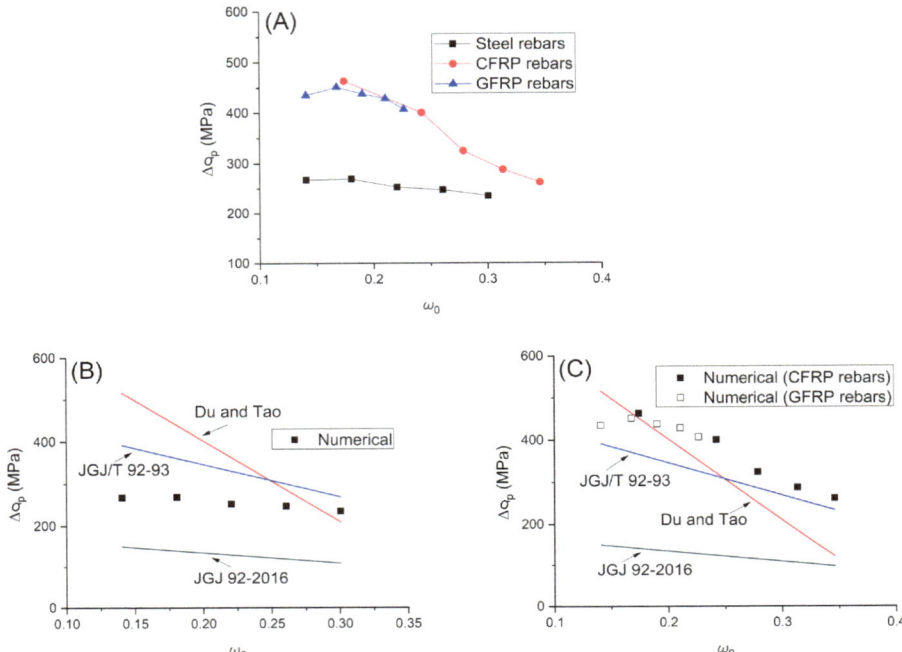

**Figure 8.7** Relationship between tendon stress increment and combined reinforcement index: (A) numerical results; (B) comparison between simplified models and numerical data for the beams with steel rebars; (C) comparison between simplified models and numerical data for the beams with fiber-reinforced polymer rebars.

JGJ/T 92−93 (JGJ/T 92-93, 1993) suggested the following equation for the computation of $\Delta\sigma_{\mathrm{p}}$:

$$\Delta\sigma_{\mathrm{p}} = 500 - 770\omega_0 \tag{8.3a}$$

for $L/d_{\mathrm{p}} \leq 35$; and

$$\Delta\sigma_{\mathrm{p}} = 250 - 380\omega_0 \tag{8.3b}$$

for $L/d_{\mathrm{p}} > 35$.

JGJ 92-2016 (2016) proposed a modification of the above equation, which is expressed as follows:

$$\Delta\sigma_{\mathrm{p}} = (240 - 335\omega_0)(0.45 + 5.5h/L_0)\frac{L_2}{L_1} \tag{8.4}$$

where $\omega_0$ is not greater than 0.4; $h$ is the cross-sectional depth; $L_0$ is the span length; $L_1$ is the tendon length between end anchorages for continuous members; and $L_2$ is the total length of loaded spans.

It should be noted that the above models were developed for the members with steel rebars. To evaluate the applicability of the models, the predictions by the simplified equations against the numerical data for the members with steel rebars are presented in Fig. 8.7B, while those for the members with FRP rebars are presented in Fig. 8.7C. It is observed that, in general, the Du and Tao model and JGJ/T 92−93 are unsafe while JGJ 92−2016 is conservative when predicting the tendon stress in the members with steel rebars. In addition, the Du and Tao model substantially overestimates the influence of $\omega_0$. For the members with FRP rebars, the effect of $\omega_0$ on the tendon stress is overestimated by the Du and Tao model while underestimated by JGJ/T 92−93 and JGJ 92−2016. Moreover, JGJ 92−2016 appears to be overly conservative.

### 8.3.2  Proposed model

According to the linear fit to the numerical data of the members with steel or FRP rebars as illustrated in Fig. 8.8, the following equation is proposed for predicting the stress increment in external tendons at ultimate:

$$\Delta\sigma_p = 303 - 220\omega_0 \tag{8.5a}$$

for the members with steel rebars; and

$$\Delta\sigma_p = 626 - 1032\omega_0 \tag{8.5b}$$

for the members with FRP rebars.

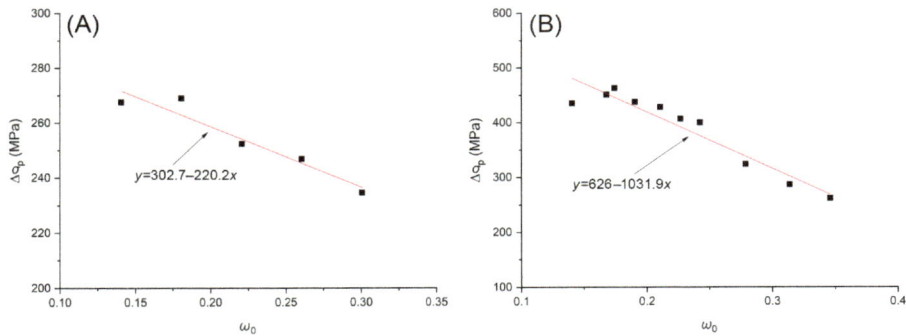

**Figure 8.8** Linear fit to the numerical data: (A) beams with steel rebars; (B) beams with fiber-reinforced polymer rebars.

It is worth mentioning that at given cross-sectional and material properties, the value of $\omega_0$ in the members with steel rebars is known (see Eq. (8.1a)) whereas that in the members FRP rebars is unknown (see Eq. (8.1b)). Consequently, the value of $\Delta\sigma_p$ in the members with steel rebars can be calculated directly by using Eq. (8.5a), while the computation of $\Delta\sigma_p$ in the members with FRP rebars needs to combine Eq. (8.5b) with the section equilibrium equation.

The axial equilibrium equation of the critical section of the members with steel rebars is

$$0.85 f_{ck} b \beta_1 c_u = A_p(\sigma_{pe} + \Delta\sigma_p) + A_r f_y - A'_r f'_y \tag{8.6}$$

where $\beta_1$ is the stress-block factor for concrete, taken equal to 0.85; $A_r$ and $A'_r$ are the area of tensile and compressive rebars, respectively. Substituting Eq. (8.5a) into Eq. (8.6) results in

$$c_u = \frac{A_p(\sigma_{pe} + 303 - 220\omega_0) + A_r f_y - A'_r f'_y}{0.85 f_{ck} b \beta_1} \tag{8.7}$$

Hence, the flexural strength for the members with steel rebars is calculated from

$$M_u = A_p(\sigma_{pe} + \Delta\sigma_p)d_e + A_r f_y d_r - A'_r f'_y d'_r - 0.85 f_{ck} b (\beta_1 c_u)^2 / 2 \tag{8.8}$$

where $d_r$ and $d'_r$ are the effective depth of tensile and compressive rebars, respectively; $d_e$ is the effective depth at ultimate of external tendons. The value of $d_e$ can be obtained by

$$d_e = R_d d_p \tag{8.9}$$

where $R_d$ the depth reduction factor as a result of second-order effects of externally prestressed concrete members, which may be calculated from ACI Committee 440 (2004)

$$R_d = 1.14 - 0.005(L/d_p) - 0.19(S_d/L) \le 1.0 \tag{8.10a}$$

for center-point loading; and

$$R_d = 1.25 - 0.01(L/d_p) - 0.38(S_d/L) \le 1.0 \tag{8.10b}$$

for third-point loading.

On the other hand, the axial equilibrium of the critical section of the members with FRP rebars is given by

$$0.85 f_{ck} b \beta_1 c_u = A_p(\sigma_{pe} + \Delta\sigma_p) + A_r \sigma_r - A'_r \sigma'_r \tag{8.11}$$

where

$$\sigma_r = E_f \varepsilon_u \left( \frac{d_r}{c_u} - 1 \right) \tag{8.12}$$

$$\sigma'_r = E'_f \varepsilon_u \left( 1 - \frac{d'_r}{c_u} \right) \tag{8.13}$$

where $E_f$ and $E'_f$ are the elastic modulus of tensile and compressive FRP rebars, respectively; $\varepsilon_u$ is taken equal to 0.003. Combining Eqs. (8.1b), (8.5b), (8.12) and (8.13) with Eq. (8.11) leads to

$$c_u = \frac{-B + \sqrt{B^2 - 4AC}}{2A} \tag{8.14}$$

where

$$A = 0.85 f_{ck} b \beta_1$$

$$B = A_r E_f \varepsilon_u (1 - 1032 \rho_p / f_{ck}) + A'_r E'_f \varepsilon_u - A_p(\sigma_{pe} + 626 - 1032 \omega_p)$$

$$C = -A_r E_f \varepsilon_u d_r (1 - 1032 \rho_p / f_{ck}) - A'_r E'_f \varepsilon_u d'_r$$

where

$$\rho_p = \frac{A_p}{b d_p} \tag{8.15}$$

$$\omega_p = \frac{A_p \sigma_{pe}}{b d_p f_{ck}} \tag{8.16}$$

Hence, the flexural strength of the members with FRP rebars is obtained by

$$M_u = A_p(\sigma_{pe} + \Delta \sigma_p) d_e + A_r \sigma_r d_r - A'_r \sigma'_r d'_r - 0.85 f_{ck} b (\beta_1 c_u)^2 / 2 \tag{8.17}$$

A comparison of the ultimate tendon stress increment and flexural strength predicted by the proposed analytical model with the numerical results for the members with steel or FRP rebars is presented in Table 8.2 and Fig. 8.9. It is seen that there is a good agreement between the analytical predictions and numerical data. The mean discrepancy for the ultimate tendon stress increment is 1.03% with a standard deviation of 4.08%, while that for the ultimate moment is −4.33% with a standard deviation of 2.32%.

**Table 8.2** Comparison of ultimate tendon stress increment and moment predicted by analytical model with numerical results.

| Rebars | $\rho_r$ (%) | $\Delta\sigma_p$ (MPa) | | | $M_u$ (kN m) | | |
|---|---|---|---|---|---|---|---|
| | | Analytical | Numerical | Error (%) | Analytical | Numerical | Error (%) |
| Steel | 0.22 | 272.05 | 267.60 | 1.66 | 654.40 | 653.12 | 0.20 |
| | 0.70 | 263.25 | 269.05 | −2.15 | 812.72 | 831.63 | −2.27 |
| | 1.19 | 254.45 | 252.44 | 0.80 | 962.98 | 997.08 | −3.42 |
| | 1.67 | 245.65 | 246.93 | −0.52 | 1105.17 | 1154.67 | −4.29 |
| | 2.16 | 236.85 | 234.63 | 0.95 | 1239.31 | 1315.42 | −5.79 |
| CFRP | 0.22 | 448.74 | 463.02 | −3.09 | 837.89 | 874.32 | −4.17 |
| | 0.70 | 382.53 | 400.19 | −4.41 | 1054.40 | 1129.98 | −6.69 |
| | 1.19 | 338.41 | 324.23 | 4.37 | 1189.31 | 1256.83 | −5.37 |
| | 1.67 | 304.47 | 287.09 | 6.05 | 1288.01 | 1366.46 | −5.74 |
| | 2.16 | 276.65 | 261.86 | 5.65 | 1365.57 | 1466.04 | −6.85 |
| GFRP | 0.22 | 483.51 | 435.02 | 11.15 | 713.43 | 706.31 | 1.01 |
| | 0.70 | 455.67 | 451.08 | 1.02 | 808.85 | 840.51 | −3.77 |
| | 1.19 | 433.69 | 437.72 | −0.92 | 882.11 | 928.20 | −4.97 |
| | 1.67 | 415.19 | 428.42 | −3.09 | 942.27 | 1004.81 | −6.22 |
| | 2.16 | 399.09 | 407.03 | −1.95 | 993.58 | 1063.79 | −6.60 |

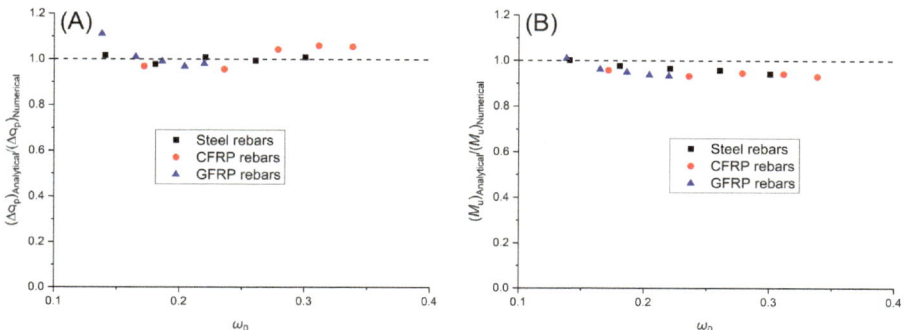

**Figure 8.9** Comparison between analytical predictions and numerical results: (A) tendon stress increment at ultimate; (B) ultimate moment.

## 8.4   Conclusions

This chapter presents a numerical and analytical work conducted on simply supported prestressed concrete members with external CFRP tendons, aimed at identifying the effect of providing FRP bonded rebars instead of steel ones. The results show that FRP (especially CFRP) rebars lead to better exploitation of noncritical sections than steel rebars, particularly notable at a low $\rho_r$ level. The crack mode is improved by providing a minimum amount of rebars, while the improvement is more effective using CFRP rebars than using GFRP or steel rebars. CFRP rebars lead to larger ultimate load and neutral axis depth while smaller ultimate curvature and tensile rebar strain than steel rebars. Such values registered by GFRP rebars could be larger or smaller than those by steel rebars, depending on the $\rho_r$ level. GFRP rebars mobilize substantially higher ultimate deflection and tendon stress increment than steel rebars. CFRP rebars lead to similar observation at a low $\rho_r$ level, while the difference between the values for the members with CFRP and steel rebars diminishes as $\rho_r$ increases.

Both JGJ/T 92−93 and JGJ 92−2016 underestimate the influence of the combined reinforcement index on the stress in external tendons at ultimate in the members with FRP rebars. Moreover, JGJ 92−2016 appears to be overly conservative in predicting the ultimate tendon stress. An analytical model is introduced for predicting the tendon stress at ultimate and flexural strength in simply supported externally prestressed concrete members with FRP or steel rebars. The model predictions are in good agreement with the numerical results.

## References

ACI Committee 318. (2019). *Building code requirements for structural concrete (ACI 318-19) and commentary (ACI 318R-19)*. Farmington Hills, MI.

ACI Committee 440. (2006). *Guide for the design and construction of structural concrete reinforced with FRP bars.* ACI 440.1R-06, Farmington Hills, MI.

ACI Committee 440. (2004). *Prestressing concrete structures with FRP tendons.* ACI 440.4R-04, Farmington Hills, MI.

Ahmad, S. (2003). Reinforcement corrosion in concrete structures, its monitoring and service life prediction—a review. *Cement and Concrete Composites, 25*(4-5), 459−471. Available from https://doi.org/10.1016/s0958-9465(02)00086-0.

Alam, M. S., & Hussein, A. (2017). Relationship between the shear capacity and the flexural cracking load of FRP reinforced concrete beams. *Construction and Building Materials, 154*, 819−828. Available from https://doi.org/10.1016/j.conbuildmat.2017.08.006.

Aziz, M. A., Abdel-Sayed, G., Ghrib, F., Grace, N. F., & Madugula, M. K. S. (2005). Analysis of concrete beams prestressed and post-tensioned with externally unbonded carbon fiber reinforced polymer tendons. *Canadian Journal of Civil Engineering, 32*(6), 1138−1151.

Barris, C., Torres, L., Vilanova, I., Miàs, C., & Llorens, M. (2017). Experimental study on crack width and crack spacing for glass-FRP reinforced concrete beams. *Engineering Structures, 131*, 231−242. Available from https://doi.org/10.1016/j.engstruct.2016.11.007.

Bennitz, A., Schmidt, J. W., Nilimaa, J., Täljsten, B., Goltermann, P., & Ravn, D. L. (2012). Reinforced concrete T-beams externally prestressed with unbonded carbon fiber-reinforced polymer tendons. *ACI Structural Journal, 109*(4), 521−530.

Du, G. C., & Tao, X. K. (1985). Ultimate stress of unbonded tendons in partially prestressed concrete beams. *PCI Journal, 30*(6), 72−91.

Dundar, C., Tanrikulu, A. K., & Frosch, R. J. (2015). Prediction of load-deflection behavior of multi-span FRP and steel reinforced concrete beams. *Composite Structures, 132*, 680−693. Available from https://doi.org/10.1016/j.compstruct.2015.06.018.

Ghallab, A., & Beeby, A. W. (2005). Factors affecting the external prestressing stress in externally strengthened prestressed concrete beams. *Cement and Concrete Composites, 27*(9-10), 945−957. Available from https://doi.org/10.1016/j.cemconcomp.2005.05.003.

Grace, N. F., & Abdel-Sayed, G. (1998). Behavior of externally draped CFRP tendons in pre-stressed concrete bridges. *PCI Journal, 43*(5), 88−101. Available from https://doi.org/10.15554/pcij.09011998.88.101.

JGJ 92-2016. (2016). *Technical specification for concrete structures prestressed with unbonded tendons.* Beijing: China Architecture & Building Press.

JGJ/T 92-93. (1993). *Technical specification for concrete structures prestressed with unbonded tendons.* Beijing: China Planning Press.

Ju, M., Park, Y., & Park, C. (2018). Cracking control comparison in the specifications of ser-viceability in cracking for FRP reinforced concrete beams. *Composite Structures, 182*, 674−684. Available from https://doi.org/10.1016/j.compstruct.2017.09.016.

Lou, T., Lopes, S. M. R., & Lopes, A. V. (2012). Numerical analysis of behaviour of con-crete beams with external FRP tendons. *Construction and Building Materials, 35*, 970−978. Available from https://doi.org/10.1016/j.conbuildmat.2012.04.055.

Mostofinejad, D., Hosseini, S. M., Nader Tehrani, B., Eftekhar, M. R., & Dyari, M. (2019). Innovative warp and woof strap (WWS) method to anchor the FRP sheets in strength-ened concrete beams. *Construction and Building Materials, 218*, 351−364. Available from https://doi.org/10.1016/j.conbuildmat.2019.05.117.

Pang, M., Li, Z., & Lou, T. (2020). Numerical study of using FRP and steel rebars in simply supported prestressed concrete beams with external FRP tendons. *Polymers, 12*(12), 2773. Available from https://doi.org/10.3390/polym12122773.

Wang, L., Kawaguchi, K., Xu, J., & Han, Q. (2019). Effects of transverse constraints on the longitudinal compressive strength of unidirectional CFRP pultruded plates and rods. *Composite Structures*, *207*, 740−751. Available from https://doi.org/10.1016/j.compstruct.2018.09.071.

Wang, X., Shi, J., Wu, G., Yang, L., & Wu, Z. (2015). Effectiveness of basalt FRP tendons for strengthening of RC beams through the external prestressing technique. *Engineering Structures*, *101*, 34−44. Available from https://doi.org/10.1016/j.engstruct.2015.06.052.

Zhou, A., Chow, C. L., & Lau, D. (2018). Structural behavior of GFRP reinforced concrete columns under the influence of chloride at casting and service stages. *Composites Part B: Engineering*, *136*, 1−9. Available from https://doi.org/10.1016/j.compositesb.2017.10.011.

Zhu, H., Dong, Z., Wu, G., Chen, H., Li, J., & Liu, Y. (2017). Experimental evaluation of bent FRP tendons for strengthening by external prestressing. *Journal of Composites for Construction*, *21*(5), 04017032. Available from https://doi.org/10.1061/(asce)cc.1943-5614.0000811.

# Using FRP rebars instead of steel rebars in continuous concrete members with external tendons

## 9.1 Introduction

The arrangement of internal rebars in externally prestressed concrete members is necessary to make sure that the members behave in a good manner. The current design codes require a minimum amount of internal rebars in such members. For example, the minimum steel rebars in externally prestressed concrete members specified in ACI Committee 318 (ACI Committee 318, 2019) are 0.4% of the area between the cross-sectional centroidal axis and flexural tension face. Corrosion of conventional steel reinforcement results in progressive deterioration in structural performance. Replacing steel reinforcement with noncorrosive fiber-reinforced polymers (FRPs) is a valid solution. In practice, carbon fiber-reinforced polymer (CFRP) and glass fiber-reinforced polymer (GFRP) are widely used, and both composite materials may be selected as a replacement for steel rebars (ACI Committee 440, 2015). However, FRP rebars are of linear stress−strain law, which might lead to unfavorable brittle failure of members. Furthermore, FRP rebars normally show a lower modulus of elasticity but a higher strength than steel rebars. Therefore, members with FRP rebars may behave differently from those with steel rebars (Pang et al., 2020).

Externally prestressed concrete members in practice are often continuous over multiple supports. Many efforts have been made to deepen the understanding of continuous externally prestressed concrete members. Laboratory tests on continuous externally prestressed concrete members were conducted by different researchers, including those by Chan and Au (2015), Tan and Tjandra (2007), Harajli et al. (2002), and Aravinthan et al. (2005). A wide range of test variables were covered in these tests, for example, tendon type, tendon profile, loading pattern, concrete strength, rebar area, casting method, and confinement condition. Numerical simulations are a valuable complement to laboratory tests, which have limitations with reference to the number and size of specimens. Comprehensive numerical works have been performed, and extensive parametric analyses were carried out to reveal the impact of various parameters on the behavior of continuous externally prestressed concrete members (Lou et al., 2013, 2014, 2017; Lou et al., 2020). Both laboratory tests and numerical simulations showed that steel tendons could be well replaced by CFRP tendons in external prestressing application (Lou et al., 2017; Tan & Tjandra, 2007) and that bonded rebars play a critical role in the performance of externally prestressed concrete members (Chan & Au, 2015; Lou et al., 2014). In

Prestressed Members with External Fiber-Reinforced Polymer (FRP) Tendons. DOI: https://doi.org/10.1016/B978-0-443-23877-2.00009-2

addition, the pattern of loading (symmetrical/unsymmetrical) was demonstrated to markedly impact the behavior of continuous externally prestressed concrete members (Aravinthan et al., 2005; Lou et al., 2013). All of the aforementioned works on continuous externally prestressed concrete members were focused on the usage of steel rebars. The study on the use of FRP rebars in continuous externally prestressed concrete members is limited.

The strain in external tendons is incompatible with that in the adjacent concrete. Therefore, the tendon stress is member-dependent and cannot be determined from section equilibrium equations. While many parameters may have an important impact on tendon stress, only one or a few critical parameters can be introduced in design equations for practical purposes. The neutral axis depth was mostly adopted in codes of practice as a key parameter in the prediction of tendon stress at ultimate (AASHTO, 1994; AASHTO, 2017; ACI committee 440, 2004), while the combined reinforcement index was employed in the Chinese code (JGJ 92-2016, 2016). However, all the available design equations, as well as other existing models for predicting tendon stress in continuous unbonded prestressed concrete members (Ghallab, 2013; Harajli, 2012; Maguire et al., 2017; Roberts-Wollmann et al., 2005; Zhou & Zheng, 2014), were valid only when steel rebars are used. In Chapter 8, a simplified model for tendon stress prediction of simply supported externally prestressed concrete members with FRP rebars was presented. This model was developed based on linear regression to data of simply supported members and, therefore, may not be applicable to continuous members. For example, this model did not account for the pattern of loading, which is a critical parameter impacting the tendon stress in continuous externally prestressed concrete members.

Moment redistribution in continuous prestressed concrete members needs to be carefully considered for an economical and safe structural design. A few works have been performed to evaluate the redistribution behavior of prestressed concrete members with external tendons. Aravinthan et al. (2005) tested 6 two-span prestressed concrete members with external steel tendons under symmetrical or unsymmetrical loading. They concluded that symmetrical loading led to a positive redistribution of moments in the support section and a negative one in the midspan section and that moment redistribution under unsymmetrical loading was insignificant. It should be noted that this conclusion has resulted from the particular reinforcement arrangement of the specimens. Moment redistribution in a critical section might be positive or negative at symmetrical loading and might be important or unimportant at unsymmetrical loading, depending on the arrangement of bonded reinforcement (Lou et al., 2013). The experimental results by Chan and Au (2015) indicated neither the neutral axis depth nor net strain in the extreme tensile reinforcement correlated well with the amount of moment redistribution in externally prestressed concrete members, confirming that moment redistribution is member-dependent. Results obtained from numerical simulations led to similar observations (Lou et al., 2014; Lou et al., 2020). All of the aforementioned works were focused on moment redistribution in externally prestressed concrete members with steel rebars. The brittleness of FRP rebars would raise concerns about the ability to redistribute moments in continuous members. The effect of adopting FRP rebars instead of steel ones on moment redistribution in continuous externally prestressed concrete members needs to be addressed.

This chapter reports the feasibility of using FRP rebars as an alternative to steel rebars in continuous externally prestressed concrete members (Lou et al., 2021, 2022), taking into account the pattern of loading. Numerical assessments on nonlinear behavior and moment redistribution in two-span externally CFRP prestressed concrete members with steel, CFRP, or GFRP rebars are presented. Moreover, several codes of practice are introduced and reasonable recommendations for quantifying the ultimate tendon stress and moment redistribution in continuous externally prestressed concrete members with FRP or steel rebars are made.

## 9.2   Global and ultimate behavior

Fig. 9.1 illustrates a sketch of a two-span externally prestressed concrete beam under symmetrical or unsymmetrical loading. In the case of symmetrical loading, $P_1 = P_2 = P$, while in the case of unsymmetrical loading, $P_1 = 2P_2 = P$, where $P$ = magnitude of the applied load; and $P_1$ and $P_2$ = concentrated loads applied on the left and right-hand spans, respectively. Because a single concentrated load at midspan is applied on each of the spans, the external tendons are deviated at midspan and center support, as illustrated in Fig. 9.1, to achieve desirable structural performance. The main variables are the type and ratio of rebars. Three rebar types (steel, CFRP, and GFRP) are selected, and their properties are the same as those used in Chapter 8 (see Table 8.1), that is, for CFRP and GFRP rebars, the rupture strengths are 1840 and 750 MPa, respectively, and the elastic moduli are 147 and 40 GPa, respectively; while the yield strength and elastic modulus of steel rebars are 450 MPa and 200 GPa, respectively. The tensile rebar ratios are variables, that is, $\rho_{r2} = 0.22\%-2.16\%$, and $\rho_{r1}/\rho_{r2} = 1.5$. The compressive rebar ratios are constant, that is, $\rho_{r3} = \rho_{r4} = 0.22\%$. The concrete properties are as follows: $f_{ck} = 60$ MPa, $f_t = 4.4$ MPa, $E_c = 39$ GPa. External CFRP tendons are used, and their related parameters are as follows: $A_p = 1000$ mm$^2$, $f_f = 1840$ MPa, $E_p = 147$ GPa, $\sigma_{p0} = 1104$ MPa.

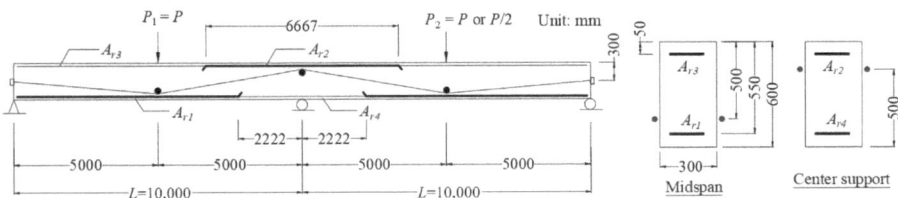

**Figure 9.1** Continuous externally prestressed concrete beams.

## 9.2.1  Failure and crack mode

Fig. 9.2A and B shows the concrete strain distribution in failure in the members under symmetrical and unsymmetrical loading, respectively. The results of the members without any rebars are also illustrated for comparison. Crushing failure is observed in all the members when concrete in the critical section reaches its ultimate compressive strain of 0.003. It is seen that the members without rebars are crushed at the critical midspan. In the case of symmetrical loading, the members with FRP rebars are crushed at the center support while crushing failure of the members with steel rebars may happen at midspan (e.g., at $\rho_{r2} = 0.22\%$), center support (e.g., at $\rho_{r2} = 2.16\%$) or both the critical sections (e.g., at $\rho_{r2} = 1.19\%$). In the case of unsymmetrical loading, the members with FRP/steel rebars fail by concrete crushing at the critical midspan. At failure, FRP rebars and tendons are far below their rupture strength.

The cracking mode can be symbolized by concrete tensile strain as illustrated in Fig. 9.2. Cracking occurs when concrete reaches its cracking strain. The members without rebars in failure exhibit significant strain concentration (or cracking concentration), that is, strains spike at the critical midspan and center support whereas they are marginal over other regions. In other words, these members remain almost uncracked over the length except at the critical midspan and support section, where a few principal cracks exist. These cracks are of very large width (i.e., concrete tensile strains in the bottom fiber are very large) at midspan but the crack width (i.e., concrete tensile strain in the top fiber) is relatively small at center support. When minimum rebars are provided ($\rho_{r2} = 0.22\%$), the crack width at the critical midspan is reduced and more cracks form over the span. The alleviation of cracking concentration is more effective by using FRP rebars, especially CFRP rebars, than by using steel rebars. This observation is attributed to that at ultimate, FRP rebars cause a higher stress in external tendons and may also experience a higher rebar stress despite a lower modulus of elasticity, when compared to steel rebars. At center support, the crack width appears to enlarge by providing minimum FRP rebars, attributed to full exploitation of the support section. As the rebar ratio increases, the crack width becomes smaller while the crack zone gets more extensive as expected.

## 9.2.2  Global behavior

The load—deflection curves of the members under symmetrical and unsymmetrical loading are shown in Fig. 9.3A and B, respectively. The members with CFRP/GFRP rebars exhibit two-stage behavior, that is, pre- and postcracking stages. In each stage, there is an approximately linear load—deflection relationship. Before yielding, the members with steel rebars show a response close to that of the members with CFRP rebars. After yielding of steel rebars, there is a marked reduction in flexural stiffness.

Fig. 9.4A and B shows the moment—curvature curves of the members under symmetrical and unsymmetrical loading, respectively. The moment—curvature relationship exhibits a bilinear curve for a section with CFRP/GFRP rebars and a

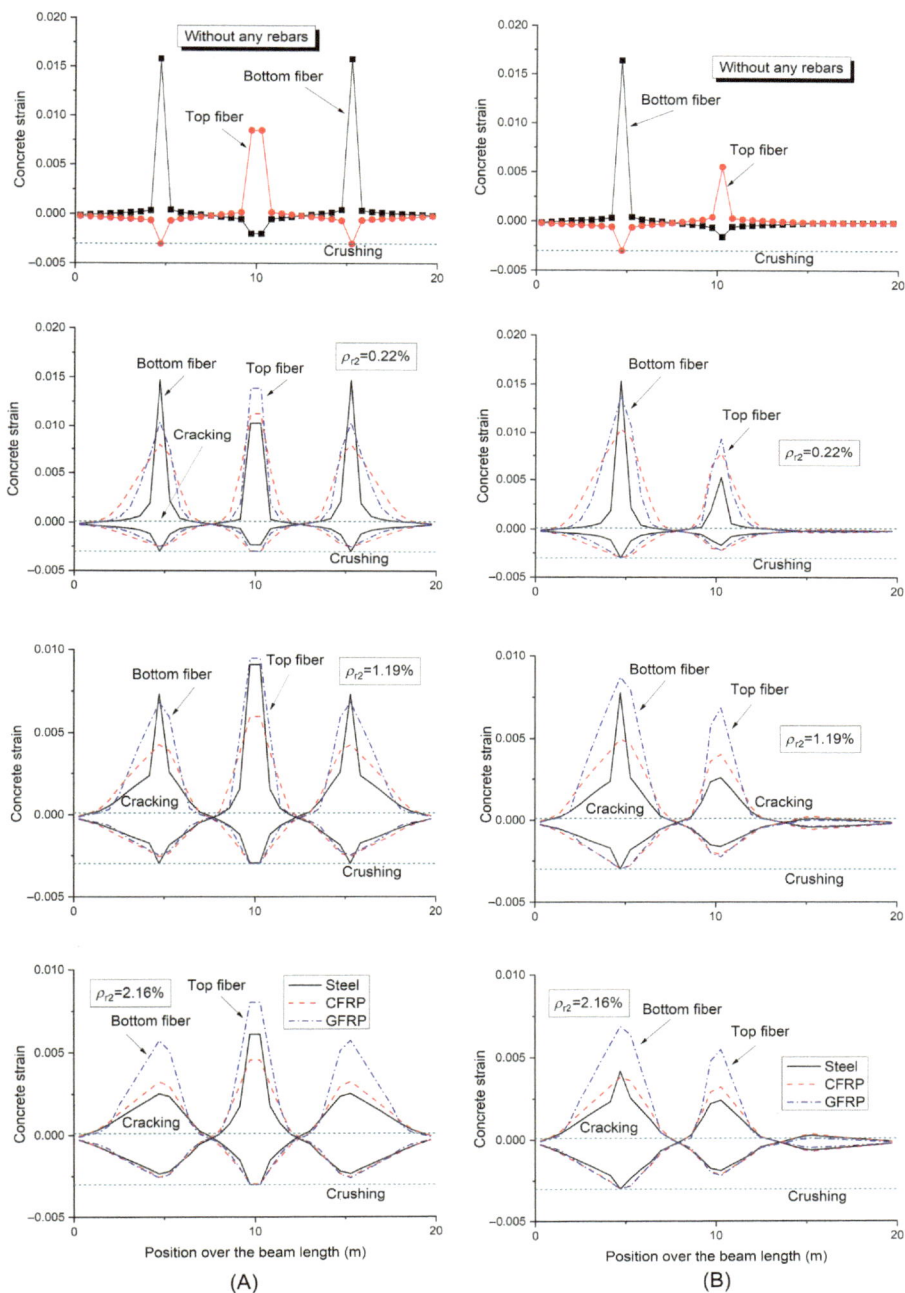

**Figure 9.2** Concrete strain distribution: (A) symmetrical loading; (B) unsymmetrical loading.

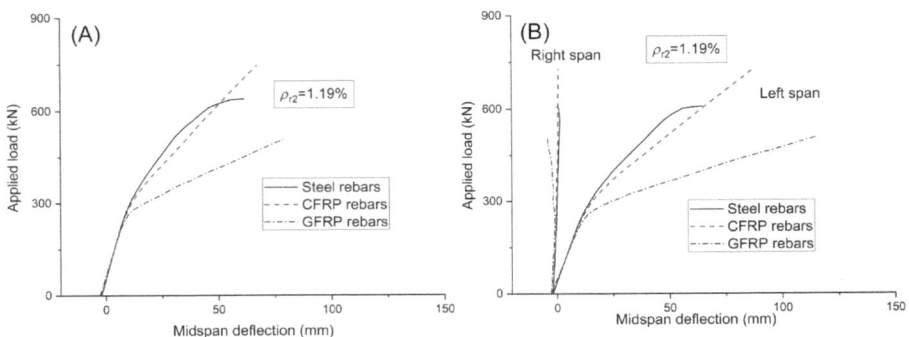

**Figure 9.3** Load—deflection behavior: (A) symmetrical loading; (B) unsymmetrical loading.

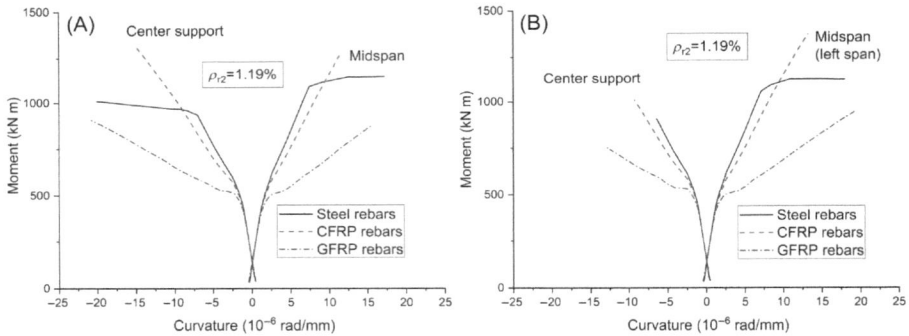

**Figure 9.4** Moment—curvature behavior: (A) symmetrical loading; (B) unsymmetrical loading.

trilinear curve for a section with steel rebars (if yielded). At unsymmetrical loading, the support section with steel rebars has not yielded in failure and, therefore, shows a bilinear behavior. The stiffness of sections with FRP/steel rebars is identical until cracking. The cracking-induced stiffness reduction of a GFRP-reinforced section is substantially more than that of a steel-reinforced section. When a section with steel bars has yielded, the moment hardly increases with the continuing increase in curvature.

Fig. 9.5A and B shows the tendon stress versus deflection curves for the members under symmetrical and unsymmetrical loading, respectively. There is a nearly linear development in tendon stress while deflecting. The slope by GFRP rebars is larger than that by CFRP/steel rebars. Meanwhile, symmetrical loading leads to a greater slope than unsymmetrical loading. The slopes by steel, CFRP, and GFRP rebars are 2.20, 2.15, and 2.62 MPa/mm, respectively, at symmetrical loading, while are 1.02, 1.08, and 1.32 MPa/mm, respectively, at unsymmetrical loading.

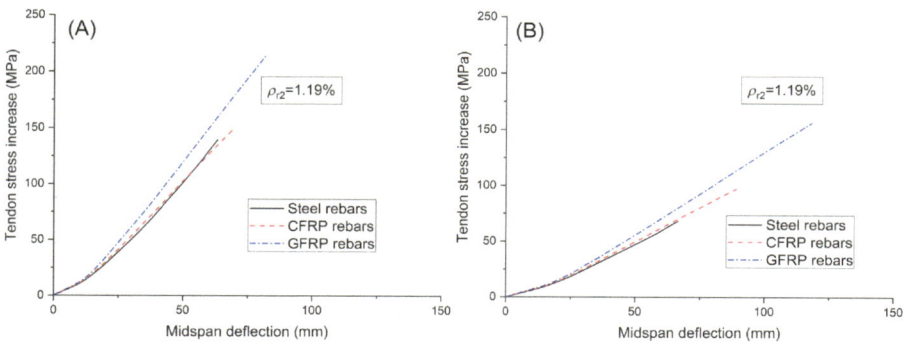

**Figure 9.5** Increase in tendon stress at deflecting: (A) symmetrical loading; (B) unsymmetrical loading.

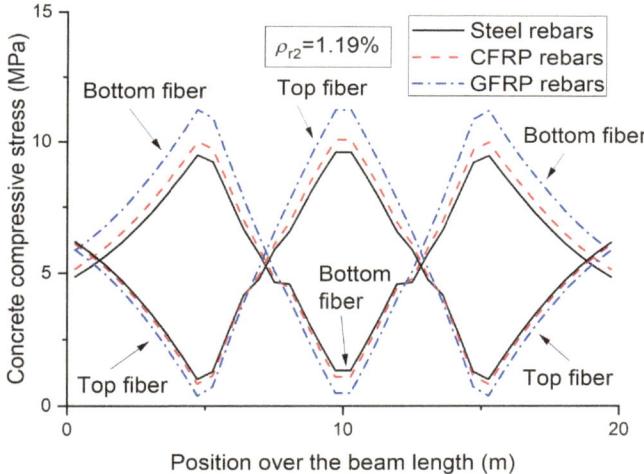

**Figure 9.6** Distribution of stresses in top and bottom fibers along the beam before load application.

Before load application (i.e., under self-weight and prestressing), the whole member is in compression (see Fig. 9.6) and there is sagging bending in center support and hogging bending in midspan. The corresponding neutral axis depths ($c$) are negative, that is, somewhere below the bottom in center support or above the top in midspan. During loading, the sagging (hogging) bending in center support (midspan) gradually disappears and the hogging (sagging) bending occurs and gradually develops. Accordingly, $c$ approaches rapidly to negative infinity, and then abruptly to positive infinity, followed by a quick decrease in the value. At $c < 600$ mm, the evolution in $c$ with the moment for the members under symmetrical and unsymmetrical loading is shown in Fig. 9.7A and B, respectively. It is seen that the decrease

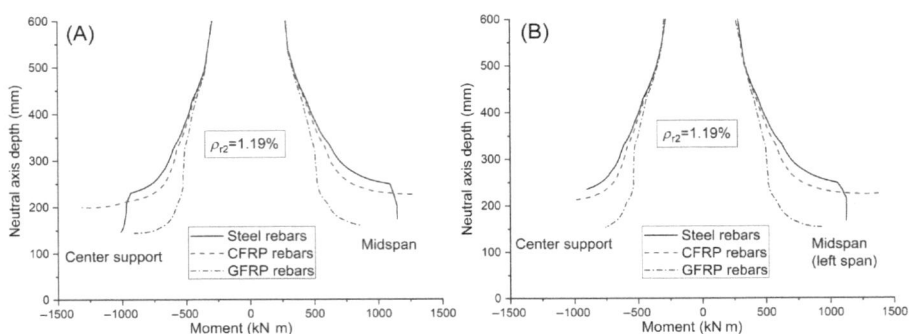

**Figure 9.7** Variation in neutral axis depth with moment: (A) symmetrical loading; (B) unsymmetrical loading.

in $c$ slows down after the cracking of the section. For the sections with steel rebars, yielding leads to resuming a quick decrease in $c$.

### 9.2.3  Ultimate behavior

The $P_u-\rho_{r2}$ and $\delta_u-\rho_{r2}$ relationships are shown in Fig. 9.8A and B, respectively, where $P_u$ is the ultimate load, and $\delta_u$ is the ultimate deflection. Fig. 9.8A shows that the increase in $P_u$ is approximately linear with increasing $\rho_{r2}$, while steel rebars cause a higher increase rate than CFRP/GFRP rebars. GFRP rebars lead to substantially lower (around 30% lower) $P_u$ than CFRP rebars. The $P_u$ of members with steel rebars is close to that of members with GFRP rebars at $\rho_{r2} = 0.22\%$ and to that of members with CFRP rebars at $\rho_{r2} = 2.16\%$. Symmetrical loading mobilizes slightly higher $P_u$ than unsymmetrical loading.

Fig. 9.8B shows that for members with steel rebars, unsymmetrical loading results in moderately higher (4%–21% higher) $\delta_u$ than symmetrical loading, as evidenced by laboratory tests (Aravinthan et al., 2005; Tan & Tjandra, 2007). It is also surprising to note that for members with FRP rebars, unsymmetrical loading leads to significantly larger (31%–46% larger) $\delta_u$ when compared to symmetrical loading. This observation can be explained by the fact that at unsymmetrical loading, members with FRP rebars exhibit greater rotation due to the lower elastic modulus of FRP rebars than members with steel rebars. The decrease in $\delta_u$ with increasing $\rho_{r2}$ appears to be more apparent by CFRP rebars than by GFRP/steel rebars. At a given load pattern, GFRP rebars lead to substantially higher (around 40% higher at symmetrical loading and around 80% higher at unsymmetrical loading) $\delta_u$ than steel rebars irrespective of the $\rho_{r2}$ level. CFRP rebars result in a significantly greater $\delta_u$ than steel rebars at $\rho_{r2} = 0.22\%$, but the difference diminishes as $\rho_{r2}$ increases.

Fig. 9.8C illustrates the $\Delta\sigma_p-\rho_{r2}$ relationship, where $\Delta\sigma_p$ is the increase in tendon stress at ultimate. It is seen that as $\rho_{r2}$ increases, the $\Delta\sigma_p$ value for members with GFRP or steel rebars tends to slightly decrease while that of members with CFRP rebars decreases quickly. Under symmetrical or unsymmetrical loading,

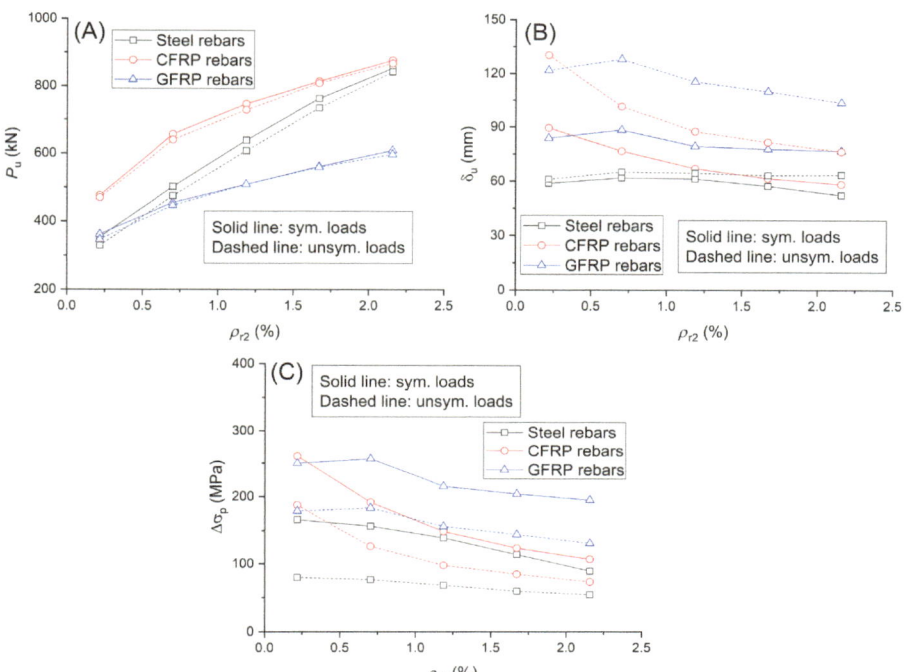

**Figure 9.8** Typical ultimate behavior: (A) variation in ultimate deflection; (B) variation in ultimate deflection; (C) variation in ultimate stress increase in tendons.

GFRP rebars result in substantially greater $\Delta\sigma_p$ than steel rebars over the entire $\rho_{r2}$ range investigated. The $\Delta\sigma_p$ values by GFRP rebars are 1.5–2.2 times those by steel rebars at symmetrical loading and around 2.4 times those by steel rebars at unsymmetrical loading. Meanwhile, members with CFRP rebars show $\Delta\sigma_p$ comparable to members with GFRP rebars at $\rho_{r2} = 0.22\%$ while to members with steel rebars at $\rho_{r2} = 2.16\%$. Due to less exploitation of plastic zone, unsymmetrical loading results in significantly lower (40%–52% lower when steel rebars are used or around 30% lower when CFRP/GFRP rebars are used) $\Delta\sigma_p$ compared to symmetrical loading.

## 9.3    Prediction of tendon stress at ultimate

### 9.3.1    Available code equations applicable to continuous members

The stress at ultimate in unbonded tendons, $\sigma_{pu}$, is conventionally written as

$$\sigma_{pu} = \sigma_{pe} + \Delta\sigma_p \tag{9.1}$$

where $\sigma_{pe}$ = effective prestress. Three typical code equations for the calculation of $\Delta\sigma_p$ in continuous members are investigated in this study.

AASHTO (1994) adopted the following expression for the prediction of $\Delta\sigma_p$:

$$\Delta\sigma_p = \Omega_u E_p \varepsilon_u \left( \frac{d_p}{c_u} - 1 \right) \frac{L_l}{L_t} \tag{9.2}$$

where $d_p$ = tendon depth; $\varepsilon_u$ = concrete ultimate cylinder compressive strain; $c_u$ = neutral axis depth at failure; $\Omega_u$ = strain reduction factor, equal to $1.5/(L/d_p)$ for one-point loading and $3.0/(L/d_p)$ for two-point or uniform loading; $L$ = span length; $L_t$ = tendon length between anchorages; and $L_l$ = length of loaded spans. Eq. (9.2) was developed based on the method assuming strain compatibility as in the case of bonded tendons and applying the factor $\Omega_u$ to consider the unbonded effect.

On the other hand, AASHTO (2017) recommended a deformation-based model, which was originally developed by Roberts-Wollmann et al. (2005). The model is expressed as follows:

$$\Delta\sigma_p = 6200 \left( \frac{d_p - c_u}{l_e} \right) \tag{9.3}$$

where

$$l_e = \frac{L_t}{1 + (N/2)} \tag{9.4}$$

where $N$ = number of support hinges required to form a mechanism crossed by tendons. It is noted that no plastic hinges would develop in externally prestressed concrete members with FRP rebars, and therefore, the application of the AASHTO, 2017 model may be limited in the case of FRP rebars.

JGJ 92-2016 (2016) proposed an equation in which the combined reinforcement index, $\omega_0$, is used as a key parameter for predicting $\Delta\sigma_p$.

$$\Delta\sigma_p = (240 - 335\omega_0)(0.45 + 5.5h/L) \frac{L_l}{L_t} \tag{9.5}$$

where $h$ = section depth; $\omega_0 = (A_p \sigma_{pe} + A_r \sigma_r)/(bd_p f_{ck})$ and $\omega_0 \leq 0.4$; $A_r$ = tensile rebar area; and $\sigma_r$ = tensile rebar stress at ultimate. When steel rebars are used, $\sigma_r = f_y$. In the case of FRP rebars, the value of $\sigma_r$ can be obtained by solving the section equilibrium equation.

## 9.3.2 Evaluation of design codes

Fig. 9.9A and B shows the numerically obtained $\Delta\sigma_p - c_u/d_p$ relationship for the members with FRP/steel rebars under symmetrical and unsymmetrical loading,

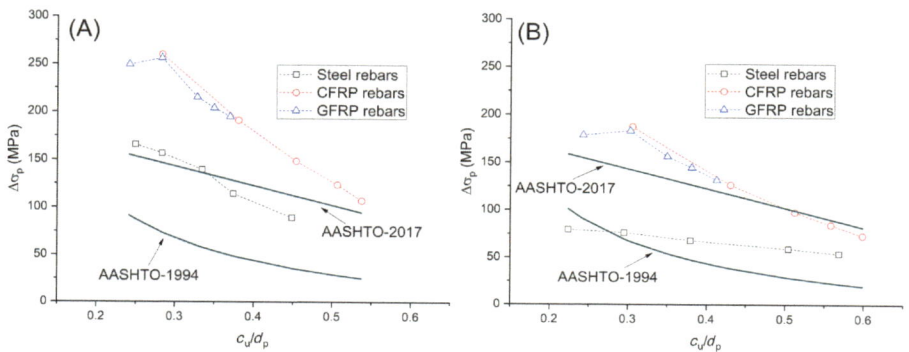

**Figure 9.9** Numerical and code predictions on $\Delta\sigma_p$ based on $c_u/d_p$: (A) symmetrical loading; (B) unsymmetrical loading.

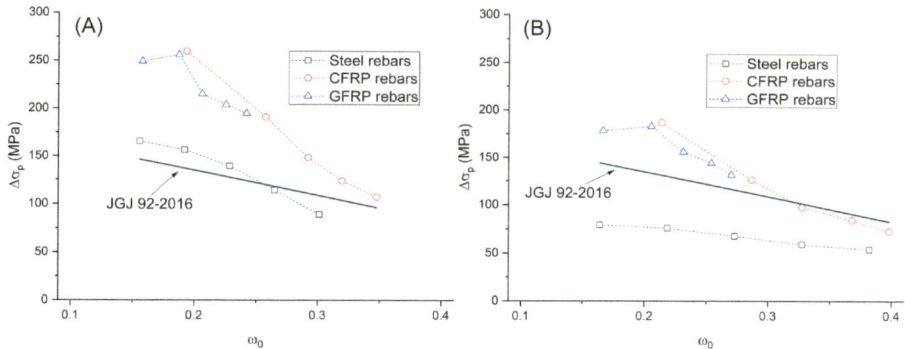

**Figure 9.10** Numerical and code predictions on $\Delta\sigma_p$ based on $\omega_0$: (A) symmetrical loading; (B) unsymmetrical loading.

respectively. The relationship curves of AASHTO, 1994; AASHTO, 2017 are also plotted in the graphs for comparison. The numerical results along with JGJ 92-2016 concerning the $\Delta\sigma_p-\omega_0$ relationship are shown in Fig. 9.10. A comparison of code predictions with numerical results is also presented in Table 9.1.

Fig. 9.9, Fig. 9.10, and Table 9.1 show that according to the numerical results, either the $\Delta\sigma_p-c_u/d_p$ or $\Delta\sigma_p-\omega_0$ relationship of the members with CFRP rebars is consistent well with that of the members with GFRP rebars. At a given $c_u/d_p$ ratio or $\omega_0$ level, FRP rebars lead to substantially higher tendon stress than steel rebars. The decrease in $\Delta\sigma_p$ with increasing $c_u/d_p$ or $\omega_0$ for the members with FRP rebars appears to be faster than that of the members with steel rebars, especially at unsymmetrical loading.

For the members with steel rebars, AASHTO, 2017 and JGJ 92-2016 are more accurate than AASHTO, 1994 in predicting the tendon stress at symmetrical loading, while the observation is opposite at unsymmetrical loading. In the case of

Table 9.1 Comparison of the $\Delta\sigma_p$ values predicted by numerical analysis and simplified equations.

| Load pattern | Rebar type | $\rho_{r2}$ (%) | $\Delta\sigma_p$ (MPa) | | | | | | $(\Delta\sigma_p)_{equation}/(\Delta\sigma_p)_{numerical}$ | | | | |
|---|---|---|---|---|---|---|---|---|---|---|---|---|---|
| | | | Nume | AO94 | AO17 | JGJ16 | Eq9.8 | Eq9.9 | AO94 | AO17 | JGJ16 | Eq9.8 | Eq9.9 |
| SL | Steel | 0.22 | 166.0 | 87.7 | 153.6 | 146.5 | 168.3 | 172.0 | 0.53 | 0.93 | 0.88 | 1.01 | 1.04 |
| | | 0.70 | 156.8 | 73.6 | 146.6 | 137.0 | 154.6 | 152.4 | 0.47 | 0.94 | 0.87 | 0.99 | 0.97 |
| | | 1.19 | 139.5 | 57.9 | 136.2 | 127.5 | 134.2 | 132.8 | 0.42 | 0.98 | 0.91 | 0.96 | 0.95 |
| | | 1.67 | 114.2 | 48.7 | 128.1 | 118.0 | 118.4 | 113.2 | 0.43 | 1.12 | 1.03 | 1.04 | 0.99 |
| | | 2.16 | 89.3 | 35.8 | 112.8 | 108.5 | 88.6 | 93.6 | 0.40 | 1.26 | 1.21 | 0.99 | 1.05 |
| | CFRP | 0.22 | 260.3 | 73.8 | 146.7 | 136.6 | 249.8 | 244.0 | 0.28 | 0.56 | 0.52 | 0.96 | 0.94 |
| | | 0.70 | 191.4 | 47.4 | 126.8 | 119.9 | 193.6 | 184.5 | 0.25 | 0.66 | 0.63 | 1.01 | 0.96 |
| | | 1.19 | 148.9 | 35.0 | 111.8 | 110.9 | 151.2 | 152.4 | 0.24 | 0.75 | 0.74 | 1.02 | 1.02 |
| | | 1.67 | 123.9 | 28.4 | 101.0 | 103.7 | 120.8 | 126.9 | 0.23 | 0.81 | 0.84 | 0.97 | 1.02 |
| | | 2.16 | 107.2 | 25.1 | 94.7 | 96.4 | 103.1 | 100.9 | 0.23 | 0.88 | 0.90 | 0.96 | 0.94 |
| | GFRP | 0.22 | 249.6 | 91.5 | 155.2 | 146.0 | 273.8 | 277.4 | 0.37 | 0.62 | 0.59 | 1.10 | 1.11 |
| | | 0.70 | 256.4 | 73.9 | 146.8 | 138.2 | 249.9 | 249.6 | 0.29 | 0.57 | 0.54 | 0.97 | 0.97 |
| | | 1.19 | 215.4 | 59.8 | 137.6 | 133.3 | 224.0 | 232.2 | 0.28 | 0.64 | 0.62 | 1.04 | 1.08 |
| | | 1.67 | 204.1 | 54.3 | 133.2 | 128.3 | 211.5 | 214.4 | 0.27 | 0.65 | 0.63 | 1.04 | 1.05 |
| | | 2.16 | 195.2 | 49.7 | 129.0 | 124.0 | 199.8 | 199.0 | 0.25 | 0.66 | 0.64 | 1.02 | 1.02 |

| Load pattern | Rebar type | $\rho_{r2}$ (%) | $\Delta\sigma_p$ (MPa) | | | | | | $(\Delta\sigma_p)_{equation}/(\Delta\sigma_p)_{numerical}$ | | | | |
|---|---|---|---|---|---|---|---|---|---|---|---|---|---|
| | | | Nume | AO94 | AO17 | JGJ16 | Eq9.8 | Eq9.9 | AO94 | AO17 | JGJ16 | Eq9.8 | Eq9.9 |
| USL | Steel | 0.22 | 79.5 | 101.1 | 158.8 | 144.4 | 84.2 | 86.0 | 1.27 | 2.00 | 1.82 | 1.06 | 1.08 |
| | | 0.70 | 76.3 | 69.5 | 144.2 | 130.1 | 77.3 | 76.2 | 0.91 | 1.89 | 1.71 | 1.01 | 1.00 |
| | | 1.19 | 68.1 | 47.6 | 127.0 | 115.9 | 67.1 | 66.4 | 0.70 | 1.86 | 1.70 | 0.99 | 0.98 |
| | | 1.67 | 59.4 | 28.6 | 101.4 | 101.6 | 59.2 | 56.6 | 0.48 | 1.71 | 1.71 | 1.00 | 0.95 |
| | | 2.16 | 54.2 | 22.1 | 88.3 | 87.4 | 44.3 | 46.8 | 0.41 | 1.63 | 1.61 | 0.82 | 0.86 |
| | CFRP | 0.22 | 187.3 | 66.2 | 142.1 | 131.4 | 124.9 | 122.0 | 0.35 | 0.76 | 0.70 | 0.67 | 0.65 |
| | | 0.70 | 126.5 | 38.5 | 116.5 | 112.3 | 96.8 | 92.3 | 0.30 | 0.92 | 0.89 | 0.77 | 0.73 |
| | | 1.19 | 97.9 | 27.7 | 99.8 | 101.6 | 75.6 | 76.2 | 0.28 | 1.02 | 1.04 | 0.77 | 0.78 |
| | | 1.67 | 84.6 | 23.1 | 90.4 | 91.0 | 60.4 | 63.4 | 0.27 | 1.07 | 1.08 | 0.71 | 0.75 |
| | | 2.16 | 73.3 | 19.5 | 82.1 | 83.2 | 51.5 | 50.5 | 0.27 | 1.12 | 1.14 | 0.70 | 0.69 |
| | GFRP | 0.22 | 178.8 | 90.8 | 154.9 | 143.7 | 136.9 | 138.7 | 0.51 | 0.87 | 0.80 | 0.77 | 0.78 |
| | | 0.70 | 183.3 | 67.1 | 142.7 | 133.5 | 125.0 | 124.8 | 0.37 | 0.78 | 0.73 | 0.68 | 0.68 |
| | | 1.19 | 156.6 | 54.2 | 133.1 | 126.8 | 112.0 | 116.1 | 0.35 | 0.85 | 0.81 | 0.72 | 0.74 |
| | | 1.67 | 144.7 | 47.4 | 126.7 | 120.9 | 105.8 | 107.2 | 0.33 | 0.88 | 0.83 | 0.72 | 0.74 |
| | | 2.16 | 131.6 | 41.4 | 120.2 | 116.6 | 99.9 | 99.5 | 0.31 | 0.91 | 0.89 | 0.76 | 0.76 |

SL, Symmetrical loading; USL, unsymmetrical loading; Nume, numerical; AO94, AASHTO, 1994; AO17, AASHTO, 2017; JGJ16, JGJ 92–2016; Eq9.8, Proposed Eq. (9.8); Eq9.9, Proposed Eq. (9.9).

symmetrical loading, AASHTO, 2017 and JGJ 92-2016 yield good predictions against numerical data with mean differences of only 4.43% and −1.64%, respectively. In contrast, AASHTO, 1994 highly underestimates $\Delta\sigma_p$. The predictions by this model are between 40.08% and 52.84% of the corresponding numerical results. On the other hand, when unsymmetrical loading is applied, AASHTO, 2017 and JGJ 92-2016 overestimate $\Delta\sigma_p$ by as high as 81.83% and 70.92% on average, respectively. By contrast, AASHTO, 1994 shows reasonable predictions although the model appears to overestimate the influence of neutral axis depth on $\Delta\sigma_p$.

For the members with FRP rebars, AASHTO, 2017 and JGJ 92-2016 appear to be better than AASHTO, 1994 in estimating tendon stress. At symmetrical loading, all the models lead to underestimations, especially for AASHTO, 1994. The predictions by AASHTO, 1994 are only 26.82% of the numerical results on average. The ratios of code to numerical predictions range from 0.56 to 0.88 for AASHTO, 2017, and from 0.52 to 0.89 for JGJ 92-2016. The predictions by AASHTO, 2017 and JGJ 92-2016 are improved as the rebar ratio increases. At unsymmetrical loading, AASHTO, 1994 also underestimates $\Delta\sigma_p$ by as high as 66.58% on average. Both AASHTO, 2017 and JGJ 92-2016 show satisfactory predictions. The average error is 8.25% with an SD of 11.32% for AASHTO, 2017, and 10.96% with an SD of 14.0% for JGJ 92-2016. However, the effect of neutral axis depth or combined reinforcement index on the tendon stress is underestimated by AASHTO, 2017 or JGJ 92-2016.

### 9.3.3  Proposed equations

According to linear fit regarding the $\Delta\sigma_p$–$c_u/d_p$ relationship illustrated in Fig. 9.11, the following expressions with the parameter $c_u/d_p$ are proposed to predict $\Delta\sigma_p$:

for continuous members with steel rebars,

$$\Delta\sigma_p = \left(268 - 400 c_u/d_p\right)\frac{L_{fl}}{L_t} \tag{9.6a}$$

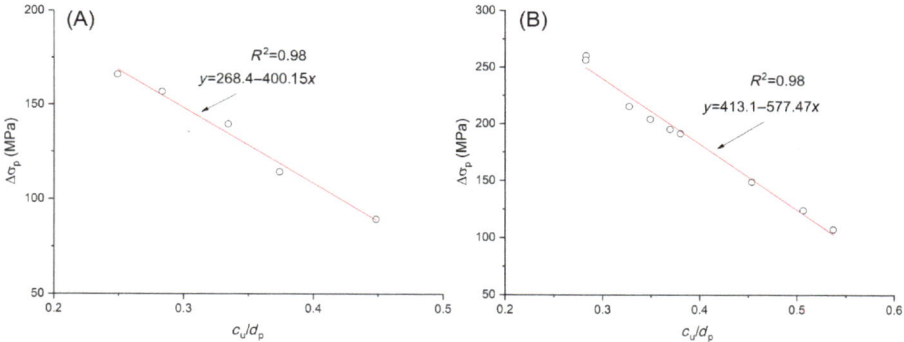

**Figure 9.11** Linear fit to numerical data regarding the $\Delta\sigma_p$–$c_u/d_p$ relationship: (A) beams with steel rebars; (B) beams with FRP rebars.

while for continuous members with FRP rebars,

$$\Delta\sigma_p = \left(413 - 577c_u/d_p\right)\frac{L_{fl}}{L_t} \tag{9.6b}$$

where $L_{fl}$ = length of fully loaded spans. $L_{fl}$ is used instead of $L_l$ to consider the load pattern because some of the spans may be partially loaded. It is evident that $L_{fl}$ leads to conservative predictions while $L_l$ results in unsafe estimations. For example, in the case of unsymmetrical loading ($P_1 = 2P_2$) shown in Fig. 9.1, the value $L_{fl}$ is $L$ while the value of $L_l$ is $2L$.

On the other hand, the fit curves to numerical data regarding the $\Delta\sigma_p-\omega_0$ relationship illustrated in Fig. 9.12 lead to the following expressions:

for continuous members with steel rebars,

$$\Delta\sigma_p = \left(256 - 539\omega_0\right)\frac{L_{fl}}{L_t} \tag{9.7a}$$

while for continuous members with FRP rebars,

$$\Delta\sigma_p = \left(424 - 930\omega_0\right)\frac{L_{fl}}{L_t} \tag{9.7b}$$

It is noted that the member with GFRP rebars having an area of 360 mm$^2$ is excluded from the numerical data used for the regression analysis because the area is far below the minimum of GFRP rebars required in externally prestressed concrete members. It should also be noted that Eq. (9.6) and Eq. (9.7) are valid only for CFRP tendons with an elastic modulus of 147 GPa used in the reference beam. Since the increase in tendon stress is directly proportional to the tendon modulus of elasticity (steel tendons are assumed to be within the elastic range), these equations

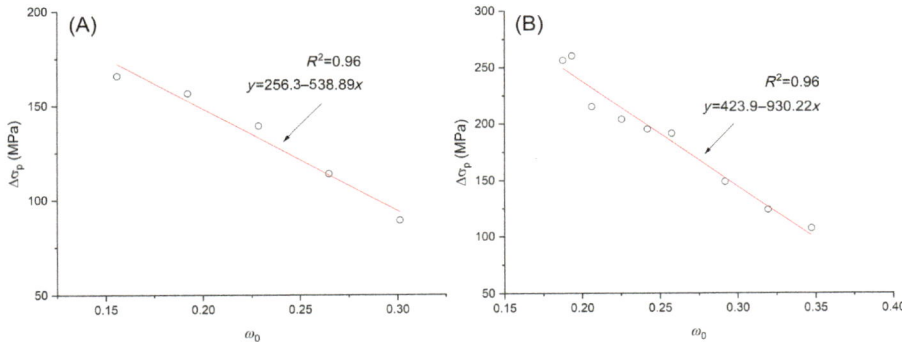

**Figure 9.12** Linear fit to numerical data regarding the $\Delta\sigma_p-\omega_0$ relationship: (A) beams with steel rebars; (B) beams with FRP rebars.

can be extended to general cases. The predictive equation extended from Eq. (9.6) is expressed by

$$\Delta\sigma_p = E_p\left(1.82 - 2.72c_u/d_p\right)\frac{L_{fl}}{L_t} \tag{9.8a}$$

for continuous members with steel rebars; and by

$$\Delta\sigma_p = E_p\left(2.81 - 3.93c_u/d_p\right)\frac{L_{fl}}{L_t} \tag{9.8b}$$

for continuous members with FRP rebars, where $E_p$ = tendon elastic modulus in GPa. On the other hand, the predictive equation extended from Eq. (9.7) is expressed by

$$\Delta\sigma_p = E_p(1.74 - 3.67\omega_0)\frac{L_{fl}}{L_t} \tag{9.9a}$$

for continuous members with steel rebars; and by

$$\Delta\sigma_p = E_p(2.88 - 6.33\omega_0)\frac{L_{fl}}{L_t} \tag{9.9b}$$

for continuous members with FRP rebars.

Fig. 9.13 and Table 9.1 show a comparison of different simplified equations for predicting $\Delta\sigma_p$ with numerical simulations. It is seen that AASHTO, 1994 leads to the poorest predictions, most of which are overly conservative. The data of AASHTO, 2017 and JGJ 92-2016 are also rather scattered, indicating unfavorable correlations with numerical results. The correlations are greatly improved by using the proposed Eq. (9.8) and Eq. (9.9), and the majority of the data are on the safe side, indicating accurate and conservative predictions of the proposed equations.

## 9.4 Moment redistribution

### 9.4.1 Support reaction and bending moment

Figs. 9.14 and 9.15 show the development of end support reactions and bending moments for the members with different types of rebars, respectively. The results are generated for $\rho_{r2}$ of 1.19% and, unless otherwise stated, the results presented in the following sections are for symmetrical loads. The elastic values obtained from the analysis assuming linearly elastic properties of materials are also plotted. The reaction or moment illustrated in the graphs consists of three components induced by dead load, applied load, and external prestressing, respectively. It is noted the external cables are a bit below their concordant line, leading to a small upward

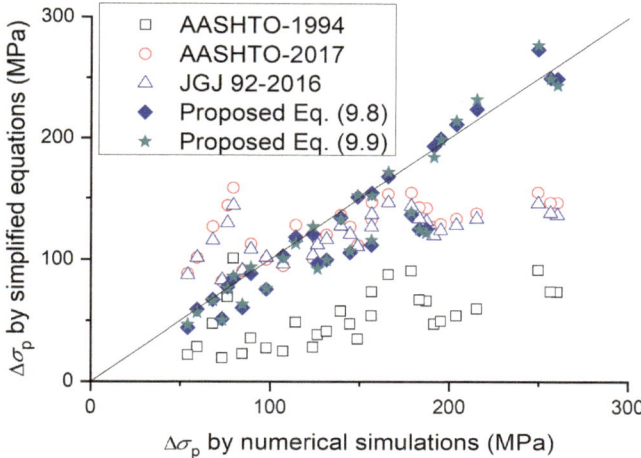

**Figure 9.13** Comparison of different simplified equations for calculating $\Delta\sigma_p$ with numerical simulations.

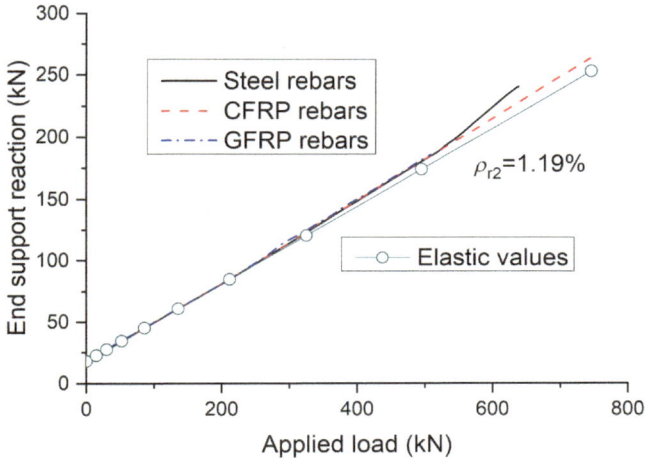

**Figure 9.14** Development of end support reactions for the beams with different types of rebars.

secondary reaction at the end support and, correspondingly, small positive secondary moments along the span.

It is seen that the actual reaction or moment does not deviate from the elastic value until the cracking load is reached. Beyond that, the load versus reaction or moment relationship for the members with FRP rebars exhibits approximately a linear manner up to failure. Before steel yielding, the load versus reaction or moment behavior for the member with steel rebars is very similar to that with FRP rebars.

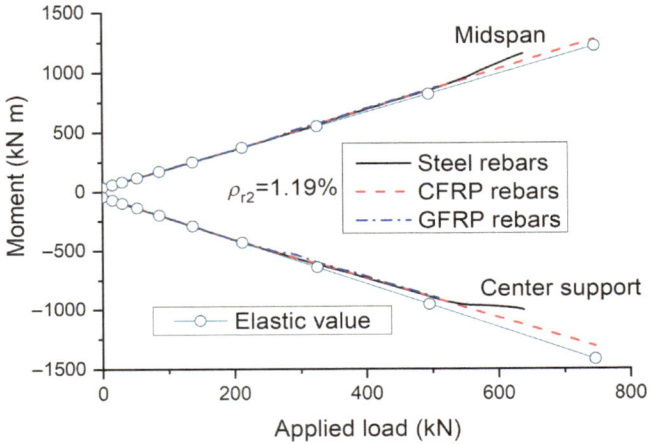

**Figure 9.15** Development of bending moments for the beams with different types of rebars.

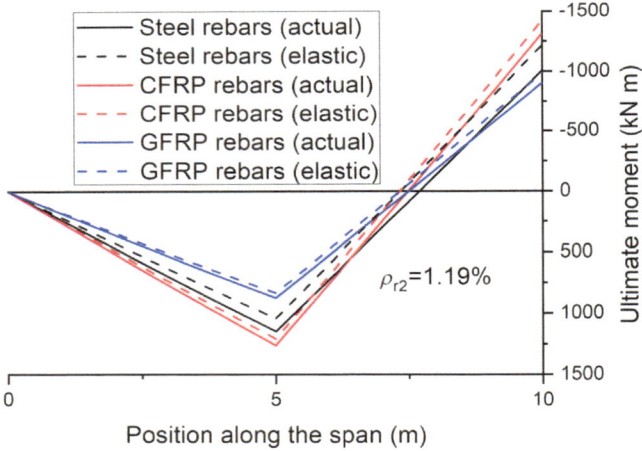

**Figure 9.16** Moment distribution for the beams with different types of rebars.

When the steel rebars at the center support begin to yield, moments are redistributed from the center support to the midspan. As a consequence, there appears a faster
increase in the reaction at the end support as shown in Fig. 9.14. Correspondingly,
the positive moment at the midspan grows quicker and the negative moment at the
center support grows slower as shown in Fig. 9.15.

Fig. 9.16 shows the moment distribution at ultimate for the members with different types of rebars ($\rho_{r2} = 1.19\%$). It is seen that at the center support, the actual
moment is smaller than the elastic moment, leading to a positive redistribution of
moments. The phenomenon is opposite at the midspan. Moreover, the difference

between the actual and elastic moments in the member with FRP rebars is slight, indicating an insignificant redistribution of moments. On the other hand, the difference in the member with steel rebars is substantial especially at the center support, indicating a notable redistribution of moments.

### 9.4.2 Reaction ratio and moment ratio

Denote by $R_1$ and $R_2$ the load-induced actual reactions at the end and center supports, respectively, by $R_{e1}$ and $R_{e2}$ the load-induced elastic reactions at the end and center supports, respectively, by $M_1$ and $M_2$ the load-induced actual moments at the midspan and center support, respectively, and by $M_{e1}$ and $M_{e2}$ the load-induced elastic moments at the midspan and center support, respectively. During the loading process, the $R_{e2}/R_{e1}$ or $M_{e2}/M_{e1}$ ratio for a continuous member remains constant according to elastic theory, while the $R_2/R_1$ or $M_2/M_1$ ratio would vary when redistribution of moments occurs.

Fig. 9.17 shows the evolution of load-induced reactions and reaction ratios for the members with different types of rebars ($\rho_{r2} = 1.19\%$), while the development of load-induced moments and moment ratios are presented in Fig. 9.18. The results confirm that the elastic reaction ratio or moment ratio keeps unchanged despite the load level. The $R_{e2}/R_{e1}$ ratios for the members with steel, CFRP, and GFRP rebars are 4.33, 4.34, and 4.38, respectively, while those of $M_{e2}/M_{e1}$ are 1.16, 1.17, and 1.19, respectively. The slight difference is attributed to the different contributions of the rebars to the transformed section. On first cracking, moments are redistributed from the center support toward the midspan, leading to slower development of $R_2$ or $M_2$ and faster development of $R_1$ or $M_1$ compared to their elastic values. Consequently, the $R_2/R_1$ or $M_2/M_1$ ratio begins to decrease. When the crack development stabilizes, the actual reaction or moment ratio for the members with FRP rebars tends to stabilize until the ultimate failure. For the members with steel rebars, the yielding of steel rebars leads to further moment redistribution from the center support to the midspan, causing a further decrease in the $R_2/R_1$ or $M_2/M_1$ ratio.

### 9.4.3 Degree of moment redistribution

Fig. 9.19 shows the development of moment redistribution with increasing load for the members with different types of rebars ($\rho_{r2} = 1.19\%$). Moment redistribution does not happen until the occurrence of the first cracking. After cracking, the degree of redistribution increases quickly. When the redistribution for the members with FRP rebars reaches a plateau, it tends to stabilize up to failure. The member with steel rebars exhibits similar redistribution behavior to that of the members with FRP rebars up to first steel yielding and thereafter resumes a quick redistribution development.

Moment redistribution relies strongly on the ductility described by either the neutral axis depth or net strain in tensile rebars. Figs. 9.20 and 9.21 show the moment redistribution versus neutral axis depth and net strain in tensile rebars

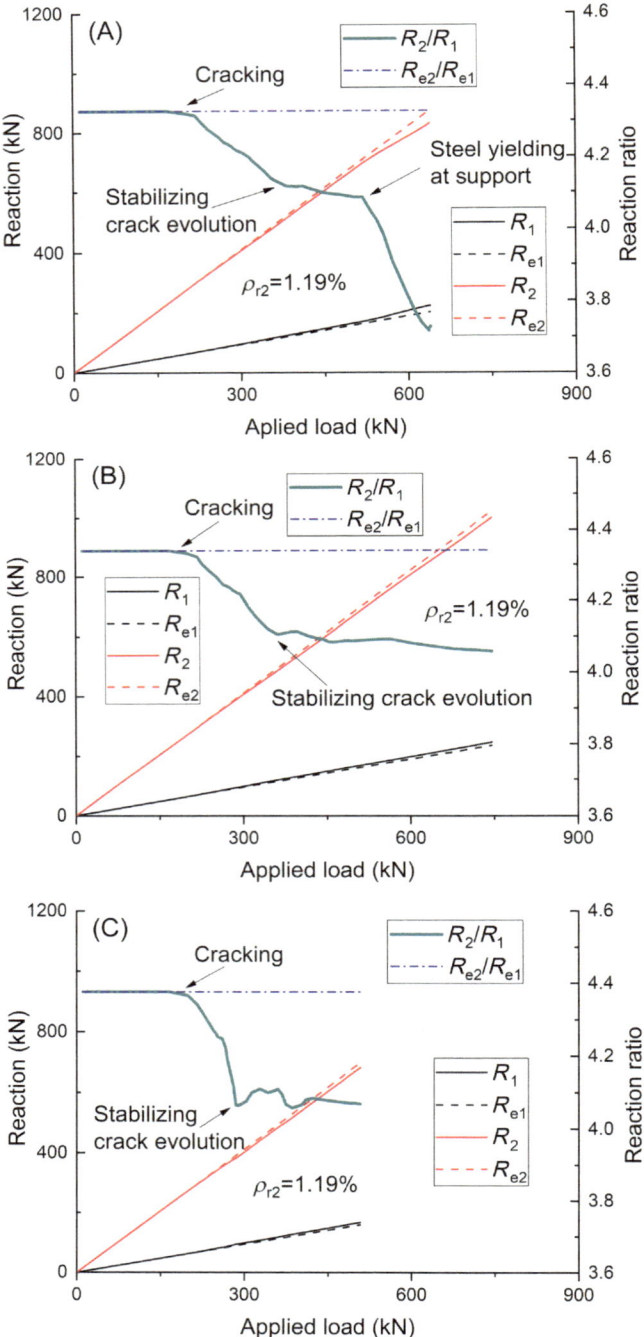

**Figure 9.17** Development of reaction ratios for the beams with different types of rebars: (A) steel rebars; (B) CFRP rebars; (C) GFRP rebars.

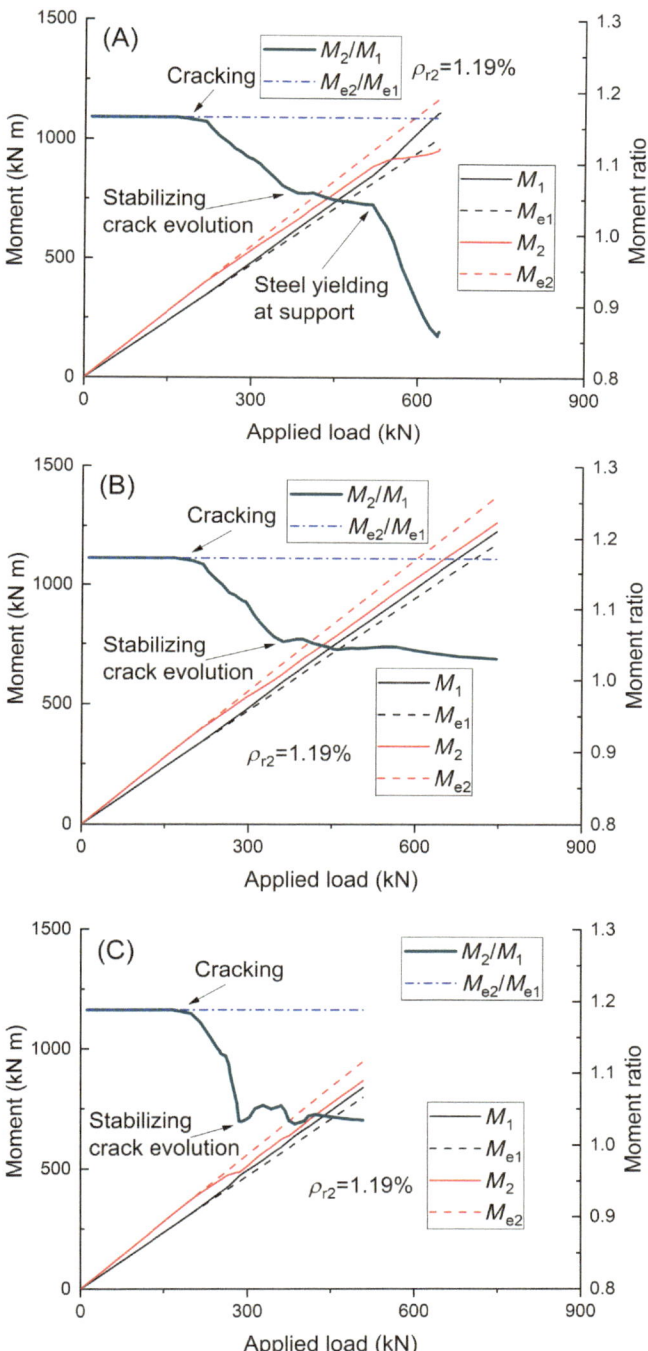

**Figure 9.18** Development of moment ratios for the beams with different types of rebars: (A) steel rebars; (B) CFRP rebars; (C) GFRP rebars.

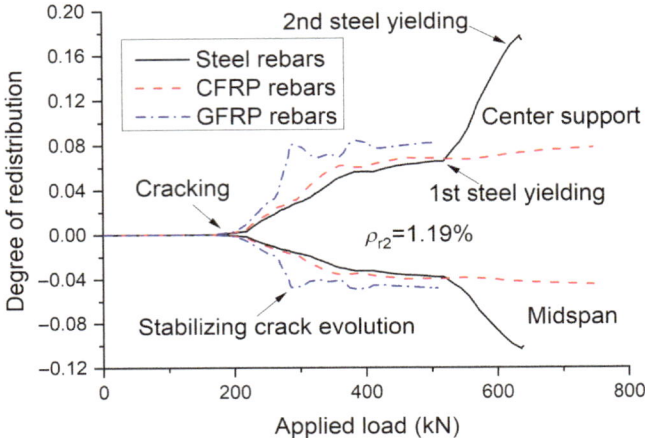

**Figure 9.19** Load versus moment redistribution curves for the beams with different types of rebars.

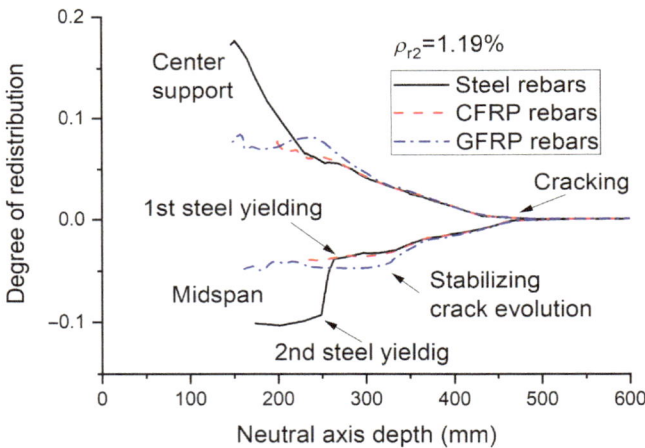

**Figure 9.20** Neutral axis depth versus moment redistribution curves for the beams with different types of rebars.

curves for the members with different types of rebars ($\rho_{r2} = 1.19\%$), respectively. The curves consist of three distinct stages for the members with FRP rebars while there are two additional stages for the members with steel rebars. The first stage corresponds to the elastic stage with zero moment redistribution. In this stage, the neutral axis shifts rapidly from infinity toward the extreme compressive fiber of the section, while the rebar strain is marginal. In the second stage, moment redistribution develops linearly with decreasing neutral axis depth or increasing rebar strain until the crack evolution stabilizes. In these two stages, the members with FRP

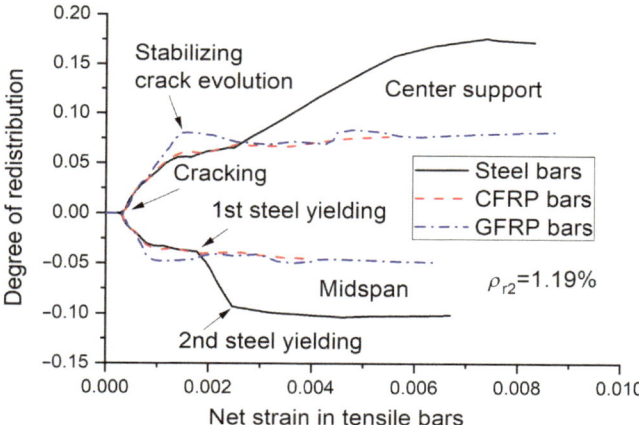

**Figure 9.21** Rebar strain versus moment redistribution curves for the beams with different types of rebars.

rebars exhibit approximately the same behavior as that of the members with steel rebars. The third stage is characterized by stabilizing redistribution. For the members with FRP rebars, this stage continues until failure, accompanied by substantial variation in neutral axis depth and rebar strain. For the members with steel rebars, the fourth stage, triggered by the yielding of steel bars at the center support, is characterized by quick development of moment redistribution with decreasing neutral axis depth or increasing rebar strain. The fifth stage, triggered by the yielding of steel bars at the midspan, is featured by stabilizing redistribution of moments with varying neutral axis depth or rebar strain up to the ultimate.

The change in the value of $\beta_u$ (degree of redistribution at ultimate) over the center support with the $\rho_{r2}$ level is displayed in Fig. 9.22. It is seen that the $\beta_u$ value for the members with GFRP rebars stabilizes around 8% with varying $\rho_{r2}$. The $\beta_u$ value for the members with CFRP rebars slightly increases with increasing $\rho_{r2}$ up to 0.7%, and thereafter it turns to decrease slightly. In general, moment redistribution in the members with CFRP rebars is very close to that in the members with GFRP rebars. For the members with steel rebars, the $\beta_u$ value increases with increasing $\rho_{r2}$ up to 1.67%. Thereafter, a higher $\rho_{r2}$ level results in a lower value of $\beta_u$. This phenomenon can be explained by the fact that the amount of rebars influences both the ductility and relative stiffness between the critical positive and negative moment zones. As $\rho_{r2}$ increases, the ductility of the center support section decreases, leading to a decrease in moment redistribution. Meanwhile, a higher $\rho_{r2}$ gives rise to a larger stiffness difference between the center support and midspan, leading to a higher degree of redistribution. Therefore, the variation in the $\beta_u$ value with varying $\rho_{r2}$ depends on the combined effects of ductility and relative stiffness.

It is also seen in Fig. 9.22 that moment redistribution in the members with steel rebars is substantially higher than that in the members with FRP rebars, attributed to the notable contribution of steel yielding. The difference between the $\beta_u$ values

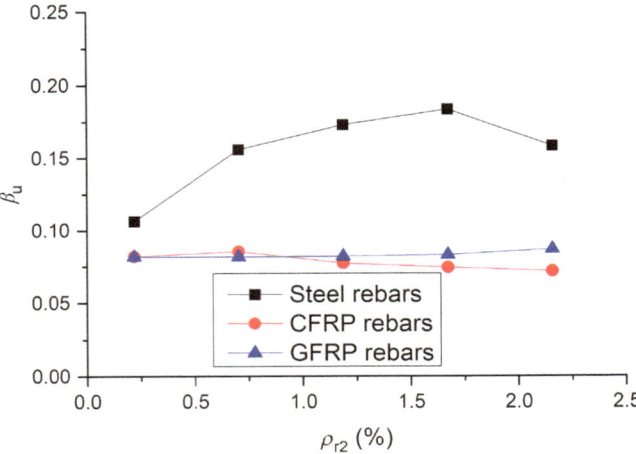

**Figure 9.22** Variation of $\beta_u$ with varying $\rho_{r2}$ for the beams with different types of rebars.

for the members with steel and FRP rebars tends to enlarge as $\rho_{r2}$ increases up to 1.67% and thereafter narrows. For $\rho_{r2} = 0.22\%$, 1.67%, and 2.16%, the redistribution mobilized by steel rebars is 1.3, 2.5, and 2.2 times, respectively, that by CFRP rebars and 1.3, 2.2, and 1.8 times, respectively, that by GFRP rebars.

# 9.5    Prediction of moment redistribution

## 9.5.1    Evaluation of design codes

While several codes or guides, for example, ACI 440.1R-15 (ACI Committee 440, 2015) and ACI 440.4R-04 (ACI Committee 440, 2004), dealing with FRP bars/tendons are available, they did not provide specific rules for moment redistribution in continuous members reinforced with FRP bars or prestressed with FRP tendons. It has been shown that prestressed concrete members with external CFRP and steel tendons exhibited similar redistribution behavior (Lou et al., 2017). On the other hand, the redistribution behavior of the members with FRP rebars is significantly different from that of the members with steel rebars, as discussed in previous sections. Therefore, it is worth checking if the current codes for reinforced or prestressed concrete members are applicable to the cases when FRP bars/tendons are used. Three codes of practice are considered, namely Eurocode 2 (CEN, 2004), CSA A23.3-04 (CSA, 2004), and ACI 318-19 (ACI Committee 318, 2019). These codes adopted either the neutral axis depth (CEN, 2004; CSA, 2004) or net strain in extreme tensile reinforcement (ACI Committee 318, 2019) as a key parameter for calculating the allowable moment redistribution in reinforced or prestressed concrete members.

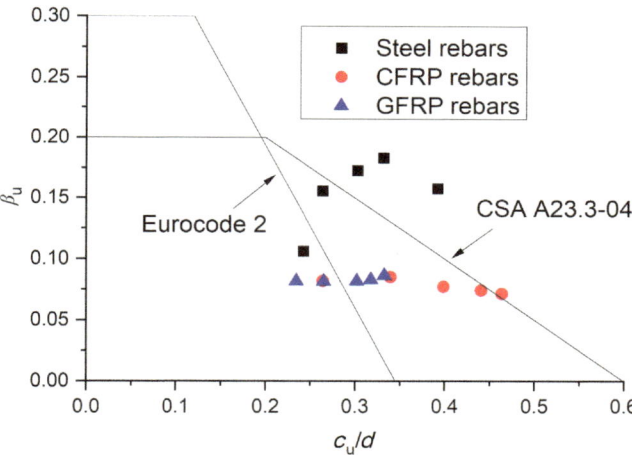

**Figure 9.23** Variation of $\beta_u$ with varying $c_u/d$ for the beams with different types of rebars along with the code curves (Eurocode 2 and CSA A23.3-04).

Fig. 9.23 shows the numerically obtained data regarding the $c_u/d-\beta_u$ relationship along with the code curves (Eurocode 2 and CSA A23.3-04). According to the numerical analysis, the $\beta_u$ value for the members with steel rebars is increased substantially by 72.45% when varying $c_u/d$ from 0.24 to 0.33. On the other hand, the members with FRP rebars exhibit stabilizing redistribution at ultimate regardless of the value of $c_u/d$. However, such observations from numerical simulations are either opposite to or inconsistent with the design codes, as the latter exhibit a trend of decrease in $\beta_u$ with increasing $c_u/d$. Therefore, both codes cannot reflect the trend regarding the variation of $\beta_u$ with varying $c_u/d$. It is also observed in the figure that most of the data lie beyond the Eurocode 2 curve while below the CSA A23.3-04 curve, indicating conservative predictions by Eurocode 2 but nonconservative predictions by CSA A23.3-04.

Fig. 9.24 illustrates the numerically obtained data regarding the $\varepsilon_t-\beta_u$ relationship along with the ACI 318-19 curve. It is seen that as far as the variation of $\beta_u$ with varying $\varepsilon_t$ is concerned, ACI 318-19 fails to reflect the actual tendency for the members with steel or FRP rebars. In addition, all of the data for the members with steel rebars are above the code curve, demonstrating safe predictions of ACI 318-19. For the members with FRP rebars, some of the data are below the code curve, implying unsafe predictions of ACI 318-19.

A comparison of the $\beta_u-\rho_{r2}$ relationship predicted by the design codes and numerical analysis for the members with different types of rebars is given in Fig. 9.25. According to the design codes, the $\beta_u$ value for the members with steel or FRP rebars consistently decreases as $\rho_{r2}$ increases. However, this is not concordant with the numerical prediction regarding the $\beta_u-\rho_{r2}$ relationship, as can be

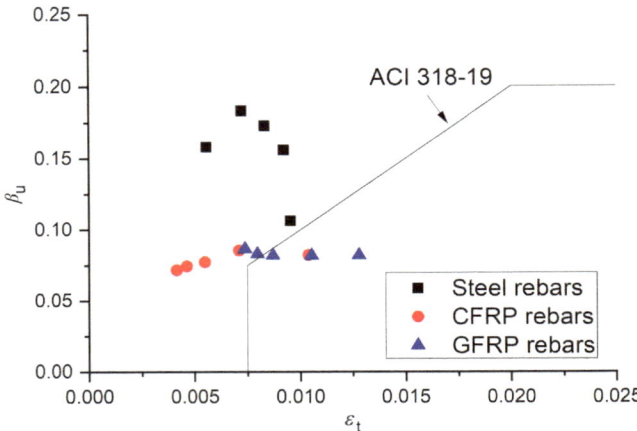

**Figure 9.24** Variation of $\beta_u$ with varying $\varepsilon_t$ for the beams with different types of rebars along with the ACI 318-19 curve.

seen in Fig. 9.25. This can be attributed to the fact that the design codes account for the section ductility only, neglecting the influence of relative stiffness, while both ductility and relative stiffness are affected by the amount of rebars. As a consequence, the influence of rebar amount on moment redistribution could not be reasonably reflected in the design codes. In addition, according to the code prediction, the redistribution for the members with CFRP rebars is substantially lower than that for the members with GFRP rebars. The redistribution values for the members with GFRP rebars are almost identical to or higher than those for the members with steel rebars. The aforementioned observations are also inconsistent with the actual (i.e., numerically predicted) influence of the type of rebars on moment redistribution. Therefore, the design codes could not reflect the influence of rebars (both the amount and type) on the moment redistribution in externally prestressed concrete members.

It can also be observed from Fig. 9.25 that Eurocode 2 is generally conservative for the members with CFRP or steel rebars but it might be nonconservative for the members with GFRP rebars at low $\rho_{r2}$ levels ($\rho_{r2}$ < 0.95%). CSA A23.3-04 is unsafe when predicting the moment redistribution in the members with FRP rebars. When steel rebars are used, CSA A23.3-04 is not safe at $\rho_{r2}$ < 0.86%. ACI 318-19 is conservative for the members with steel rebars, whereas it appears to be unsafe for the members with CFRP rebars at $\rho_{r2}$ < 0.52% and for the members with GFRP rebars at $\rho_{r2}$ < 1.49%.

## 9.5.2  Recommended equation

As illustrated in Fig. 9.22, the $\beta_u$ value for the members with FRP rebars is around 8% regardless of the type and amount of FRP rebars. Therefore, a redistribution

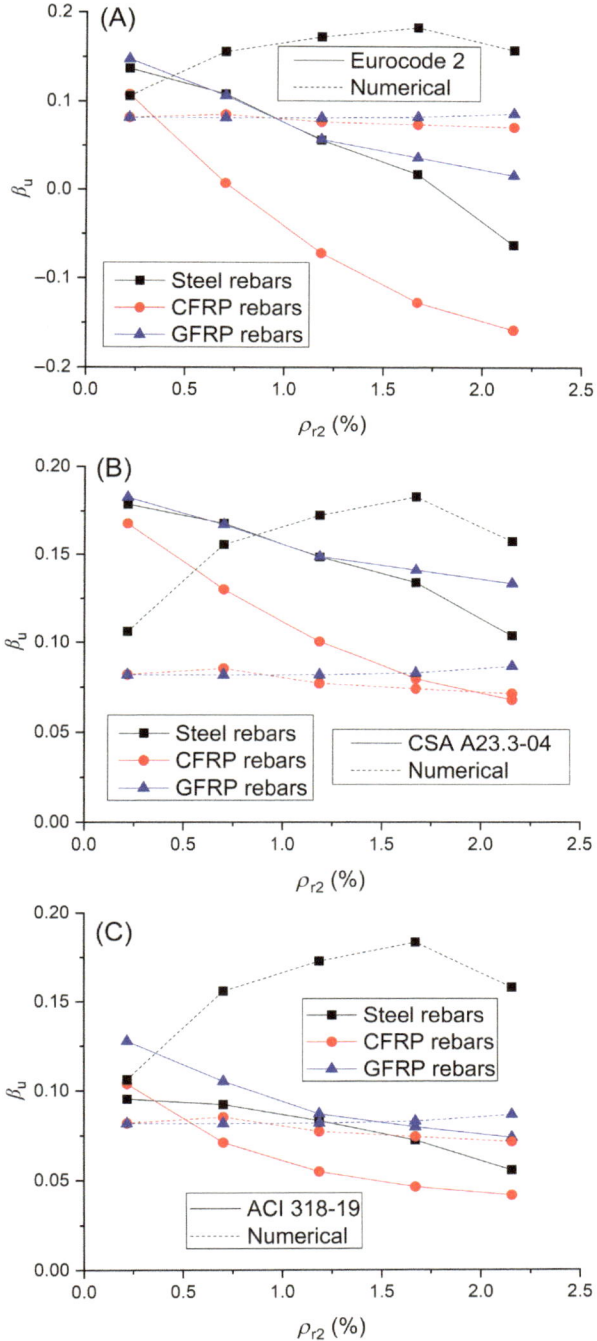

**Figure 9.25** Comparison of numerically predicted $\beta_u$ values and code predictions: (A) Eurocode 2; (B) CSA A23.3-04; (C) ACI 318-19.

value of 8% may be used in the design of continuous externally prestressed concrete members with FRP rebars. For the members with steel rebars, a modified CSA A23.3-04 equation with a new coefficient $\lambda_{csa}$ reflecting the relative stiffness of critical sections (Lou et al., 2020) may be adopted (see Chapter 5). Hence, the following equation is recommended to predict the moment redistribution at ultimate in externally prestressed concrete members with steel and FRP rebars.

$$\beta_u = \begin{cases} \lambda_{csa}(0.3 - 0.5c_u/d) & \text{for steel rebars} \\ 8\% & \text{for FRP rebars} \end{cases} \tag{9.10}$$

where

$$\lambda_{csa} = 0.43 + 2.71\ln(\omega_1/\omega_2) - 0.84\ln^2(\omega_1/\omega_2) \tag{9.11}$$

where $\omega$ is the combined reinforcement index; the subscripts 1 and 2 represent the midspan and center support, respectively.

Fig. 9.26 shows the variation of $(\beta_u)_{sim}/(\beta_u)_{act}$ against $c_u/d$ for the 15 members with different types of rebars, where $(\beta_u)_{sim}$ represents the moment redistribution calculated from the simplified equations (i.e., CSA A23.3-04 and recommended), and $(\beta_u)_{act}$ represents the actual moment redistribution generated by the numerical analysis. It is seen that the recommended equation is substantially more accurate than CSA A23.3-04 for quantifying moment redistribution in these members. In addition, the data by the recommended equation are mostly on the safe side (i.e., $(\beta_u)_{sim}/(\beta_u)_{act} < 1$). By contrast, most of the predictions by CSA A23.3-04 are unsafe (i.e., $(\beta_u)_{sim}/(\beta_u)_{act} > 1$).

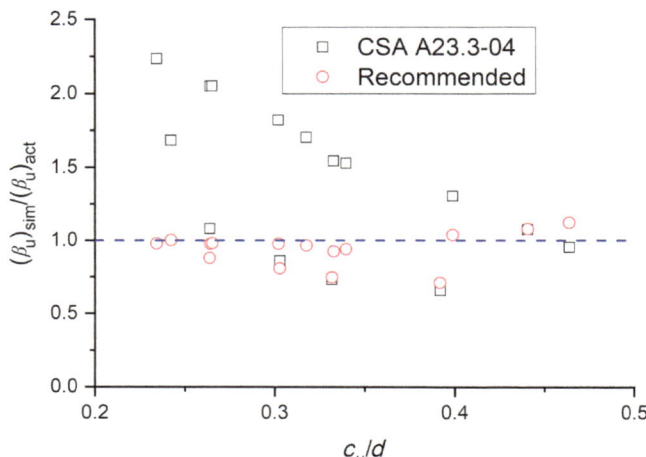

**Figure 9.26** Values of $(\beta_u)_{sim}/(\beta_u)_{act}$ against $c_u/d$ according to CSA A23.3-04 and recommended equations.

## 9.6 Conclusions

This chapter reports the use of FRP rebars as a replacement for steel rebars in two-span continuous externally prestressed concrete members. Comprehensive performance assessments are presented. Several code equations for calculating the ultimate tendon stress and moment redistribution are introduced, and practical simplified equations are recommended.

The results show that at symmetrical loading, crushing failure of members with FRP rebars occurs at center support while members with steel rebars may be crushed at midspan or center support, depending on the rebar area. Providing minimum FRP rebars appears to be more effective in alleviating the cracking concentration than providing steel rebars. FRP rebars lead to a slower increase in ultimate load with increasing $\rho_{r2}$ than steel rebars. Members with steel rebars show an ultimate load close to members with GFRP rebars at $\rho_{r2} = 0.22\%$ but to members with CFRP rebars at $\rho_{r2} = 2.16\%$. The effect of load pattern on the ultimate deflection is more significant for using FRP rebars than using steel rebars. Unsymmetrical loading leads to $4\%-21\%$ and $31\%-46\%$ higher ultimate deflection for members with steel and FRP rebars, respectively, when compared to symmetrical loading. The effect of load pattern on $\Delta\sigma_p$ in members with FRP rebars is not as significant as that of members with steel rebars. Unsymmetrical loading results in $40\%-52\%$ lower $\Delta\sigma_p$ when steel rebars are used or around $30\%$ lower $\Delta\sigma_p$ when CFRP/GFRP rebars are used in comparison to symmetrical loading.

At symmetrical loading, the code equations (JGJ 92-2016, 2016; AASHTO, 1994; AASHTO, 2017) substantially underestimate $\Delta\sigma_p$ in members with FRP rebars. At unsymmetrical loading, the design codes except for AASHTO, 1994 are unsafe when predicting $\Delta\sigma_p$ in members with steel rebars. Two simplified equations, one based on the parameter $c_u/d_p$ and the other on the parameter $\omega_0$, are proposed to calculate $\Delta\sigma_p$ in continuous externally prestressed concrete members with FRP/steel rebars. The proposed equations show much better correlations with numerical simulations than the code equations.

Moment redistribution in members with FRP rebars is contributed by concrete cracking, and it tends to stabilize after the stabilization of crack evolution. For members with steel rebars, apart from the contribution by concrete cracking, steel yielding leads to further development of moment redistribution. Steel rebars lead to significantly higher redistribution of moments than FRP rebars. The redistribution difference between the members with steel and FRP rebars enlarges with increasing $\rho_{r2}$ up to $1.67\%$ and decreases thereafter.

The current codes of practice (Eurocode 2, CSA A23.3-04, and ACI 318-19) could not reflect the influence of both the amount and type of rebars on moment redistribution in prestressed concrete members with external tendons. In addition, the codes may lead to unsafe predictions in moment redistribution in members with FRP rebars. A simplified equation is recommended to predict moment redistribution in externally prestressed concrete members with steel or FRP rebars. It is shown that the recommended equation yields accurate and conservative predictions.

# References

AASHTO. (1994). *LRFD bridge design specifications.* Washington, DC.

AASHTO. (2017). *LRFD bridge design specifications* (8th ed.). Washington, DC.

ACI Committee 318. (2019). *Building code requirements for structural concrete (ACI 318-19) and commentary (ACI 318R-19).* Farmington Hills, MI.

ACI committee 440. (2004). *Prestressing concrete structures with FRP tendons.* ACI 440.4R-04, Farmington Hills, MI.

ACI committee 440. (2015). *Guide for the design and construction of structural concrete reinforced with FRP bars.* ACI 440.1R-15, Farmington Hills, MI.

Aravinthan, T., Witchukreangkrai, E., & Mutsuyoshi, H. (2005). Flexural behavior of two-span continuous prestressed concrete girders with highly eccentric external tendons. *ACI Structural Journal, 102*(3), 402−411.

CEN. (2004). *Eurocode 2 (EC2): Design of concrete structures-Part 1-1: General rules and rules for buildings.* EN 1992-1-1, Brussels, Belgium.

Chan, K. H. E., & Au, F. T. K. (2015). Behaviour of continuous prestressed concrete beams with external tendons. *Structural Engineering and Mechanics, 55*(6), 1099−1120. Available from https://doi.org/10.12989/sem.2015.55.6.1099.

CSA. (2004). *Design of concrete structures.* A23.3-04, Mississauga, Ontario, Canada.

Ghallab, A. (2013). Calculating ultimate tendon stress in externally prestressed continuous concrete beams using simplified formulas. *Engineering Structures, 46*, 417−430. Available from https://doi.org/10.1016/j.engstruct.2012.07.018.

Harajli, M. H. (2012). Tendon stress at ultimate in continuous unbonded post-tensioned members: Proposed modification of ACI 318, Eq. (18-4) and (18-5). *ACI Structural Journal., 109*(2), 183−192. Available from https://doi.org/10.14359/51683629.

Harajli, M. H., Mabsout, M. E., & Al-Hajj, J. A. (2002). Response of externally post-tensioned continuous members. *ACI Structural Journal, 99*(5), 671−680.

JGJ 92-2016. (2016). *Technical specification for concrete structures prestressed with unbonded tendons.* Beijing: China Architecture & Building Press.

Lou, T., Li, Z., & Pang, M. (2021). Moment redistribution in continuous externally CFRP prestressed beams with steel and FRP rebars. *Polymers, 13*(8), 1181. Available from https://doi.org/10.3390/polym13081181.

Lou, T., Li, Z., & Pang, M. (2022). Behavior of externally prestressed continuous beams with FRP/steel rebars under symmetrical/unsymmetrical loading: Numerical study. *Case Studies in Construction Materials, 17*, e01196.

Lou, T., Lopes, S. M. R., & Lopes, A. V. (2013). Flexural response of continuous concrete beams prestressed with external tendons. *Journal of Bridge Engineering, 18*(6), 525−537. Available from https://doi.org/10.1061/(ASCE)BE.1943-5592.0000392.

Lou, T., Lopes, S. M. R., & Lopes, A. V. (2014). Factors affecting moment redistribution at ultimate in continuous beams prestressed with external CFRP tendons. *Composites Part B: Engineering, 66*, 136−146. Available from https://doi.org/10.1016/j.compositesb.2014.05.007.

Lou, T., Lopes, S. M. R., & Lopes, A. V. (2017). Effect of linear transformation on nonlinear behavior of continuous prestressed beams with external FRP cables. *Engineering Structures, 147*, 410−424. Available from https://doi.org/10.1016/j.engstruct.2017.06.029.

Lou, T., Peng, C., Karavasilis, T. L., Min, D., & Sun, W. (2020). Moment redistribution versus neutral axis depth in continuous PSC beams with external CFRP tendons. *Engineering Structures, 209*, 109927. Available from https://doi.org/10.1016/j.engstruct.2019.109927.

Maguire, M., Chang, M., Collins, W. N., & Sun, Y. (2017). Stress increase of unbonded tendons in continuous posttensioned members. *Journal of Bridge Engineering*, *22*(2), 04016115. Available from https://doi.org/10.1061/(asce)be.1943-5592.0000991.

Pang, M., Li, Z., & Lou, T. (2020). Numerical study of using FRP and steel rebars in simply supported prestressed concrete beams with external FRP tendons. *Polymers*, *12*(12), 2773. Available from https://doi.org/10.3390/polym12122773.

Roberts-Wollmann, C. L., Kreger, M. E., Rogowsky, D. M., & Breen, J. E. (2005). Stresses in external tendons at ultimate. *ACI Structural Journal*, *102*(2), 206−213.

Tan, K. H., & Tjandra, R. A. (2007). Strengthening of RC continuous beams by external prestressing. *Journal of Structural Engineering*, *133*(2), 195−204. Available from https://doi.org/10.1061/(ASCE)0733-9445(2007)133:2(195).

Zhou, W., & Zheng, W. (2014). Unbonded tendon stresses in continuous post-tensioned beams. *ACI Structural Journal*, *111*(3), 525−536. Available from https://doi.org/10.14359/51686569.

# Externally prestressed steel–concrete composite girders  **10**

## 10.1  Introduction

Steel–concrete composite girders are widely used in bridge infrastructure. This structural typology offers efficient and economical construction due to its higher stiffness and strength compared to those of steel or concrete beams (CEN, 2004a, 2004b). Despite their major advantages, steel–concrete composite girders show structural deterioration over their lifespan due to various reasons such as traffic loads higher than those considered in their initial design and steel corrosion. Using external tendons for prestressing has been recognized as an effective technique for strengthening existing structures (Nordin, 2005). The same technique is also considered an attractive means for the construction of new bridges (Brockmann & Rogenhofer, 2000). The noteworthy advantages offered by external prestressing include the enhancement of flexural capacities, the reduction in deflections and steel stresses at service loads, and the improvement of crack and fatigue resistance (El-Zohairy, Salim, & Saucier, 2019).

In externally prestressed steel–concrete composite girders, the tendons interact with the composite girder through deviators and anchorages. The tendon profile between adjacent deviators/anchorages remains rectilinear throughout the loading history. As a consequence, the tendon eccentricities or effective depths vary with the member deformation, leading to second-order effects in prestressed composite girders. The second-order effects are one of the key characteristics of externally posttensioned members. While many works on prestressed composite girders have been conducted (Abu-Sena et al., 2020; Chen & Jia, 2010; El-Zohairy & Salim, 2017; Moscoso, Tamayo, & Morsch, 2017; Zhang & Fu, 2009), little information on second-order effects in these girders has been reported. Dall'Asta et al. (2006) evaluated the performance of prestressed composite girders with different deviators (0 deviators, 1 deviator at midspan, and 2 deviators at third points) by a finite element analysis considering or neglecting second-order effects. Their study showed that compared to the composite girder without second-order effects, the composite girders with 0, 1, and 2 deviators exhibit 25.2%, 1.7%, and 9.5% lower ultimate load, respectively. In other words, two deviators at third points are less efficient than one deviator at midspan when minimizing the second-order effect. This finding is consistent with the study on externally prestressed concrete members (Lou & Xiang, 2010).

As external tendons are unbonded to adjacent components of composite girders, the tendon strain or stress depends on the whole member deformation. The estimation of ultimate tendon stress is a major task for the flexural strength design of prestressed composite girders. The available empirical equations (He & Liu, 2010; Pang, Li, & Lou, 2020; Roberts-Wollmann et al., 2005), as well as those

**Prestressed Members with External Fiber-Reinforced Polymer (FRP) Tendons.** DOI: https://doi.org/10.1016/B978-0-443-23877-2.00010-9

recommended in design codes (e.g., ACI committee 440 (ACI committee 440, 2004)), for predicting the ultimate tendon stress were developed for externally prestressed concrete members, and therefore, may not be applicable to prestressed composite girders. In addition, most of these equations did not account for the second-order effects, which have been demonstrated to have an important effect on tendon stress at the ultimate limit state (Harajli, Khairallah, & Nassif, 1999; Lou & Xiang, 2010). It is essential to propose a rational equation for predicting the ultimate tendon stress in prestressed composite girders for practical purposes.

The behavior of simply supported prestressed composite girders has been well addressed. In practice, composite girders, which are continuous over multiple spans, are a common structural typology. The hyperstatic restraints of continuous prestressed composite girders result in a more complicated behavior compared to that of simply supported ones. However, relatively few research works have been conducted on continuous prestressed composite girders (Chen, Wang, & Jia, 2009; Nie et al., 2009). The nonlinear behavior of continuous prestressed composite girders needs to be adequately addressed.

This chapter describes the flexural behavior of externally prestressed steel—concrete composite girders (Lou & Karavasilis, 2019a; Lou, Karavasilis, & Chen, 2021). An evaluation of second-order effects in simply supported prestressed composite girders with external tendons is conducted. An equation for calculating the ultimate stress in external tendons is proposed, taking into account the second-order effects in these girders. Additionally, the effectiveness of strengthening the continuous steel—concrete composite girder using external tendons is evaluated, and secondary moments in continuous prestressed composite girders having different tendon layouts and load patterns are analyzed. The effect of the span ratio is also discussed. Conclusions relevant to the design and performance of prestressed composite girders with external tendons are drawn.

## 10.2    Assessment of second-order effects

Simply supported externally prestressed steel—concrete composite girders having a span of 20.0 m, as shown in Fig. 10.1, are used. The tendons are of rectilinear profile, having an initial depth of 745 mm. The main variables are the deviator spacing represented by $S_d/L$ (ratio of deviator spacing to span length) and load type (one-point load at midspan, two-point loads each being $L/8$ away from the midspan, and uniform load). Both variables are closely linked to the second-order effects. The tendon area $A_p$, original prestress $\sigma_{p0}$, and area of steel bars $A_s$ along with the material properties are presented in Fig. 10.1.

### 10.2.1    Failure and cracking mode

As illustrated in Fig. 10.2, when the concrete strain reaches the specified value of 0.003, the slab is crushed, resulting in the failure of the composite girders. Crushing

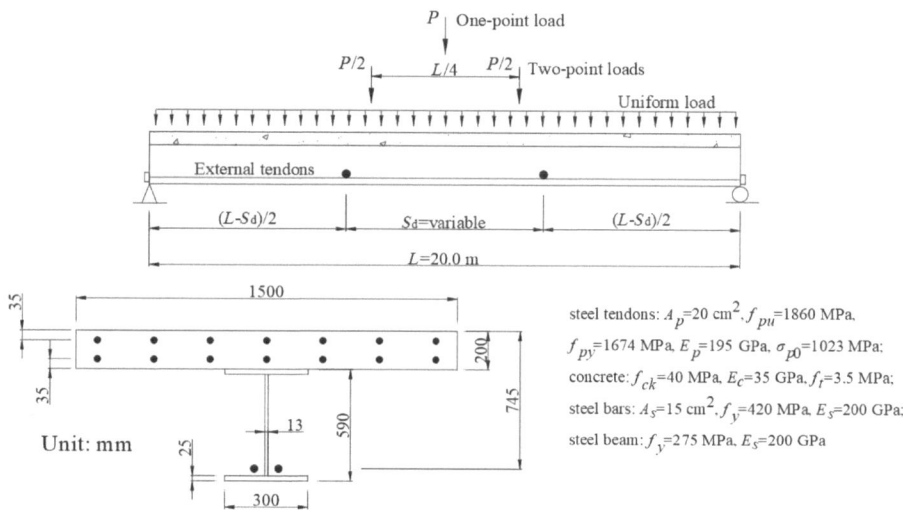

**Figure 10.1** Details of simply supported prestressed composite girders.

failure occurs at midspan except for the girder with $S_d/L = 1.0$ under two-point or uniform loading. In the case of $S_d/L = 1.0$, there are reductions in tendon effective depth along the span except at midspan. When uniform loading is applied, failure occurs at the weakest sections near the midspan. On the other hand, for two-point loading, failure happens at the sections where the loads are applied.

For single-span composite girders used in this section, cracking in a concrete slab is not significant. This is confirmed by the small tensile strain in concrete as shown in Fig. 10.2. Cracking takes place when concrete gets to its cracking strain, which is $100 \ \mu\varepsilon$. It is seen in Fig. 10.2 that the crack zone for one-point loading is smaller than that for uniform or two-point loading while the maximum crack widths are almost identical for different types of loading. The deviator spacing appears to have no noticeable influence on the crack mode. Apart from the main cracking in the bottom slab fiber in the zone around the midspan, a few cracks form in the top slab fiber near the support.

## 10.2.2   Displacement and tendon effective depth

Fig. 10.3 shows the influence of deviator spacing on the change in tendon effective depths according to different loading types. The maximum reduction in tendon depth is at midspan except for $S_d/L=0$, where it is located at points that are $L/6$ away from the midspan. The loss in tendon depth (i.e., second-order effects) is not important for $S_d/L=0$ and 0.25 but becomes increasingly significant as $S_d/L$ increases. One-point loading leads to a substantially smaller reduction in tendon depth (i.e., second-order effects) than uniform or two-point loading.

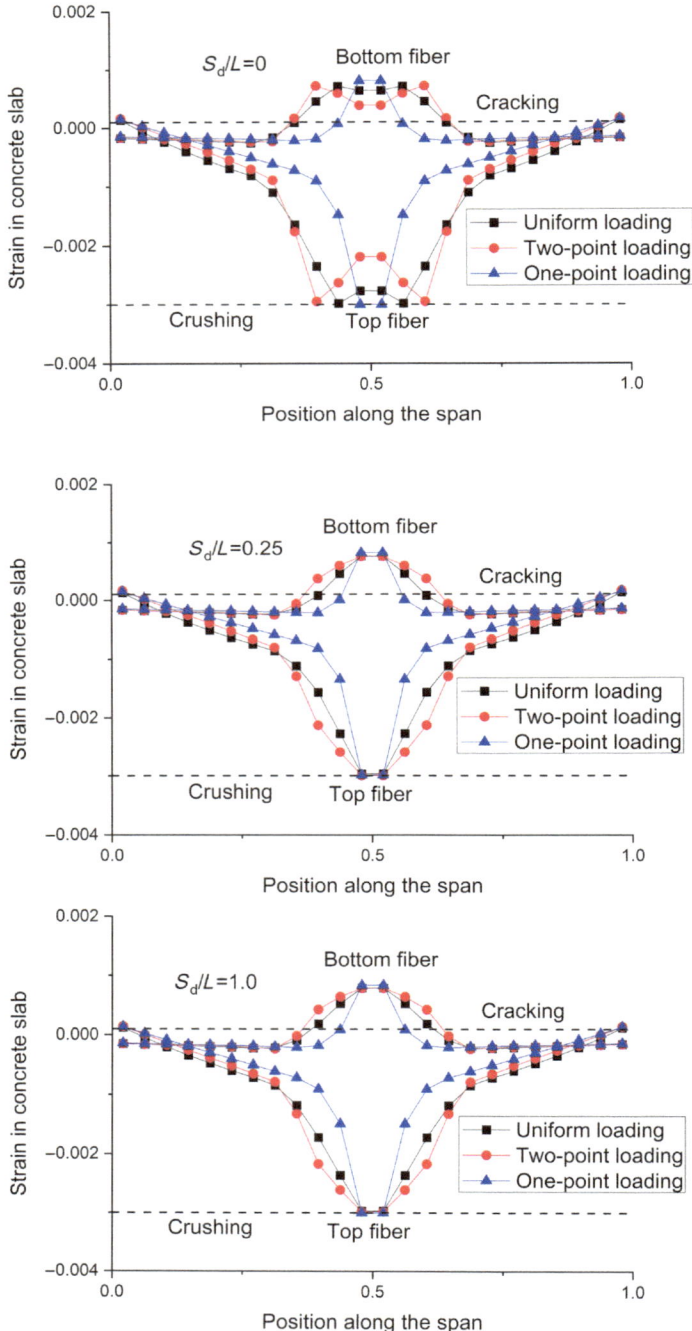

**Figure 10.2** Strain distribution in concrete slab.

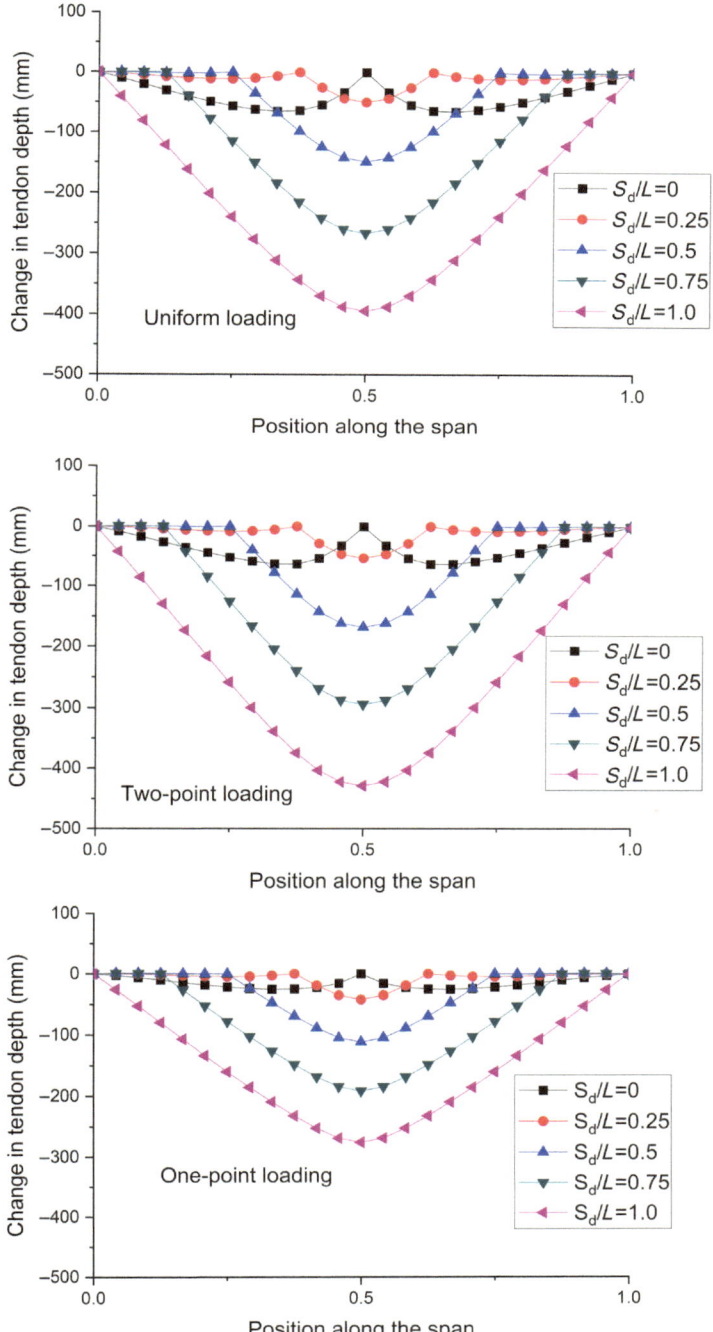

**Figure 10.3** Change in effective depths of external tendons over the span (in failure).

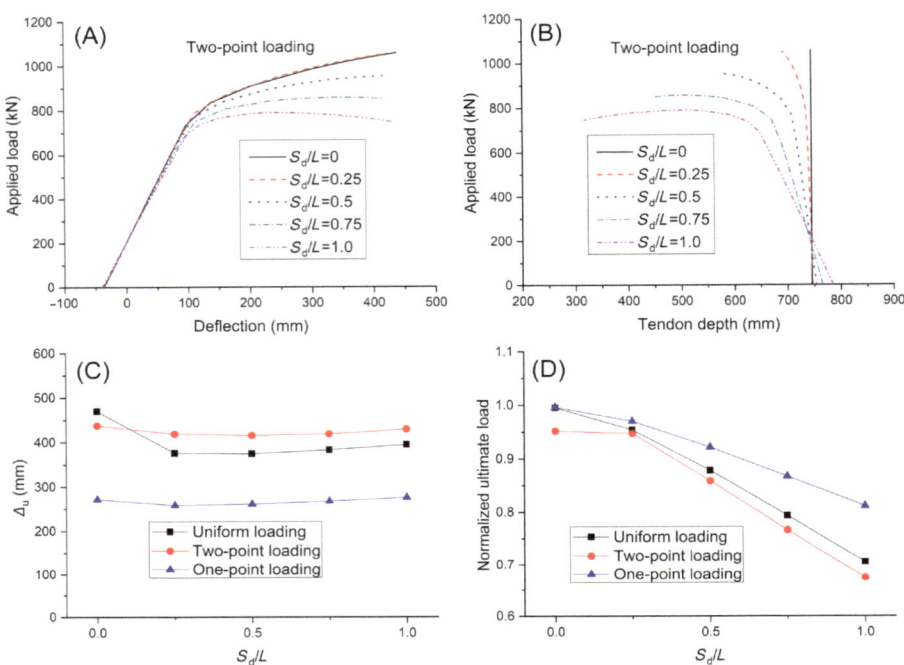

**Figure 10.4** Load versus displacement: (A) deflection development at midspan with applied load; (B) change in tendon effective depth at midspan with applied load; (C) variation of ultimate deflection with varying deviator spacing and load type; (D) variation of ultimate load with varying deviator spacing and load type.

Fig. 10.4A and B shows the influence of deviator spacing on the load versus deflection and tendon effective depth curves for composite girders under two-point loading, respectively. The load−deflection curve consists of two distinct stages. The transition is caused by the yielding of the steel beam. Concrete cracking and reinforcement yielding appear to have negligible effects on the load−deflection behavior of composite girders. Before yielding, the deflection development is small (see Fig. 10.4A) and the reduction in tendon effective depth is insignificant (see Fig. 10.4B). Therefore the girders with different $S_d/L$ ratios exhibit nearly the same load−deflection behavior in the preyielding stage due to the negligible second-order effects. After yielding, the deflection develops quickly (see Fig. 10.4A) and accordingly the second-order effects become increasingly notable if not minimized (see Fig. 10.4B). As a consequence, the postyielding load−deflection curves differ for the girders with different $S_d/L$ ratios. For $S_d/L=0$ and 0.25, the second-order effects are minimized, leading to higher postyielding flexural stiffness compared to other deviator configurations. For $S_d/L=1.0$, the second-order effects are extremely significant, resulting in a decrease in applied load with increasing deflection before collapse.

The variation of the ultimate deflection ($\Delta_u$) with varying $S_d/L$ for the composite girders under different load types is shown in Fig. 10.4C. For a given type of loading, the ultimate deflection remains almost unchanged regardless of the $S_d/L$ ratio, indicating that the ultimate deflection is independent of the second-order effects. However, when uniform loading is applied, the girder with $S_d/L=0$ exhibits 24% higher ultimate deflection than those with other $S_d/L$ ratios. This observation is consistent with a previous study by Dall'Asta et al. (2006). At a given $S_d/L$ ratio, one-point loading leads to considerably lower ultimate deflection than uniform or two-point loading. This can be attributed to the fact that the plastic zone developed by one-point loading is smaller than that by uniform or two-point loading, as demonstrated in Fig. 10.2.

The variation of the ultimate load (normalized) with varying $S_d/L$ for the composite girders under different load types is shown in Fig. 10.4D. The normalized ultimate load is defined as the ratio of the ultimate load of a prestressed composite girder to that neglecting second-order effects. It is seen that the ultimate load decreases with increasing deviator spacing, attributed to increasing second-order effects. For $S_d/L=0$, the ultimate load by uniform or one-point loading is very close to 1.0, indicating negligible second-order effects. The ultimate load by two-point loading is 0.95, attributed to the reduction in tendon effective depth in failure sections where the point loads are applied. For $S_d/L \geq 0.25$, there is approximately a linear relationship between the deviator spacing and the ultimate load. Because of the smaller second-order effects, the decrease in ultimate load for one-point loading is slower than that for uniform or two-point loading. As a result, the discrepancy of the ultimate loads by different types of loading becomes increasingly notable as $S_d/L$ increases.

### 10.2.3 Stress in external tendons

Fig. 10.5A and B shows the increase in tendon stress versus deflection curves for different $S_d/L$ ratios and types of loading, respectively. When the second-order effects are minimized (i.e., $S_d/L=0$ and 0.25), the stress—deflection relationship is approximately linear. For $S_d/L > 0.25$, the relationship exhibits a nonlinear manner, with a degree of nonlinearity depending on the second-order effects. At a given $S_d/L$ ratio, the increase in tendon stress by one-point loading appears to be slower than that for uniform or two-point loading.

At ulimate, the variation of the stress increment ($\Delta\sigma_{ps}$) in external tendons with varying $S_d/L$ for the composite girders under different load types is shown in Fig. 10.5C. Due to second-order effects, the $\Delta\sigma_{ps}$ value decreases as $S_d/L$ increases. The rate of decrease is lower for one-point loading than for uniform or two-point loading. This can be explained by the lower deflection and hence smaller

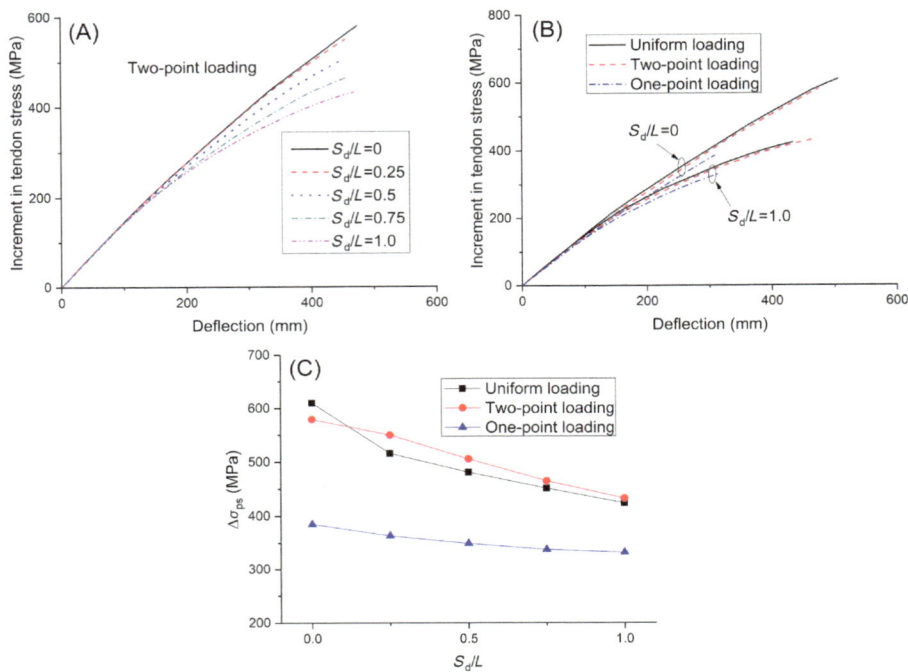

**Figure 10.5** Stress increase in external tendons: (A) development of midspan deflection against increment in tendon stress for different $S_d/L$ ratios; (B) development of midspan deflection against increment in tendon stress for different load types; (C) variation of ultimate tendon stress increment with varying deviator spacing and load type.

second-order effects produced by one-point loading. At a given $S_d/L$ ratio, one-point loading leads to substantially lower $\Delta\sigma_{ps}$ than uniform or two-point loading, attributed to smaller plastic zones developed at ultimate.

## 10.2.4   Curvature ($\kappa$) versus neutral axis depth (c)

Initially, there are hogging curvatures in prestressed composite girders under self-weight load only. The corresponding neutral axis depths in the critical section vary between 26.7 mm ($S_d/L=0.25$) and 48.1 mm ($S_d/L=1.0$), that is, the initial neutral axes lie within the slab. When the girders are subjected to monotonic loading, the hogging curvatures gradually disappear and then the sagging curvatures occur and develop. Accordingly, the neutral axes move rapidly toward upward infinity, and then jump to downward infinity and rise thereafter. Fig. 10.6A and B shows the neutral axis depth versus curvature ($c-\kappa$) relationship (when $c < 600$ mm) for the girders with different $S_d/L$ values and loading types. It is observed that both the

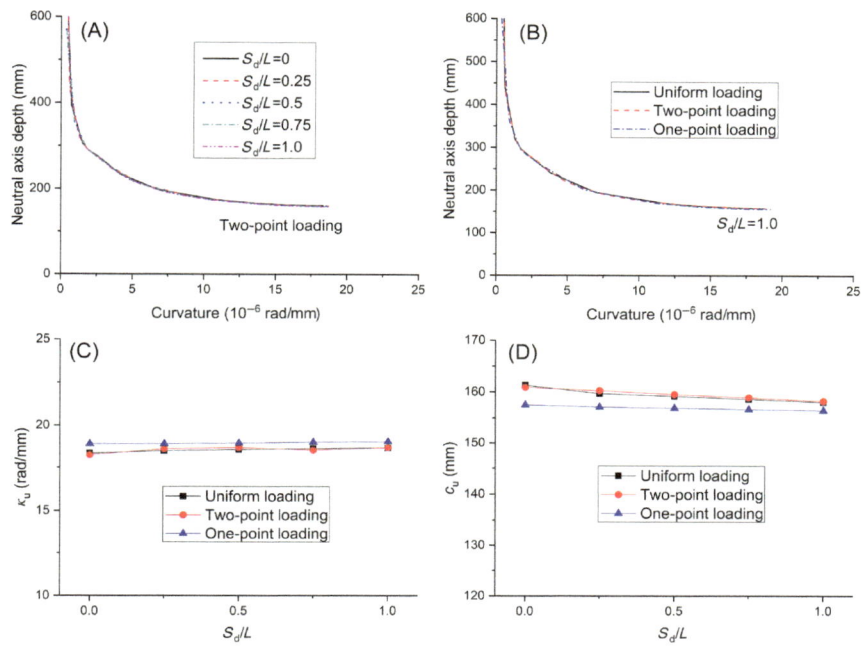

**Figure 10.6** Curvature versus neutral axis depth: (A) development of curvature against neutral axis depth in critical section for different $S_d/L$ ratios; (B) development of curvature against neutral axis depth in critical section for different load types; (C) variation of ultimate curvature with varying deviator spacing and load type; (D) variation of neutral axis depth at ultimate with varying deviator spacing and load type.

deviator spacing and load type have practically no influence on the $c-\kappa$ behavior, indicating that this behavior is independent of the second-order effects. According to the plane section assumption, the strain in any fiber in a composite section can be expressed as a function of $c$ and $\kappa$. Therefore at a specified $c$ or $\kappa$, the strains (and hence stresses) in concrete, steel bars, and steel beams are independent of the deviator spacing and load type.

At ultimate, the variations in curvature ($\kappa_u$) and neutral axis depth ($c_u$) with varying $S_d/L$ are illustrated in Fig. 10.6C and D, respectively. It is seen that as $S_d/L$ increases, the $\kappa_u$ value remains almost unchanged while the value of $c_u$ decreases slightly. At a given $S_d/L$ ratio, one-point loading leads to a slightly higher $\kappa_u$ value and lower $c_u$ value compared to uniform or two-point loading. According to the above observations and the plane section assumption, it can be inferred that the influence of both deviator spacing and tendon type on the strains (and hence stresses) at ultimate in the critical section is negligible or very limited. This statement is confirmed by the results generated by the proposed model, which are not presented in this chapter due to limited space.

## 10.3 Proposed equation for predicting ultimate tendon stress

ACI 440.4R-04 (ACI committee 440, 2004) recommended the following equation, which employs strain compatibility in combination with a bond reduction factor, to predict the ultimate stress increase, $\Delta\sigma_p$, in unbonded tendons:

$$\Delta\sigma_p = \Omega_u E_p \varepsilon_{cu} \left( \frac{d_p}{c_u} - 1 \right) \tag{10.1}$$

where $\Omega_u$ is the bond reduction factor. Eq. (10.1) is applicable to both steel and FRP tendons. For uniform or two-point loading, the value of $\Omega_u$ in flexural strength design is expressed as

$$\Omega_u = \frac{3.0}{L/d_p} \tag{10.2a}$$

while for one-point loading as

$$\Omega_u = \frac{1.5}{L/d_p} \tag{10.2b}$$

Note that Eq. (10.2) was developed based on the regression analysis on test data of prestressed concrete members with unbonded or external tendons. To check if the expression is applied to the case of prestressed composite girders, the $\Omega_u$ versus $L/d_p$ relationship by numerical simulations is compared to that by ACI 440.4R-04 in Fig. 10.7. The results are generated for the prestressed composite girders with deviators all along the span (i.e., no second-order effects). According to ACI 440.4R-04, the $\Omega_u$ value decreases quickly with increasing $L/d_p$. This phenomenon,

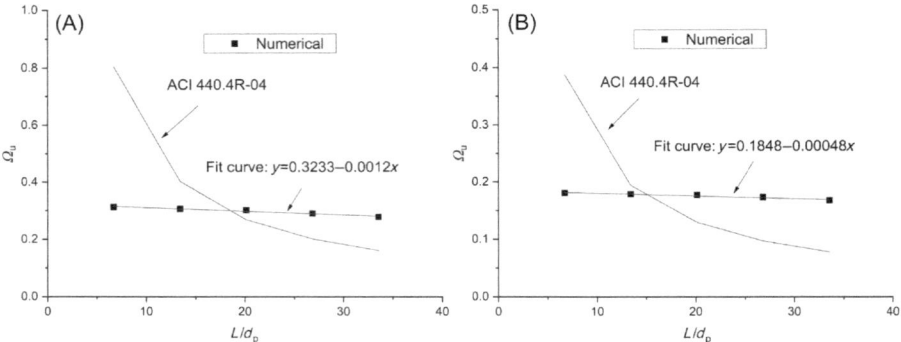

**Figure 10.7** Variation of $\Omega_u$ with $L/d_p$ according to ACI 440.4R-04 and numerical simulations: (A) two-point loading; (B) one-point loading.

however, is not consistent with numerical simulations, which show a slight change in $\Omega_u$ with varying $L/d_p$. In addition, at a low $L/d_p$ ratio, ACI 440.4R-04 leads to significantly unsafe predictions. According to the linear fit on numerical data, the following recommendation on $\Omega_u$ for prestressed composite girders neglecting second-order effects is made.

For uniform or two-point loading,

$$\Omega_u = 0.32 - 0.0012L/d_p \tag{10.3a}$$

while for one-point loading,

$$\Omega_u = 0.18 - 0.00048L/d_p \tag{10.3b}$$

As illustrated previously, the second-order effects play an important role in developing the stress increase in external tendons. Therefore such an effect needs to be included in the predictive equation for $\Delta\sigma_p$ in prestressed composite girders. A reduction factor due to second-order effects ($R_{sec}$) is introduced.

$$\Delta\sigma_p = R_{sec}\Omega_u E_p \varepsilon_{cu}\left(\frac{d_p}{c_u} - 1\right) \tag{10.4}$$

$$R_{sec} = \frac{\Delta\sigma_p}{(\Delta\sigma_p)_{nsoe}} \tag{10.5}$$

where $(\Delta\sigma_p)_{nsoe}$ denotes the ultimate tendon stress increment in prestressed composite girders neglecting second-order effects. Fig. 10.8 shows the variation of $R_{sec}$ with varying deviator spacing for two-point loading and one-point loading, respectively. According to the fit curve, the expression of $R_{sec}$ for uniform or two-point loading is given by

$$R_{sec} = 0.95 - 0.248S_d/L \tag{10.6a}$$

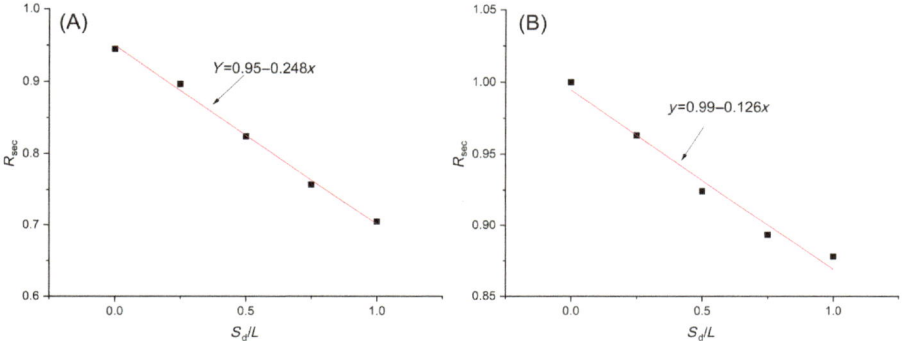

**Figure 10.8** Relationship between $R_{sec}$ and $S_d/L$: (A) two-point loading; (B) one-point loading.

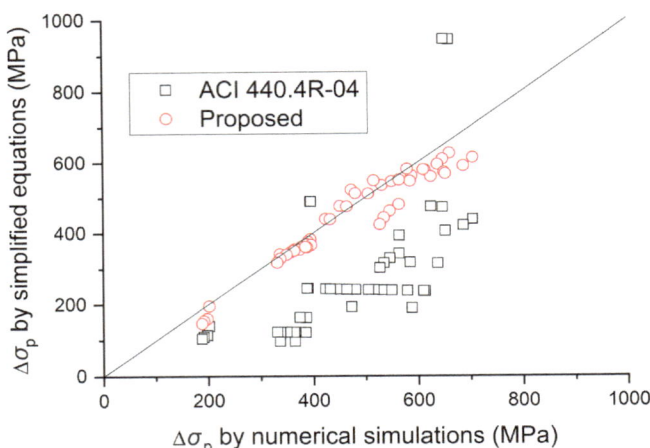

**Figure 10.9** Correlation of $\Delta\sigma_p$ by simplified equations and numerical simulations.

while for one-point loading by

$$R_{\text{sec}} = 0.99 - 0.126 S_d / L \tag{10.6b}$$

It is worth mentioning that the above equation is developed based on the data produced for very slender composite girders (i.e., $L/d_p = 26.8$). Therefore the $R_{\text{sec}}$ value expressed by Eq. (10.6) is conservative as the second-order effects are also related to the slenderness of prestressed composite girders.

The $\Delta\sigma_p$ values predicted by simplified equations are correlated to the numerical simulations in Fig. 10.9. A total of 47 specimens are used for the correlation. These include 15 specimens with different $S_d / L$ ratios and load types presented in the previous section, 20 specimens with different $L/d_p$ ratios, having deviators at third points or all along the span, and under one-point or two-point loading used in this section, and 12 specimens with different tendon types and prestress levels presented in Lou and Karavasilis (2019b). It is seen that ACI 440.4R-04 leads to poor correlation. By using a modified $\Omega_u$ and introducing a new coefficient $R_{\text{sec}}$, the proposed equation leads to a much better prediction than ACI 440.4R-04.

## 10.4　General behavior of continuous girders

The continuous prestressed composite girders are 12 m long with two identical spans each of which is subjected to a concentrated load at the midspan as shown in Fig. 10.10. The depth of external tendons from the bottom flange of the steel beam is 295 mm at the end support, 45 mm at the midspan, and 545 mm at the center support. The tendon area, $A_p$, is a variable and takes values from 0 to 20 cm². The initial prestress, $\sigma_{p0}$, is 930 MPa. Top and bottom steel rebars are equally provided in

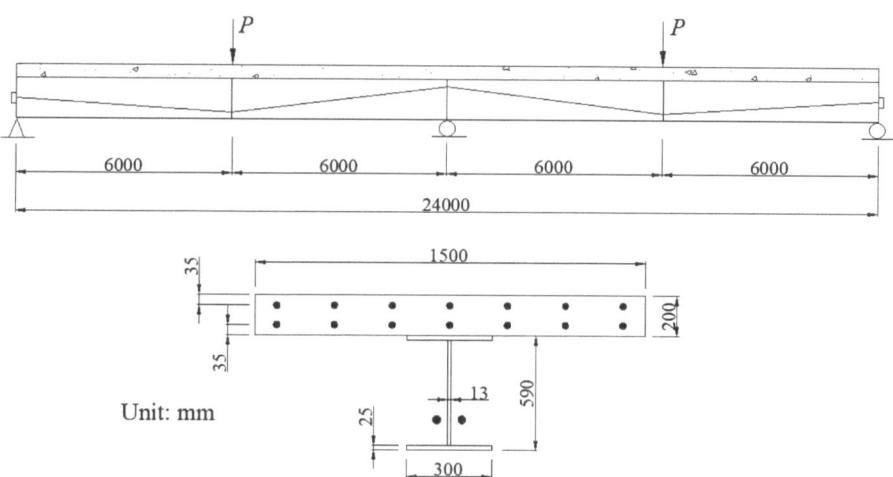

**Figure 10.10** Structure and section details of continuous prestressed composite girders.

the concrete slab. Their total area is 15 cm². The material parameters for steel tendons, steel beam, steel rebars, and concrete are the same as those used in the previous sections (see Fig. 10.1).

### 10.4.1 Moment—curvature and load—deflection behavior

All the girders have failed due to the crushing of the concrete slab at the midspan. Before the failure, some typical phases may be experienced during the loading process, such as concrete cracking at the center support, yielding of reinforcing steel bars at the critical sections, and yielding of the steel beam at the critical sections.

Fig. 10.11 shows the moment—curvature and load—deflection curves of the prestressed composite girders with various tendon areas. The response is mainly influenced by the first yielding of the steel beam. The deformation development is very slow in the preyielding stage but becomes very fast in the postyielding stage. Cracking of concrete and yielding of reinforcing bars do not have a noticeable influence on the structural response of composite girders. The curvature over the center support develops faster than the curvature over the midspan. As the tendon area increases, the ultimate midspan moment and load-carrying capacity increase notably, while the ultimate midspan curvature and deflection decrease. Strengthening the steel—concrete composite girder by external prestressing leads to an increase in the ultimate load by 13.04% for $A_p=5$ cm², by 23.87% for $A_p=10$ cm², by 33.17% for $A_p=15$ cm², and by 42.15% for $A_p=20$ cm². However, the moment—curvature response at the center support appears to be hardly influenced by the tendon area.

It is also observed that the ultimate moment over the center support is lower than the one over the midspan and the difference enlarges as the tendon area increases. This observation can be partly attributed to the influence of secondary

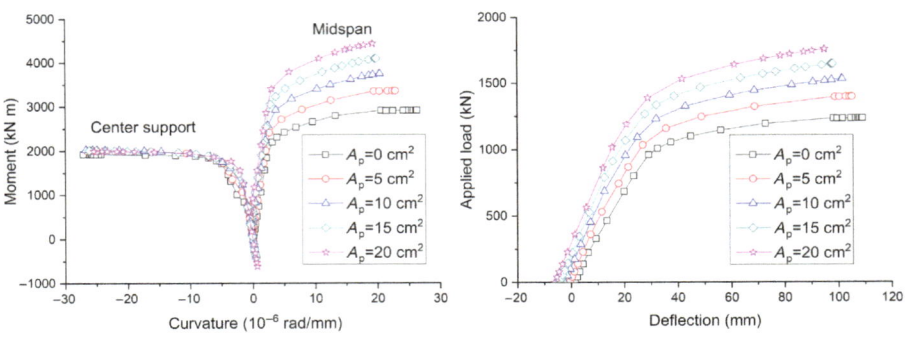

**Figure 10.11** Moment−curvature and load−deflection curves.

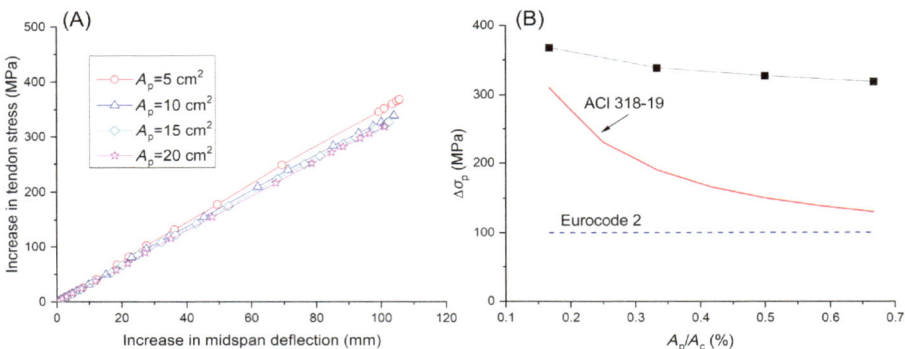

**Figure 10.12** Stress increase in external tendons: (A) relationship between tendon stress and deflection; (B) variation of $\Delta\sigma_p$ with $A_p/A_c$

moments induced by prestressing. The secondary moments, which are higher for a larger tendon area, increase the midspan moment and decrease the center support moment. The evaluation of secondary moments is examined in detail in Section 10.5.

### 10.4.2 Stress increase in external tendons

Fig. 10.12A shows the stress increase in external tendons against deflection. There is almost a linear relationship between tendon stress and deflection. A larger tendon area leads to a slightly lower rate of stress increase in external tendons. The slopes for $A_p$=5, 10, 15, and 20 cm$^2$ are 3.48, 3.26, 3.19, and 3.14 MPa/mm, respectively. At the ultimate phase of the behavior, considerable increments in tendon stress (around 350 MPa) are developed. It should be noted that the computed data are highly dependent on the specified value of $\varepsilon_u$ and that a conservative value of 0.003 for $\varepsilon_u$ is used in this analysis.

The variation of $\Delta\sigma_p$ (increase in tendon stress at ultimate) with $A_p/A_c$ (ratio of the tendon area to the concrete slab area) is shown in Fig. 10.12B. It is seen that $\Delta\sigma_p$ gradually decreases as $A_p$ increases. ACI 318−19 (ACI Committee 318, 2019) recommends an equation to calculate the value of $\Delta\sigma_p$ for normal span-to-depth ratio, that is, $\Delta\sigma_p = 70 + f_{ck}/(100\rho_p)$, where $\rho_p$ is the tendon ratio, assumed equal to $A_p/A_c$. In Eurocode 2 (CEN, 2004a), a value of 100 MPa is recommended for $\Delta\sigma_p$. It is observed that ACI 318-19 conservatively overestimates the influence of $A_p$ when calculating $\Delta\sigma_p$ in prestressed composite girders, especially at high $A_p/A_c$ levels. Eurocode 2 is more conservative when predicting $\Delta\sigma_p$ in prestressed composite girders.

### 10.4.3 Behavior of reinforced concrete slab

The development of strains in concrete slab at the midspan and center support is shown in Fig. 10.13. It is seen that at the midspan, the compressive strain in the top fiber develops faster than the tensile strain in the bottom fiber. When the compressive strain reaches 0.003 the concrete slab is crushed. At failure, the tensile strain in the bottom fiber is smaller than 0.003, indicating that the neutral axis is located at the lower half of the slab section. In addition, a larger tendon area leads to a smaller tensile strain in the bottom fiber, indicating a bigger neutral axis depth (distance between neutral axis and concrete top fiber) at ultimate. On the other hand, over the center support, tensile strains develop in concrete in both the top and bottom fibers at failure, although the concrete strain in the bottom fiber is significantly smaller than in the top fiber, thus indicating that the neutral axis lies in the steel beam. In addition, a larger tendon area leads to a higher ratio of the top fiber strain to the bottom fiber strain at ultimate, indicating a larger neutral axis depth (distance between neutral axis and steel bottom flange).

Fig. 10.14 shows the stress development in reinforcing bars at the midspan and center support. Over the center support, both the top and bottom bars develop tensile stress with the applied load. The stress development is initially slow but becomes rapid after cracking. Over the midspan, the top bars develop compressive

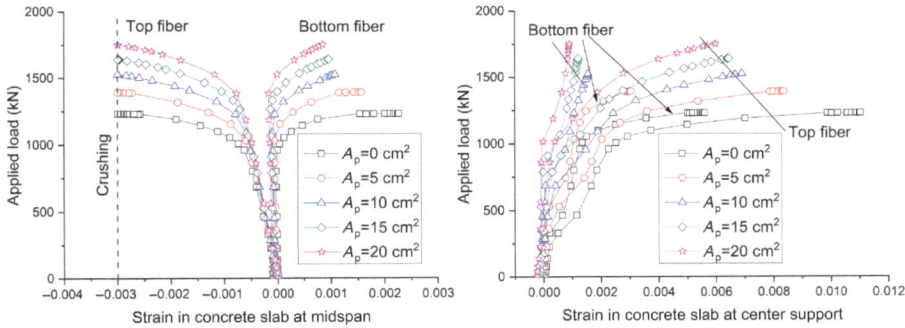

**Figure 10.13** Development of strains in concrete slab.

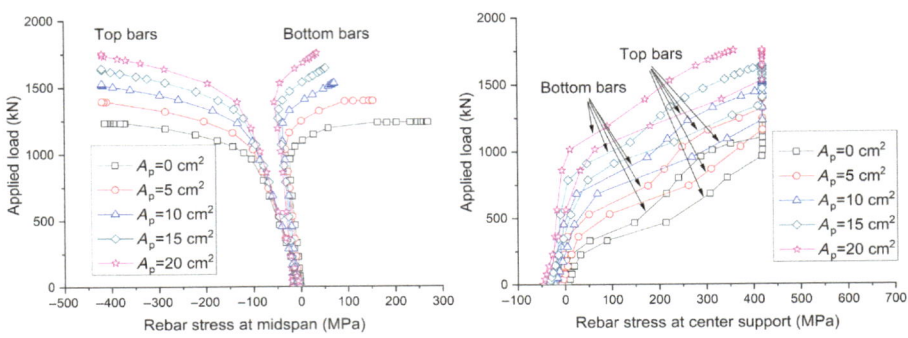

**Figure 10.14** Development of stresses in reinforcing steel bars.

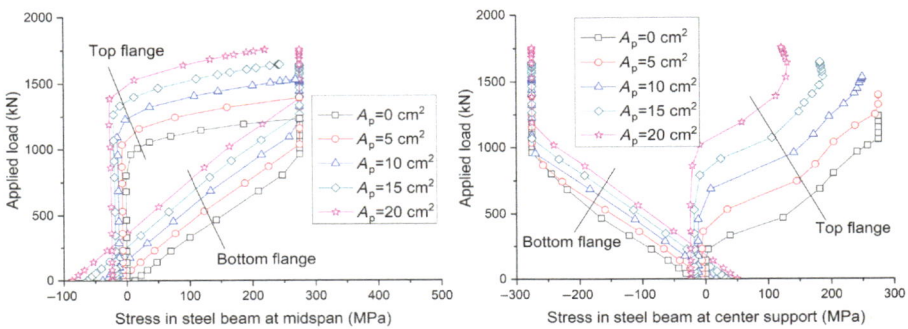

**Figure 10.15** Development of stresses in the steel beam.

stress, while the bottom bars tend to develop tensile stress as the applied loads increase. The stress development is not significant until the yielding of the steel beam. At service loads, the precompressive stress introduced by external prestressing in the bars over the center support effectively reduces the stress level. However, for the bars over the midspan, the stress difference between nonprestressed and prestressed composite girders is not apparent at service conditions.

### 10.4.4　Behavior of steel beam

Fig. 10.15 shows the stress development of the steel beam at the midspan and center support. Before load application, external prestressing induces precompressive stress over the midspan and pretensile stress over the center support in the bottom flange. After load application, the tensile stress over the midspan and the compressive stress over the center support develop linearly with increasing load up to yielding. When compared to the nonprestressed composite girder, the prestress in the bottom flange of prestressed composite girders significantly reduces the stress level

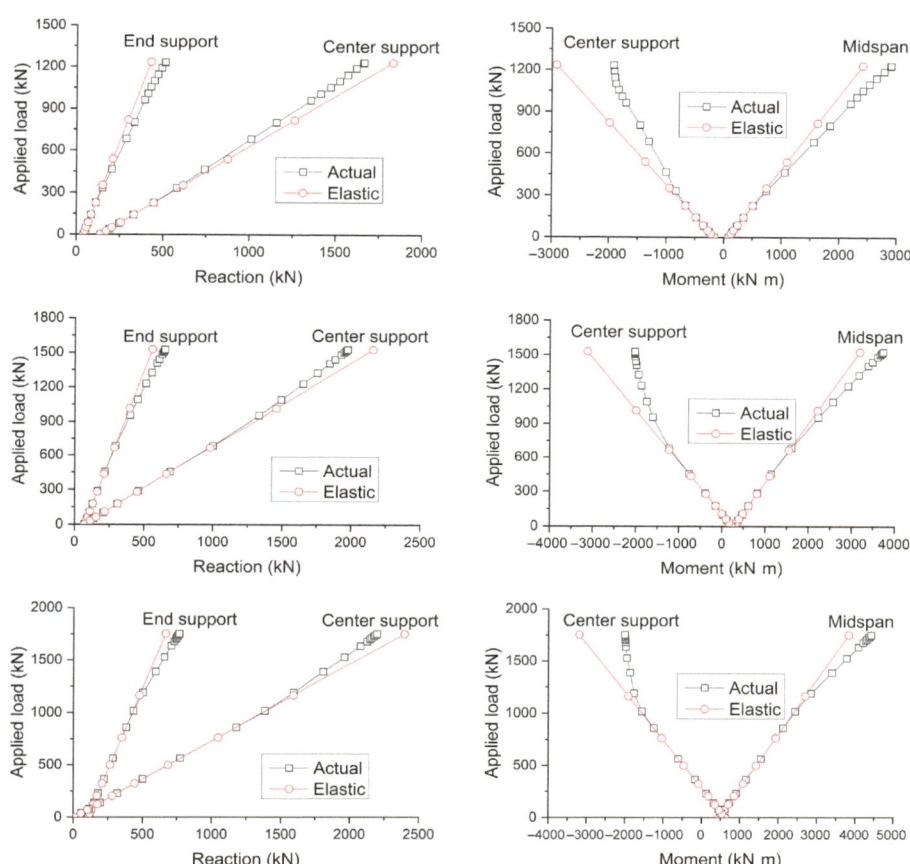

**Figure 10.16** Development of support reactions and bending moments.

at service loads. The stress reduction due to prestressing is more effective over the midspan than over the center support.

In the initial loading stage, the stresses in the top flange remain almost unchanged both over the midspan and center support. A quick increase in center support stress is observed after cracking, while the midspan stress develops rapidly after yielding in the bottom flange. The stress in the top flange may or may not reach its yield strength, depending on the tendon area.

### 10.4.5 Moment redistribution

Fig. 10.16 shows the load–reaction and load–moment behavior for the composite girders with various amounts of external tendons. The actual values are exactly the same as the elastic ones until the first cracking. Thereafter, the actual values begin to deviate from the elastic ones due to the redistribution of moments, and such deviation becomes more apparent as the load increases. As the first cracking takes place over the center support, moments tend to be redistributed from the negative moment region to the positive moment region. This leads to higher-end reaction and midspan moment and correspondingly lower center support reaction and moment when compared to the elastic values.

Fig. 10.17A shows the load versus the degree of moment redistribution curves for the composite girders with various amounts of external tendons. It is seen that moment redistribution does not take place until the crack load is reached. During the inelastic range of loading, the degree of moment redistribution for the prestressed composite girders develops almost linearly with the applied load. For the nonprestressed composite girder, the moment redistribution development is nonlinear due to the influence of different structural limit states such as the yielding of the steel beam. It is also noted that for the nonprestressed composite girder, moment redistribution contributed by cracking of the concrete slab is notable, accounting for 75% of the total redistribution. The contribution of concrete cracking to moment redistribution is significantly reduced when strengthening the composite girder by external prestressing. In the prestressed composite girders, cracking contributes to 60%, 41%, 29%, and 13% of the total redistribution for $A_p = 5$, 10, 15, and 20 cm$^2$, respectively.

The variation of $\beta_u$ (degree of moment redistribution at ultimate) against $A_p/A_c$ is illustrated in Fig. 10.17B. It is interesting to observe that a larger tendon area leads to a higher $\beta_u$ value. This finding is totally different from the common-sense knowledge about moment redistribution in reinforced concrete members strengthening by external prestressing. The presence of external prestressing was reported to significantly reduce the moment redistribution capacity of concrete members partly due to a reduction in flexural ductility (Lou, Lopes, & Lopes, 2014). However, according to the present analysis, strengthening of steel–concrete composite girders by external prestressing can improve the ability to redistribute the bending moments. The $\beta_u$ values for the composite girders obtained from the current analysis are close to the limit of Eurocode 4 (CEN, 2004b) but are significantly higher

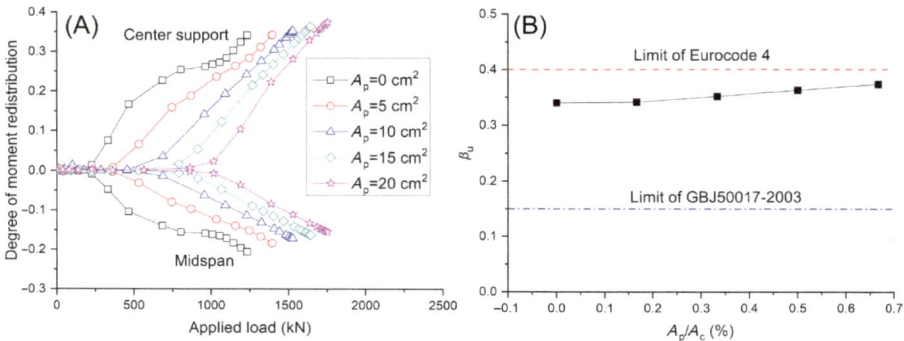

**Figure 10.17** Variation in the degree of moment redistribution: (A) load versus moment redistribution curves; (B) variation of $\beta_u$ with $A_p/A_c$.

than the limit of GB 50017-2003 (GB 50017—2003, 2003). Therefore the latter code seems to be conservative regarding the allowable moment redistribution in composite girders.

## 10.5  Secondary moments

An approach based on linear transformation was presented in Chapter 6 to compute the secondary moments in continuous prestressed concrete members. This approach is extended herein to the case of continuous prestressed composite girders. In Chapter 6, it has been numerically and experimentally demonstrated that linear transformation would not alter the general behavior of continuous prestressed concrete members. This conclusion about linear transformation holds true for continuous prestressed composite girders. To prove this statement, three different tendon profiles are considered, as shown in Fig. 10.18. Profile 1 is the same as the tendon layout illustrated in Fig. 10.10. Profiles 2 and 3 are linear transformations of Profile 1 by moving the tendon line by $-250$ and 445.9 mm, respectively, over the center support and correspondingly by $-125$ and 222.95 mm, respectively, over the midspan. Profile 3 is a concordant tendon profile producing no secondary moments. It should be noted that Profile 3 is not practical for prestressed composite girders. This profile is used to illustrate the influence of linear transformation and the quantification of secondary moments. Both symmetrical and unsymmetrical loads are used. For symmetrical loads, the concentrated load acting on the right span is the same as that acting on the left span ($P_1 = P_2 = P$). For unsymmetrical loads, the concentrated load acting on the left span is twice that acting on the right span ($P_1 = 2P_2 = P$). The tendon area is 20 cm$^2$. All the other geometric and material properties are the same as the composite girders used in Section 10.4.

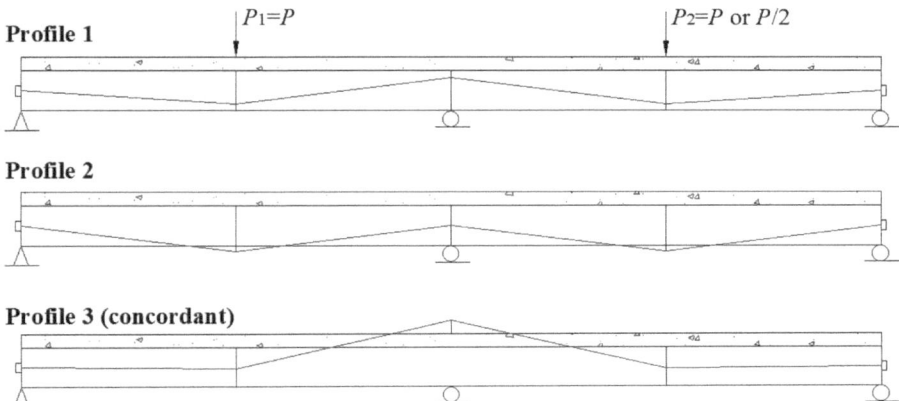

**Figure 10.18** Three linearly transformed tendon profiles.

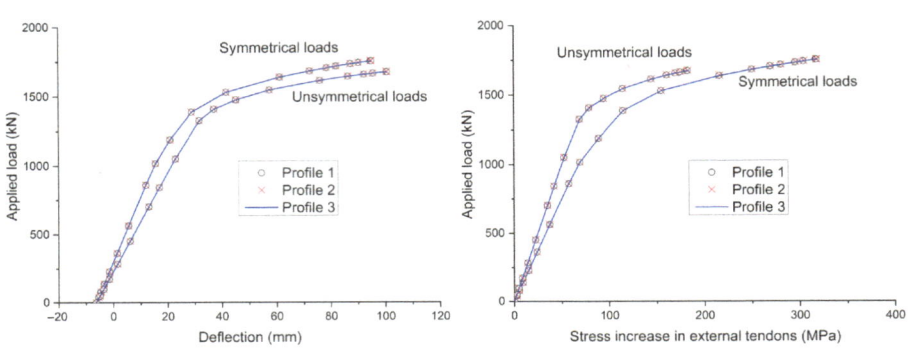

**Figure 10.19** Load versus midspan deflection and tendon stress increase curves for prestressed composite girders with different linearly transformed tendon profiles

Fig. 10.19 shows the influence of linear transformation on the load−deflection behavior and stress increase in external tendons for the prestressed composite girders under symmetrical and unsymmetrical loading. It is observed that unsymmetrical loads lead to larger deflection at the critical midspan, significantly lower tendon stress increment, and lower ultimate load-carrying capacity when compared to symmetrical loads. In addition, the full-range response curves for different linearly transformed tendon profiles are exactly identical, confirming that linear transformation does not change the flexure of prestressed composite girders.

Bending moments in a continuous prestressed composite girder consists of two components:

$$M = M_{\text{load}} + M_{\text{sec}} \tag{10.7}$$

where $M_{\text{load}}$ is the moment induced by dead and live loads; and $M_{\text{sec}}$ the secondary moment induced by prestressing.

As mentioned previously, linear transformation does not change the basic behavior of prestressed composite girders and the concordant tendon profile does not induce secondary moments. Therefore the load-induced moment for a nonconcordant tendon profile ($M_{\text{load}}$) is equal to the moment for its linearly transformed concordant tendon profile ($M_{\text{con}}$), that is, $M_{\text{load}} = M_{\text{con}}$. Substituting it into Eq. (10.7) yields the predictive equation for secondary moments in prestressed composite girders:

$$M_{\text{sec}} = M - M_{\text{con}} \tag{10.8}$$

Figs. 10.20 and 10.21 show the moment development for the prestressed composite girders under symmetrical and unsymmetrical loads, respectively. The bending moments $M$ and $M_{\text{con}}$ are obtained by the finite element analysis and the secondary moments are calculated according to Eq. (10.8). During the postelastic ranges of loading, the load−moment relationship exhibits nonlinear behavior due to moment redistribution. When compared to the case of symmetrical loads, the

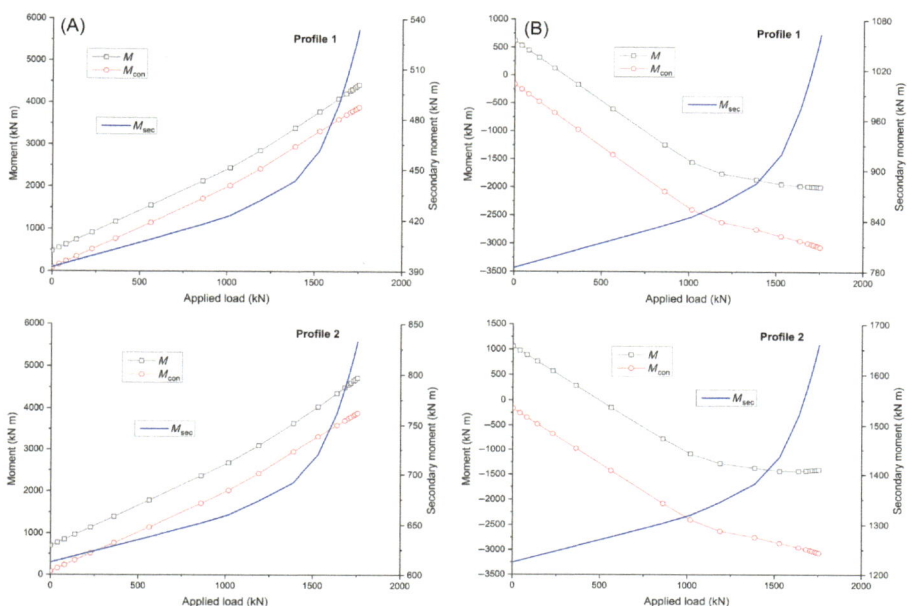

**Figure 10.20** Load versus moment and secondary moment curves at symmetrical loading: (A) midspan; (B) center support.

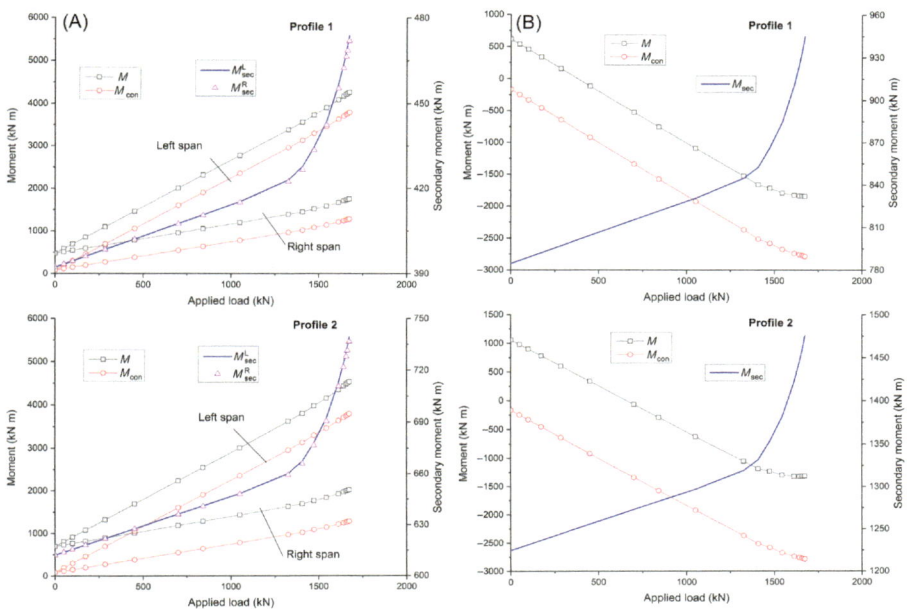

**Figure 10.21** Load versus moment and secondary moment curves at unsymmetrical loading: (A) midspan; (B) center support.

nonlinear behavior for unsymmetrical loads is less apparent, indicating lower redistribution of moments. The secondary moments increase slowly with increasing loads up to the yielding of the steel beam. Thereafter, the increase in secondary moments becomes faster, which is attributed to the rapid increase in tendon force. It is also seen that the secondary moment over the center support is twice that over the midspan. In addition, for unsymmetrical loads, the secondary moment over the left midspan ($M_{\text{sec}}^{L}$) is equal to that over the right midspan ($M_{\text{sec}}^{R}$). The results are coincident with the secondary moment law of continuous prestressed members.

Fig. 10.22 shows the secondary moment versus tendon stress curves for different tendon layouts and load patterns. It is seen that there is a linear relationship between the secondary moment and tendon stress. The slope is dependent on the tendon layout but independent of the pattern of loading. Over the midspan, the slopes for Profiles 1 and 2 are 0.44 and 0.69 kN m/MPa, respectively, while the slopes over the center support are twice those over the midspan. At ultimate, symmetrical loads lead to significantly higher secondary moments as a result of a higher tendon stress when compared to unsymmetrical loads.

External tendons used for strengthening continuous reinforced concrete members can be designed to be concordant to purposely eliminate the secondary moments (Harajli, Mabsout, & Al-Hajj, 2002). For continuous prestressed composite girders, on the other hand, the design of a concordant tendon profile is not practical. Typically, the tendon profile of prestressed composite girders is well below the concordant profile (see Fig. 10.18). As a consequence, external prestressing would produce significant secondary moments in these girders as can be observed in Figs. 10.20–10.22. In addition, secondary moments are present throughout the loading process including the yielding and ultimate stages. The latter statement contradicts a popular viewpoint, that is, secondary moments are no longer existent after

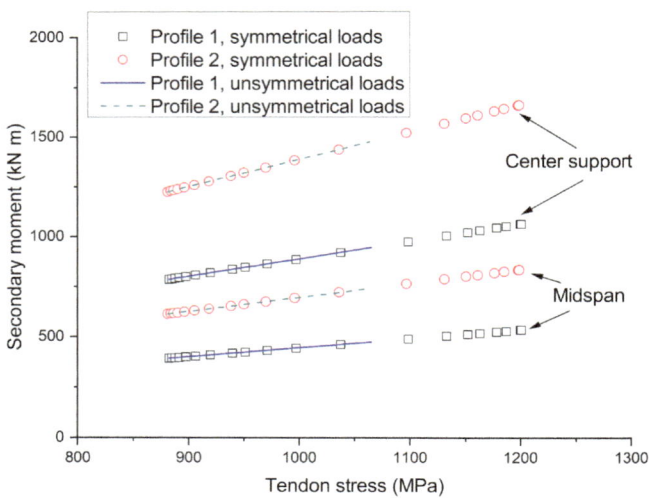

**Figure 10.22** Relationship between secondary moment and tendon stress.

the formation of plastic hinges. Indeed, as illustrated in Figs. 10.20 and 10.21, yielding of the steel beam leads to a faster development of secondary moments. Therefore secondary moments should be taken into account in the strength design of continuous prestressed composite girders.

## 10.6  Influence of span ratio

The afore-described results are produced by using steel—concrete composite girders with two equal spans. In engineering practice, continuous composite girders may be designed to have unequal spans, leading to different behavior from that of equal-span composite girders. In this section, the influence of span ratio on the behavior of two-span prestressed composite girders is examined. The reference composite girder presented in Section 10.4 is also used here, with varied length of the left span $L_1$ and fixed length of the right span $L_2$, as shown in Fig. 10.23A. Three

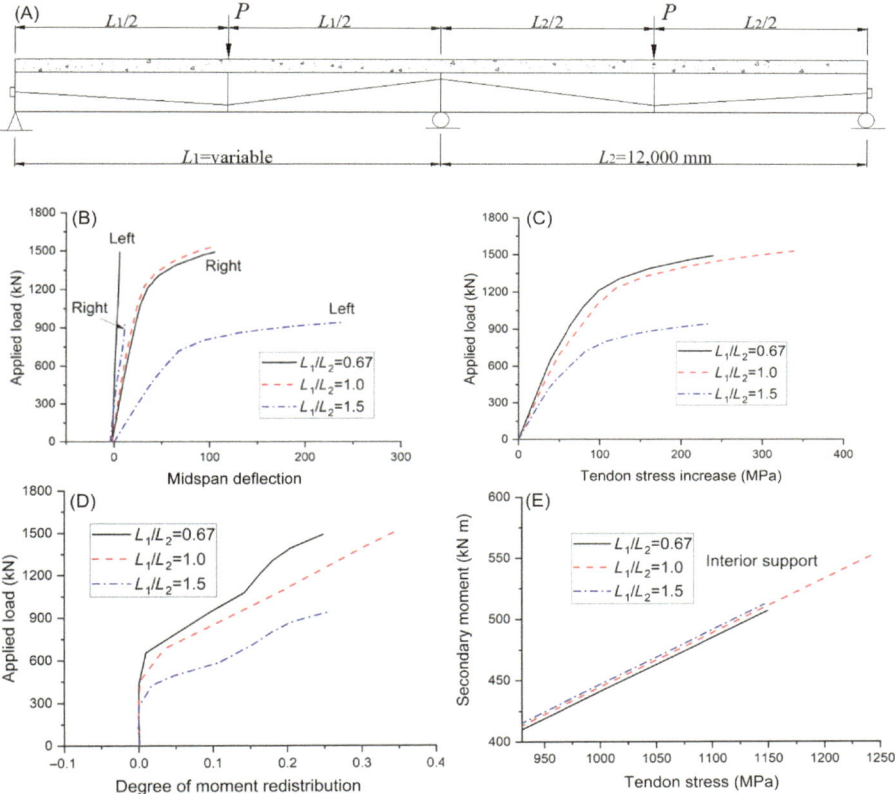

**Figure 10.23** Influence of span ratio on the behavior of two-span prestressed composite girders: (A) sketch of girders; (B) load against deflection; (C) load against tendon stress increase; (D) load against moment redistribution; (E) secondary moment against tendon stress.

different values of $L_1$ (8, 12, and 18 m) are selected for generating different $L_1/L_2$ ratios, namely 0.67, 1.0, and 1.5. The tendon area is taken equal to 10 cm$^2$.

According to the numerical analysis, the composite girder having equal spans ($L_1/L_2 = 1.0$) fails with three plastic hinges formed at the critical sections (one at the interior support and two at the left and right midspan). On the other hand, for the composite girder having unequal spans ($L_1/L_2 = 0.67$ and 1.5), only two plastic hinges have formed (one at the interior support and one at the critical midspan) when crushing failure occurs at the left (for $L_1/L_2 = 1.5$) or right midspan (for $L_1/L_2 = 0.67$). Meanwhile, the noncritical midspan is still in the elastic range, as confirmed by the load–deflection behavior shown in Fig. 10.23B, that is, slight deflection at the non-critical midspan against substantial deflection at the critical midspan. Because of fewer plastic hinges developed at failure, the composite girders with unequal spans exhibit significantly lower stress increase in external tendons and degree of moment redistribution than that with equal spans, as shown in Fig. 10.23C and D. Consequently, the composite girder with $L_1/L_2 = 0.67$ shows a bit lower ultimate load despite of shorter length while the composite girder with $L_1/L_2 = 1.5$ shows a significantly lower ultimate load when compared to that with equal spans. Fig. 10.23E shows that at a given tendon stress, $L_1/L_2 = 0.67$ or 1.5 leads to a bit lower or higher secondary moment than $L_1/L_2 = 1.0$. This can be attributed to the difference in the tendon profiles for different span ratios. In addition, because of significantly lower tendon stress at ultimate, the composite girders with unequal spans have substantially lower secondary moments than those with equal spans.

## 10.7 Conclusions

The second-order effects in externally prestressed composite girders are negligible when $S_d/L < 0.25$. The ultimate load decreases linearly with increasing $S_d/L$ (when $S_d/L \geq 0.25$) as a consequence of second-order effects. The ultimate tendon stress decreases as $S_d/L$ increases. One-point loading leads to substantially lower ultimate tendon stress than two-point or uniform loading, and the difference is particularly notable at a small $S_d/L$ ratio at which the second-order effects are not significant. Composite girders under one-point loading experience significantly smaller ultimate deflection than composite girders under two-point or uniform loading. As a result, one-point loading results in less second-order effects compared to two-point or uniform loading. Both deviator spacing and load type do not produce noticeable influence on the neutral axis depth and curvature at the critical section. Therefore it can be inferred that the stresses or strains in concrete, steel bars, and structural steel in the critical section are independent of the second-order effects.

The equation recommended in ACI 440.4R-04 for predicting the ultimate tendon stress is modified so that it can be applicable to prestressed composite girders. In the proposed equation, a modified bond reduction factor $\Omega_u$ is used, and a new parameter $R_{sec}$ (reduction factor due to second-order effects) associated with deviator spacing and load type is introduced. The proposed equation correlates well with the numerical simulations.

Strengthening of continuous steel−concrete composite girders by external tendons leads to a significant increase in the ultimate load-carrying capacity and flexural strength at the midspan, but hardly increases the ultimate moment at the center support. A higher tendon area leads to a larger difference between the midspan and center support moments partly due to the influence of secondary moments. A higher tendon area leads to a slightly smaller stress increment in external tendons and a larger neutral axis depth at ultimate. At service loads, external prestressing effectively reduces the stress level in reinforcing bars at the center support but not at the midspan. Moreover, external prestressing is more effective in reducing the stress in the bottom flange of the steel beam at the midspan compared to the center support. In contrast to common-sense knowledge that external prestressing significantly reduces moment redistribution in concrete members, strengthening of steel−concrete composite girders by external tendons leads to an increase in the degree of moment redistribution. Moment redistribution in prestressed composite girders is shown to be close to the limit of Eurocode 4 and significantly higher than the limit of GB 50017-2003.

Linear transformation does not influence the general nonlinear behavior of continuous prestressed composite girders. A previously developed approach for quantifying secondary moments in prestressed concrete members was found applicable to prestressed composite girders. Secondary moments are present in continuous prestressed composite girders at elastic, yielding, and ultimate phases of the response. Due to the large deviation from its linearly transformed concordant line and the high tendon stress increment developed at ultimate, secondary moments should be considered in the design of continuous prestressed composite girders. Due to fewer plastic hinges formed at failure, the prestressed composite girders with unequal spans ($L_1/L_2 = 0.67$ and 1.5) exhibit significantly lower ultimate tendon stress, moment redistribution, and secondary moment than those with equal spans ($L_1/L_2 = 1.0$).

# References

Abu-Sena, A. B., Shaaban, I. G., Soliman, M. S., & Gharib, K. A. M. (2020). Effect of geometrical properties on strength of externally prestressed steel-concrete composite beams. *Proceedings of the Institution of Civil Engineers: Structures and Buildings, 173*(1), 42−62. Available from https://doi.org/10.1680/jstbu.17.00172.

ACI Committee 318. (2019). *Building code requirements for structural concrete (ACI 318-19) and commentary (ACI 318R-19).* Farmington Hills, MI.

Brockmann, C., & Rogenhofer, H. (2000). Bang Na expressway, Bangkok, Thailand - World's longest bridge and largest precasting operation. *PCI Journal, 45*(1), 26−38. Available from https://doi.org/10.15554/pcij.01012000.26.38.

ACI committee 440. (2004). *Prestressing concrete structures with FRP tendons.* ACI 440.4R-04, Farmington Hills, MI.

CEN. (2004a). *Eurocode 2 (EC2): Design of concrete structures. Part 1-1: General rules and rules for buildings.* EN 1992-1-1, Brussels, Belgium.

CEN. (2004b). *Eurocode 4 (EC4): Design of composite steel and concrete structures. Part 1-1: General rules and rules for buildings.* EN 1994-1-1, Brussels, Belgium.

Chen, S., & Jia, Y. (2010). Numerical investigation of inelastic buckling of steel−concrete composite beams prestressed with external tendons. *Thin-Walled Structures, 48*(3), 233−242. Available from https://doi.org/10.1016/j.tws.2009.10.009.

Chen, S., Wang, X., & Jia, Y. (2009). A comparative study of continuous steel−concrete composite beams prestressed with external tendons: Experimental investigation. *Journal of Constructional Steel Research, 65*(7), 1480−1489. Available from https://doi.org/10.1016/j.jcsr.2009.03.005.

Dall'Asta, A., Ragni, L., & Zona, A. (2006). Steel-concrete composite beams prestressed by external tendons: Effects of material and geometric nonlinearities. *Advanced Steel Construction, 2*(1), 53−70.

El-Zohairy, A., & Salim, H. (2017). Parametric study for post-tensioned composite beams with external tendons. *Advances in Structural Engineering, 20*(10), 1433−1450. Available from https://doi.org/10.1177/1369433216684352.

El-Zohairy, A., Salim, H., & Saucier, A. (2019). Steel-concrete composite beams strengthened with externally post-tensioned tendons under fatigue. *ASCE Journal of Bridge Engineering, 24*(5), 04019027. Available from https://doi.org/10.1061/(ASCE)BE.1943-5592.0001390.

GB 50017-2003. (2003). *Code for design of steel structures.* Beijing: China Planning Press.

Harajli, M., Khairallah, N., & Nassif, H. (1999). Externally prestressed members: Evaluation of second-order effects. *Journal of Structural Engineering, 125*(10), 1151−1161. Available from https://doi.org/10.1061/(ASCE)0733-9445(1999)125:10(1151).

Harajli, M. H., Mabsout, M. E., & Al-Hajj, J. A. (2002). Response of externally post-tensioned continuous members. *ACI Structural Journal, 99*(5), 671−680.

He, Z. Q., & Liu, Z. (2010). Stresses in external and internal unbonded tendons: Unified methodology and design equations. *Journal of Structural Engineering, 136*(9), 1055−1065. Available from https://doi.org/10.1061/(ASCE)ST.1943-541X.0000202.

Lou, T., Karavasilis, T. L., & Chen, B. (2021). Assessment of second-order effect in externally prestressed steel−concrete composite beams. *Journal of Bridge Engineering, 26* (6), 04021024. Available from https://doi.org/10.1061/(ASCE)BE.1943-5592.0001718.

Lou, T., Lopes, S. M. R., & Lopes, A. V. (2014). Factors affecting moment redistribution at ultimate in continuous beams prestressed with external CFRP tendons. *Composites Part B: Engineering, 66*, 136−146. Available from https://doi.org/10.1016/j.compositesb.2014.05.007.

Lou, T., & Karavasilis, T. L. (2019a). Numerical assessment of the nonlinear behavior of continuous prestressed steel-concrete composite beams. *Engineering Structures, 190*, 116−127. Available from https://doi.org/10.1016/j.engstruct.2019.04.031.

Lou, T., & Karavasilis, T. L. (2019b). Numerical evaluation of prestressed steel-concrete composite girders with external FRP or steel tendons. *Journal of Constructional Steel Research, 162*, 105698. Available from https://doi.org/10.1016/j.jcsr.2019.105698.

Lou, T., & Xiang, Y. (2010). Numerical analysis of second-order effects of externally prestressed concrete beams. *Structural Engineering and Mechanics, 35*(5), 631−643. Available from https://doi.org/10.12989/sem.2010.35.5.631.

Moscoso, A. M., Tamayo, J. L. P., & Morsch, I. B. (2017). Numerical simulation of external pre-stressed steel-concrete composite beams. *Computers and Concrete, 19*(2), 191−201. Available from https://doi.org/10.12989/cac.2017.19.2.191.

Nie, J., Tao, M., Cai, C. S., & Li, S. (2009). Deformation analysis of prestressed continuous steel-concrete composite beams. *Journal of Structural Engineering*, *135*(11), 1377−1389. Available from https://doi.org/10.1061/(asce)st.1943-541x.0000067.

Nordin, H. (2005). Strengthening structures with externally prestressed tendons. *Literature review, Luleå University of Technology Technical Report*, 6.

Pang, M., Li, Z., & Lou, T. (2020). Numerical study of using FRP and steel rebars in simply supported prestressed concrete beams with external FRP tendons. *Polymers*, *12*(12), 2773. Available from https://doi.org/10.3390/polym12122773.

Roberts-Wollmann, C. L., Kreger, M. E., Rogowsky, D. M., & Breen, J. E. (2005). Stresses in external tendons at ultimate. *ACI Structural Journal*, *102*(2), 206−213.

Zhang, N., & Fu, C. C. (2009). Experimental and theoretical studies on composite steel-concrete box beams with external tendons. *Engineering Structures*, *31*(2), 275−283. Available from https://doi.org/10.1016/j.engstruct.2008.08.004.

# Using FRP reinforcement in steel−concrete composite girders  **11**

## 11.1  Introduction

In steel−concrete composite girders, the concrete slab over negative moment regions may experience significant cracking, even at service conditions. Therefore a reasonable arrangement of rebars in the slab is necessary to ensure favorable performance. Different types of rebars can be employed to reinforce the slab. While conventional steel rebars are prone to corrosive damage, noncorrosive fiber-reinforced polymer (FRP) composites are an alternative option (ACI Committee 440, 2015). Previous works on FRP-reinforced concrete structures generally demonstrated that the brittleness of FRP rebars would lead to undesirable behavior related to ductility and moment redistribution (Ju et al., 2017; Pang et al., 2021; Santos et al., 2013). On the other hand, few works have been reported on the use of FRP rebars instead of steel ones in steel−concrete composite girders.

In externally prestressed steel−concrete composite girders, the steel tendons are commonly wrapped with grease-filled plastic sheaths to minimize the corrosion problem. This protection technique, however, will increase the cost and also add the difficulty in inspecting the tendons. The most efficient way to address the corrosion issue is to use tendons made of FRP composites. These composite materials, characterized by noncorrosive properties and a high strength−weight ratio, are gaining popularity for structural rehabilitation (Al-Saidy et al., 2007; Fam et al., 2009; Siwowski & Siwowska, 2018; Teng et al., 2012; Zhao & Zhang, 2007). Typical composite tendons include aramid FRP (AFRP) and carbon FRP (CFRP), which exhibit favorable creep performance (ACI Committee 440, 2004). Glass FRP (GFRP), in general, is not suitable for prestressing applications because of creep rupture under a low sustained load level. Both AFRP and CFRP tendons possess high tensile strength, comparable to that of prestressing steel tendons. AFRP tendons show substantially lower elastic modulus than steel tendons. The CFRP modulus of elasticity covers a large range, varying between 0.4 and 2.5 times the steel one. The relaxation of AFRP tendons is considerably more pronounced than that of steel tendons, while CFRP tendons exhibit superior relaxation resistance, comparable to or even better than low-relaxation prestressing steel (ACI Committee 440, 2004). Although the initial cost of FRP (especially CFRP) is higher than that of steel, the life span of FRP-strengthened systems may be cost-effective because of the superior characteristics of FRP materials. A major concern on the use of FRP tendons is related to anchorages, which are rather sophisticated and need to be specially designed. Common anchorages for steel tendons are not applicable to FRP tendons. Extensive studies have been conducted to develop strong anchorage systems for FRP tendons (Schmidt et al., 2012).

Prestressed Members with External Fiber-Reinforced Polymer (FRP) Tendons. DOI: https://doi.org/10.1016/B978-0-443-23877-2.00011-0

Many studies (Deng et al., 2011; Ghafoori & Motavalli, 2015; Tavakkolizadeh & Saadatmanesh, 2003; Aly. & El-Hacha, 2009) on steel–concrete composite or steel members strengthened by prestressing FRP sheets/plates have been carried out. The retrofit effectiveness of using such an FRP-prestressing technique was demonstrated in these studies. When FRP tendons are adopted instead of conventional steel ones, some challenges in the design of prestressed composite members may arise due to the nonstandard mechanical properties of the FRP material. The use of external FRP tendons for strengthening steel–concrete composite members needs to be further revealed.

This chapter reports the effect of using FRP reinforcement instead of steel one in steel–concrete composite girders (Lou & Karavasilis, 2019; Lou et al., 2022). A numerical work on the inelastic behavior of continuous steel–concrete composite girders with CFRP, GFRP, or steel rebars in the slab is presented. Then, single-span and two-span prestressed steel–concrete composite girders with external CFRP, AFRP, or steel tendons at different prestress levels (i.e., from 0% to 60%) are analyzed. With respect to continuous prestressed steel–concrete composite girders, emphasis is given to evaluating the secondary moments and moment redistribution.

## 11.2   Behavior of composite girders with FRP rebars

Fig. 11.1 shows a reference steel–concrete composite girder with two equal spans. Each span is 12-m long and under center-point loading. The slab, 1500 mm in width and 200 mm in thickness, is reinforced with two-layer (top and bottom) deformed reinforcement along the entire length, made of CFRP, GFRP, or steel bars with properties given in Table 11.1. The mechanical properties of rebars used are in compliance with ACI 440.1R-15 (ACI Committee 440, 2015). The top rebar area is

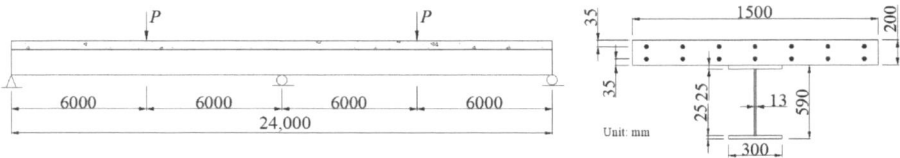

**Figure 11.1** Details of reference composite girder.

**Table 11.1** Mechanical properties of rebars.

| Rebar | Ultimate strength (MPa) | Yield strength (MPa) | Elastic modulus (GPa) | Rupture strain (%) |
|-------|-------------------------|----------------------|-----------------------|--------------------|
| Steel | 420 | 420 | 200 | >3.5 |
| CFRP | 1840 | – | 150 | 1.23 |
| GFRP | 750 | – | 40 | 1.88 |

equal to the bottom one. The rebar ratio is defined by $\rho_r = A_r/A_c$, where $A_r$ is the sum of the top and bottom rebar areas and $A_c$ is the area of the slab. $A_r$ is a variable ranging between 0 and 15,000 mm$^2$, so $\rho_r$ varies from 0% to 5.0%. The steel profile, made from HEA 600, has a yield strength of 275 MPa and an elastic modulus of 200 GPa. The concrete cylinder compressive strength is 60 MPa.

## 11.2.1 Cracking mode

Fig. 11.2 shows that when the concrete reaches its ultimate compressive strain of 0.003, crushing failure occurs (top fiber of midspan section for all girders). Before the failure, the structural steel in the critical sections yielded either tension or compression.

When concrete reaches its cracking strain of 113 $\mu\varepsilon$, cracks begin to form. The cracking at the midspan appears to be unimportant, as demonstrated by the trivial tensile strain at the bottom of the slab. In addition, both the area and type of rebars do not impact the cracking mode (crack width and zone) over the midspan region. This is attributed to that the behavior of a composite section under positive bending is controlled by concrete compression and structural steel tension. On the other hand, the crack mode over the center support region is markedly impacted by the rebars. At $\rho_r = 0$, significant cracking is observed over the center support, that is, the crack width is very large at the top of the slab, and meanwhile, the bottom of the slab is also significantly cracked. The presence of rebars effectively reduces the crack width over the center support. The reduction is more effective for providing CFRP rebars than for providing steel or GFRP rebars, as demonstrated in Fig. 11.2.

## 11.2.2 Load—deflection response

Fig. 11.3A and B shows that composite girders exhibit an approximately linear load—deflection response up to the yielding of structural steel. Beyond that, the load—deflection behavior is substantially softened as expected. These composite girders exhibit approximately the same response in the initial loading range. After cracking in the center support, the contribution of rebars becomes more important. Consequently, the composite girder with GFRP rebars exhibits smaller postcracking flexural stiffness than the composite girder with steel or CFRP rebars due to a lower elastic modulus of GFRP rebars, as shown in Fig. 11.3A. Meanwhile, a higher $\rho_r$ level results in a higher postcracking flexural stiffness of composite girders as expected (Fig. 11.3B). However, the difference in the load—deflection curves between the composite girders with different types or ratios of rebars is not so notable, indicating the limited contribution of rebars to the flexural stiffness of composite girders.

Fig. 11.3C shows that as $\rho_r$ increases from 0% to 1.0%, the ultimate deflection is almost unchanged when steel rebars are provided while it is reduced by around 4.0% when CFRP or GFRP rebars are used. With a continuing increase in $\rho_r$, the ultimate deflection for the composite girders with CFRP or steel rebars gradually increases while it tends to stabilize when GFRP rebars are arranged. As a consequence, despite a lower elastic modulus, FRP rebars lead to smaller ultimate

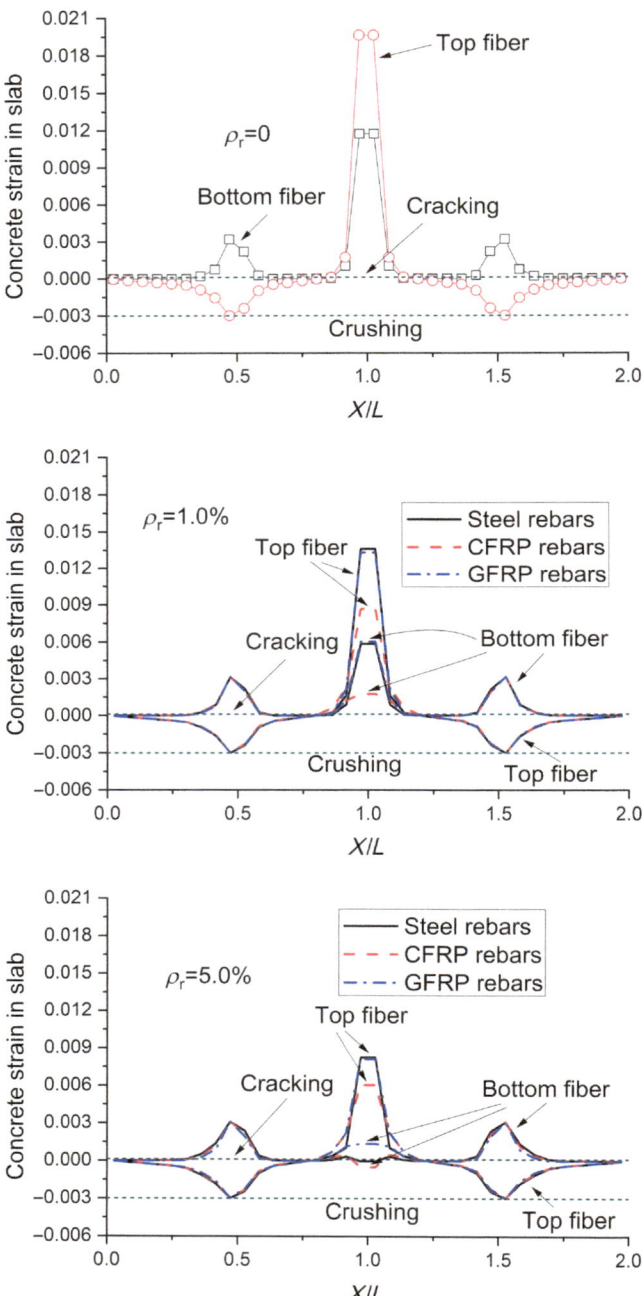

**Figure 11.2** Strain distribution in concrete slab at failure ($X/L$ = ratio of the distance from the end support to the span length).

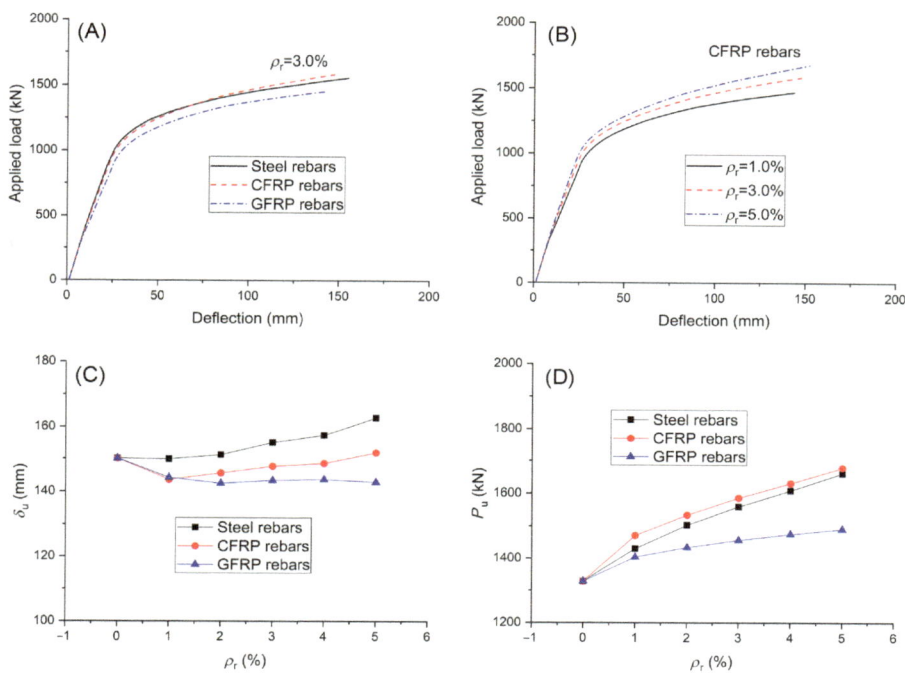

**Figure 11.3** Load versus midspan deflection: (A) load—deflection curves of girders with different rebar types; (B) load—deflection curves of girders with different rebar ratios; (C) effect of rebar ratio/type on $\delta_u$ (ultimate deflection); (D) effect of rebar ratio/type on $P_u$ (ultimate load).

deflection than steel rebars at a given $\rho_r$ level. Fig. 11.3D shows that as $\rho_r$ increases, the growth of the ultimate load is much slower for using GFRP rebars than for using steel or CFRP rebars. Composite girders with CFRP rebars show a bit higher ultimate load than composite girders with steel rebars. Providing $\rho_r = 1.0\%$, 2.0%, 3.0%, 4.0%, and 5.0% in the slab increases the ultimate load by 5.7%, 8.0%, 9.8%, 11.2%, and 12.3%, respectively, for composite girders with GFRP rebars, by 10.8%, 15.5%, 19.5%, 22.8%, and 26.4%, respectively, for composite girders with CFRP rebars, and by 7.7%, 13.2%, 17.4%, 21.1%, and 25.1%, respectively, for composite girders with steel rebars.

## 11.2.3 Curvature

Fig. 11.4A and B displays the moment—curvature curves of composite girders with different rebar types and ratios, respectively. Over the midspan, the moment—curvature evolution is primarily influenced by the yielding of structural steel at the bottom and top flanges; both the type and ratio of rebars have limited impact on the moment—curvature behavior. Over the center support, the moment—curvature

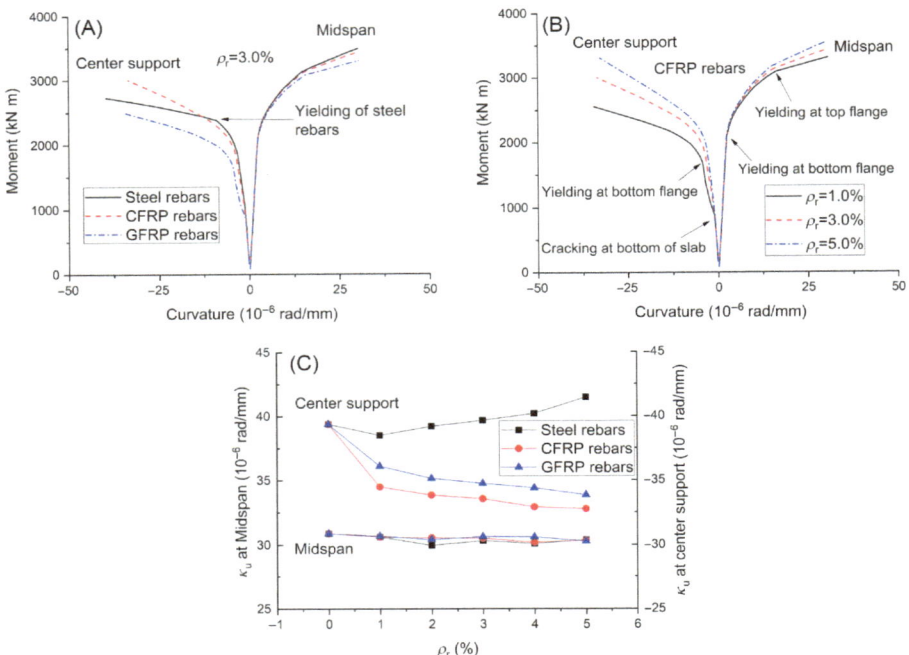

**Figure 11.4** Curvature development: (A) moment−curvature curves of girders with different rebar types; (B) moment−curvature curves of girders with different rebar ratios; (C) effect of rebar ratio/type on $\kappa_u$ (ultimate curvature).

evolution is typically influenced by cracking at the bottom of the slab and by yielding at the bottom steel flange, especially notable for the composite girder with GFRP rebars or a low rebar ratio. For the composite girder with steel rebars, a reduction in cross-sectional rigidity is observed on steel rebar yielding, as shown in Fig. 11.4A.

Fig. 11.4C shows that the ultimate curvatures at the midspan are around $30 \times 10^{-6}$ rad/mm, irrespective of the type and ratio of rebars. Over the center support, as $\rho_r$ increases, the ultimate curvature tends to increase when steel rebars are used while it decreases when CFRP or GFRP rebars are used. Consequently, at a given $\rho_r$ level, steel rebars lead to higher ultimate curvature than FRP rebars, and the difference is increasingly notable with increasing $\rho_r$.

## 11.2.4 Neutral axis

Fig. 11.5A and B shows the neutral axis shift in composite girders with different rebar types and ratios, respectively. The elastic neutral axis is located about 200 mm below the top fiber of the slab. Both the type and ratio of rebars have a slight influence on the position of the elastic neutral axis because of their different contributions to the transformed section. Over the midspan, the neutral axis remains unchanged until the structural steel yields. Thereafter, the neutral axis moves quickly toward the

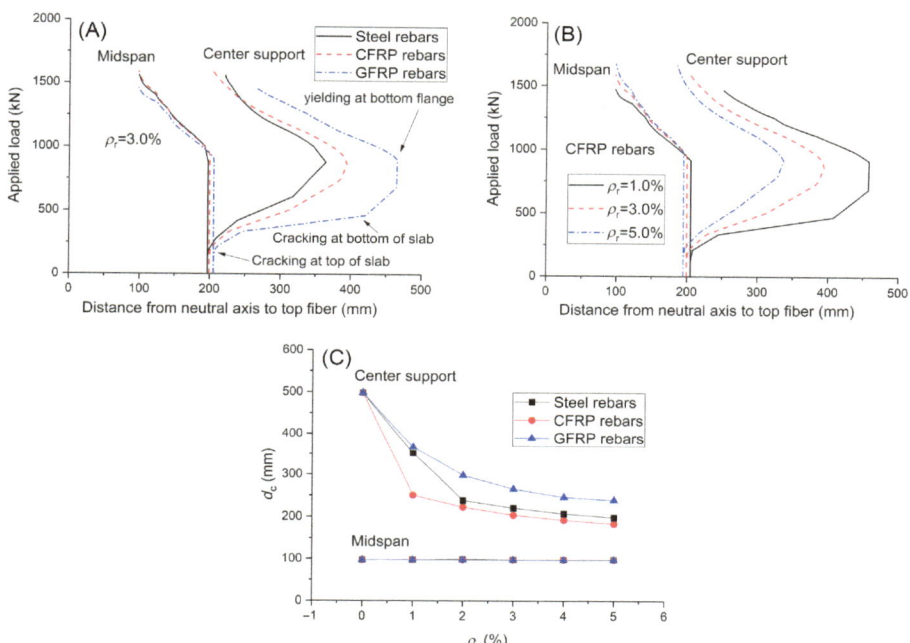

**Figure 11.5** Neutral axis evolution: (A) load versus neutral axis for girders with different rebar types; (B) load versus neutral axis for girders with different rebar ratios; (C) effect of rebar ratio/type on $d_c$ (distance from neutral axis to top fiber) at failure.

top fiber of the slab. At a given load level, the difference between the neutral axis depths by different types and ratios of rebars is insignificant. Over the center support, on cracking at the top of the slab, the neutral axis moves rapidly from its elastic position toward the bottom of the composite section. The movement slows down after cracking at the bottom of the slab. When the structural steel yields, the neutral axis turns to move upward. At a given load level, the inelastic neutral axis for the composite girder with GFRP rebars is below that with CFRP or steel rebars, as illustrated in Fig. 11.5A. Meanwhile, a higher rebar ratio leads to a smaller distance, $d_c$, from the neutral axis to the top fiber, as illustrated in Fig. 11.5B.

The variation of $d_c$ (at failure) with varying rebar ratio/type is illustrated in Fig. 11.5C. The $d_c$ values are around 100 mm at midspan, independently of the type and ratio of rebars. Over the center support, the value of $d_c$ decreases as $\rho_r$ increases. At a given $\rho_r$ level, GFRP rebars lead to a higher $d_c$ while CFRP rebars result in a smaller $d_c$ when compared to steel rebars.

## 11.2.5 Stress in rebars

Fig. 11.6A and B shows the stress development in FRP/steel rebars in composite girders. Over the midspan (Fig. 11.6A), the compressive stress in the top rebars

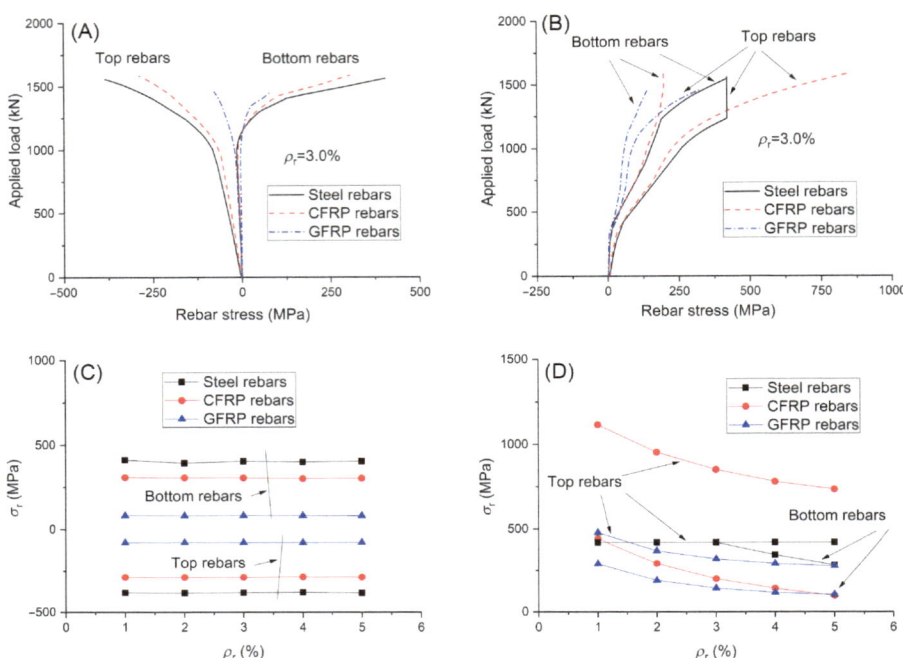

**Figure 11.6** Stress development in FRP/steel rebars: (A) load versus rebar stress at midspan; (B) load versus rebar stress at center support; (C) effect of rebar ratio/type on $\sigma_r$ (ultimate rebar stress) at midspan; (D) effect of rebar ratio/type on $\sigma_r$ at center support.

gradually develops while the bottom rebars hardly develop any stress with increasing load up to yielding at the bottom steel flange. Thereafter, the compressive stress in top rebars or the tensile stress in bottom rebars quickly develops. Over the center support (Fig. 11.6B), the stress in rebars is negligible before cracking. After that, the tensile stress in top rebars develops quickly, and the stress development is much faster than that in bottom rebars, as expected.

Fig. 11.6C and D shows the variation of the ultimate stress in rebars with varying rebar ratios/types. At ultimate, both the top and bottom steel rebars over the midspan have not yielded. The rebar stresses over the midspan appear to be independent of the $\rho_r$ level (see Fig. 11.6C). The stresses in top steel, CFRP, and GFRP rebars are around $-385$, $-288$, and $-77$ MPa, respectively, and the stresses in bottom steel, CFRP, and GFRP rebars are around 400, 300, and 80 MPa, respectively. The ratio of FRP rebar stress to steel one is exactly equal to the elastic modulus ratio. Over the center support, at failure, the top steel rebars have reached their yield strength of 420 MPa, while the bottom steel rebars may or may not yield, depending on the $\rho_r$ level (see Fig. 11.6D). The stress in FRP rebars decreases as $\rho_r$ increases. The top CFRP rebars develop substantially higher stress than the top steel or GFRP rebars but are still far below their rupture strength.

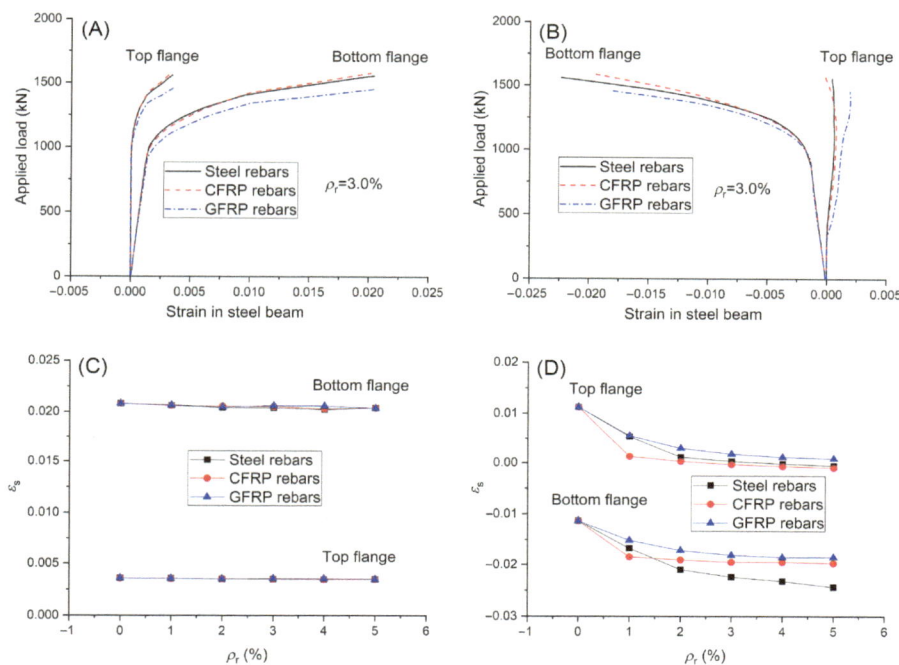

**Figure 11.7** Strain development in steel beam: (A) load versus structural steel strain at midspan; (B) load versus structural steel strain at center support; (C) effect of rebar ratio/type on $\varepsilon_s$ (ultimate strain in structural steel) at midspan; (D) effect of rebar ratio/type on $\varepsilon_s$ at center support.

## 11.2.6 Strain in structural steel

Fig. 11.7A and B shows the strain development in structural steel in composite girders with different rebar types. Over the midspan (Fig. 11.7A), before yielding the tensile strain in the bottom steel flange develops slowly while there is approximately zero strain in the top steel flange. When the bottom flange begins to yield, the strain in the bottom and top rebars develops quickly, and the responses of the composite girders with FRP and steel rebars differ. Over the center support (Fig. 11.7B), the compressive strain in the bottom steel flange gradually develops until yielding and then it turns to develop very quickly. In the top steel flange, the strain remains zero until cracking, followed by a slight development in tensile strain. The composite girders with different types of rebars assume different behaviors regarding the strain development in the top flange after cracking and regarding the strain development in the bottom flange after yielding.

Fig. 11.7C shows that the ultimate strains in the top and bottom flanges over the midspan are not impacted by the type and ratio of rebars. The ultimate strains are around 0.02 in the bottom flange and around 0.0035 in the top flange. Fig. 11.7D demonstrates that over the center support, the provision of rebars leads to a higher

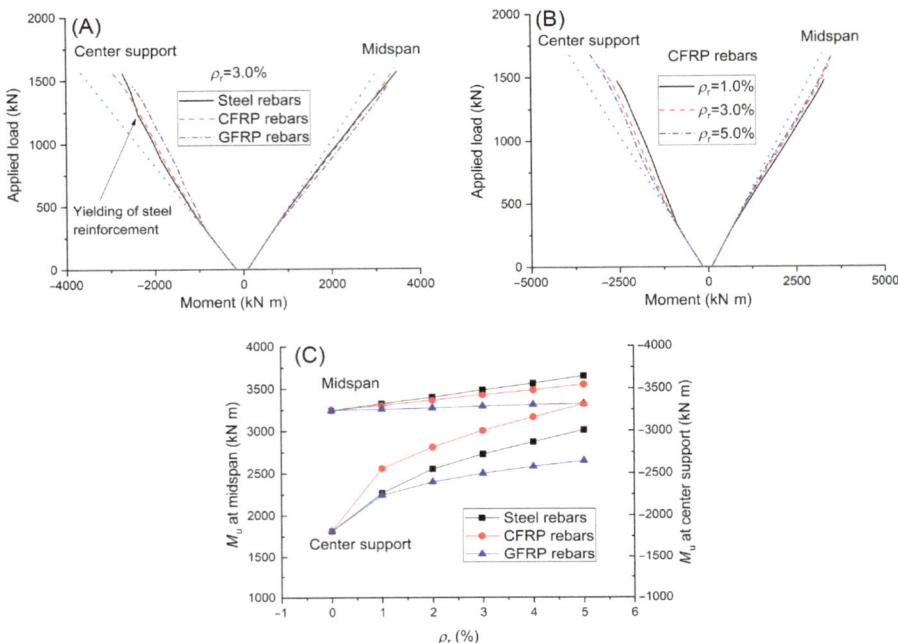

**Figure 11.8** Moment development: (A) load—moment curves of girders with different rebar types; (B) load—moment curves of girders with different rebar ratios; (C) effect of rebar ratio/type on $M_u$ (ultimate moment).

ultimate compressive strain in the bottom flange while a lower ultimate tensile strain in the top flange. The ultimate strain in structural steel changes substantially as $\rho_r$ increases from 0% to 1.0% for FRP rebars or from 0% to 2.0% for steel rebars. The change in steel strain is insignificant with the continuing increase in $\rho_r$. At $\rho_r = 1.0\%$, CFRP rebars cause a higher ultimate compressive strain in the bottom flange and a lower ultimate tensile strain in the top flange than GFRP or steel rebars. When $\rho_r$ is increased to 2.0% and above, the ultimate strain in the top flange for composite girders with CFRP rebars is comparable to that with steel rebars but is lower than that with GFRP rebars. Meanwhile, the compressive strain in the bottom flange for composite girders with CFRP rebars is comparable to that with GFRP rebars, but substantially smaller than that with steel rebars.

## 11.2.7 Moment development

Fig. 11.8A and B shows the load—moment curves of composite girders with different rebar types and ratios, respectively. The moment obtained by the proposed analysis deviates from the elastic moment after cracking. The growth of the actual support moment becomes slower than that of the elastic one, while it is opposite over the midspan. At a given load level, the deviation is more apparent for the composite girder with GFRP rebars than for the composite girder with CFRP or steel rebars (Fig. 11.8A) and that a higher rebar ratio leads to a smaller deviation (Fig. 11.8B). For the composite girder with steel rebars, reinforcement yielding leads to further deviation between the actual and elastic moments.

The variation of the ultimate moment with varying rebar type/ratio is displayed in Fig. 11.8C and Table 11.2. The midspan moment is larger than the support moment, partly attributed to the fact that composite girders tend to redistribute bending moments from the center support toward the midspan. In addition, as $\rho_r$ increases, the increase in the ultimate moment at the midspan is not as apparent as that at the center support. At a given $\rho_r$ level, CFRP rebars lead to higher support moments than steel rebars, due to higher stress developed in top rebars at ultimate. GFRP rebars lead to lower ultimate moments than steel rebars due to lower rebar stress.

### 11.2.8 Moment redistribution

The load versus moment redistribution curves of composite girders with different rebar types and ratios are illustrated in Fig. 11.9A and B, respectively. Redistribution of moments does not occur until cracking. Cracks initiate and progressively develop at the center support, resulting in a weaker support section relative to the midspan section. As a consequence, bending moments are progressively redistributed from the support to the midspan, causing progressive development of positive redistribution at the support and negative redistribution at the midspan. GFRP rebars lead to a greater development of moment redistribution than steel or CFRP rebars (Fig. 11.9A), while the redistribution development is less notable for a higher rebar ratio (Fig. 11.9B). For the composite girder with steel rebars, the yielding of rebars at the center support leads to further redistribution of moments from the support to the midspan.

At failure, the variation in moment redistribution with varying rebar ratio/type is shown in Fig. 11.9C and Table 11.2. Moment redistribution at ultimate decreases as $\rho_r$ increases. The decrease in redistribution by CFRP rebars is faster than that by GFRP or steel rebars. When $\rho_r$ is increased from 0% to 5.0%, the moment redistribution for composite girders with steel, CFRP, and GFRP rebars is reduced by 47%, 64%, and 42%, respectively. At a given $\rho_r$ level, CFRP rebars lead to much lower moment redistribution than steel rebars, while GFRP rebars cause comparable moment redistribution to steel rebars.

## 11.3　Behavior of simply supported prestressed composite girders with external FRP tendons

Fig. 11.10 shows single-span prestressed composite girders with 12.0 m in span length and subjected to two concentrated loads. The distance between the two point loads is 3.0 m. Deviators are used at the third points. The distances between the

**Table 11.2** List of bending moments and moment redistribution.

| Rebar | $\rho_r$ (%) | $M_u$ (kN·m) | | $M_e$ (kN·m) | | $\beta_u$ (%) | |
|---|---|---|---|---|---|---|---|
| | | Midspan | Support | Midspan | Support | Midspan | Support |
| – | 0.0 | 3244.8 | −1813.2 | 2585.5 | −3131.6 | −25.5 | 42.1 |
| Steel | 1.0 | 3324.3 | −2267.4 | 2778.0 | −3359.9 | −19.7 | 32.5 |
| | 2.0 | 3401.2 | −2552.3 | 2915.8 | −3523.1 | −16.6 | 27.6 |
| | 3.0 | 3482.9 | −2726.2 | 3021.8 | −3648.5 | −15.3 | 25.3 |
| | 4.0 | 3559.5 | −2869.2 | 3114.8 | −3758.6 | −14.3 | 23.7 |
| | 5.0 | 3646.7 | −3008.5 | 3214.3 | −3873.4 | −13.5 | 22.3 |
| CFRP | 1.0 | 3303.0 | −2559.0 | 2856.1 | −3452.9 | −15.6 | 25.9 |
| | 2.0 | 3364.2 | −2810.0 | 2973.3 | −3591.8 | −13.1 | 21.8 |
| | 3.0 | 3425.3 | −3003.0 | 3072.3 | −3709.0 | −11.5 | 19.0 |
| | 4.0 | 3480.2 | −3160.7 | 3157.2 | −3806.7 | −10.2 | 17.0 |
| | 5.0 | 3545.3 | −3313.6 | 3246.1 | −3912.0 | −9.2 | 15.3 |
| GFRP | 1.0 | 3258.5 | −2241.6 | 2728.5 | −3301.5 | −19.4 | 32.1 |
| | 2.0 | 3272.4 | −2397.9 | 2786.3 | −3370.1 | −17.4 | 28.8 |
| | 3.0 | 3291.1 | −2498.3 | 2829.5 | −3421.4 | −16.3 | 27.0 |
| | 4.0 | 3307.1 | −2577.6 | 2864.5 | −3462.8 | −15.5 | 25.6 |
| | 5.0 | 3319.5 | −2646.6 | 2894.0 | −3497.8 | −14.7 | 24.3 |

*Note: $M_u$, ultimate moment; $M_e$, elastic moment; $\beta_u = 1 - (M_u/M_e)$.*

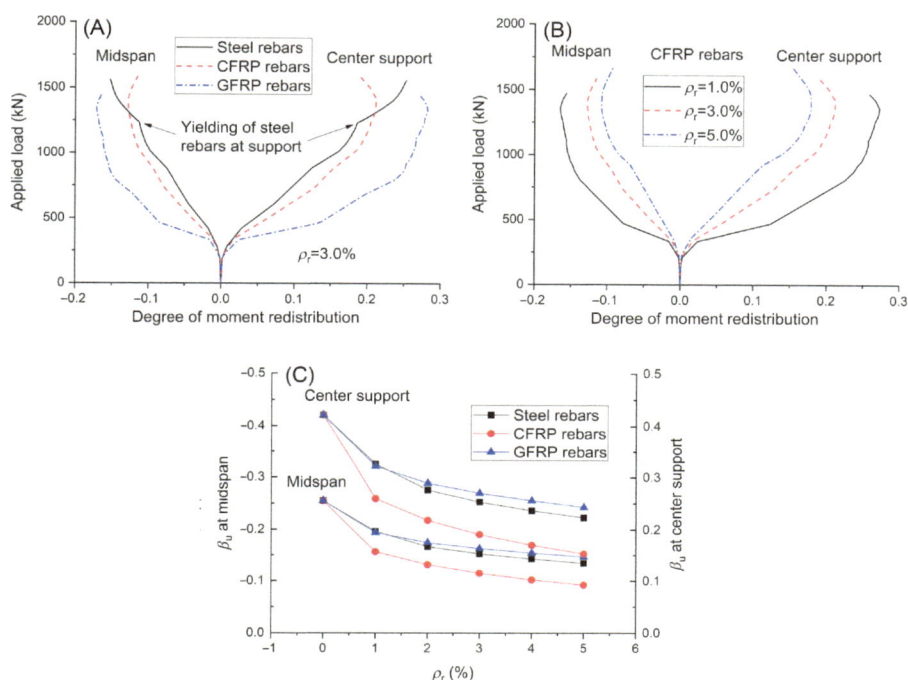

**Figure 11.9** Development of moment redistribution: (A) load−redistribution curves of girders with different rebar types; (B) load−redistribution curves of girders with different rebar ratios; (C) effect of rebar ratio/type on $\beta_u$ (moment redistribution at ultimate).

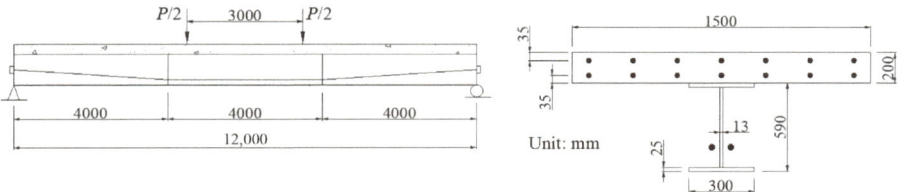

**Figure 11.10** Simply supported prestressed composite girders.

centroid of external tendons and the steel bottom flange are 295 and 45 mm at the support and deviator points, respectively. The external tendons are made of AFRP, CFRP, or prestressing steel having an elastic modulus of 50, 150, or 195 GPa, respectively. The FRP and steel tendons are assumed to have an identical tensile strength of 1860 MPa. The initial prestress level, $\sigma_{p0}/f_{pu}$, varies from 0% to 60%. The tendon area is 20 cm$^2$. The area of the top or bottom steel rebars is 7.5 cm$^2$. The yield strength and elastic modulus of steel rebar are 420 MPa and 200 GPa, respectively, and those of the steel beam are 275 MPa and 200 GPa, respectively.

The cylinder compressive strength, tensile strength, and elastic modulus of concrete are 40 MPa, 3.5 MPa, and 35 GPa, respectively.

## 11.3.1 Load—deformation behavior

Fig. 11.11A and B illustrates the effect of tendon type and prestress level on the load—deformation behavior of prestressed composite girders. The girders exhibit linear-elastic behavior up to the yielding of the steel beam, followed by a significant reduction in flexural stiffness. In the elastic stage, the increase in the stress of external tendons is small, and therefore, the tendon type has negligible influence on the elastic response characteristics. After yielding of the steel beam, AFRP tendons result in substantially lower flexural stiffness than CFRP or steel ones because of their significantly lower $E_p$. The postyielding behavior of prestressed composite girders with external CFRP and steel tendons is similar, with small discrepancies attributed to the difference between their moduli of elasticity. As expected, the prestress level significantly influences the structural behavior, that is, a higher prestress level leads to much stiffer behavior and higher yield and ultimate loads (or moments).

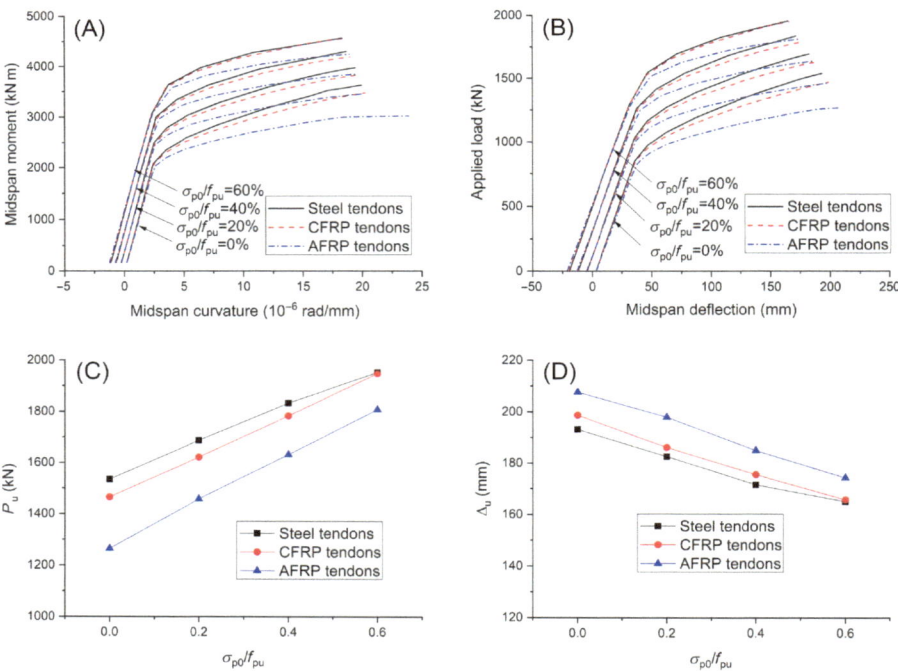

**Figure 11.11** Load—deformation characteristics: (A) moment—curvature curves; (B) load—deflection curves; (C) variation of ultimate load with the prestress level; (D) variation of ultimate deflection with the prestress level.

The variations of the ultimate load and deflection against the prestress level for AFRP, CFRP, and steel tendons are illustrated in Fig. 11.11C and D, respectively. Increasing the prestress level from 0% to 60% results in an increase in the ultimate load by 27.2%, 32.9%, and 42.9% for steel, CFRP, and AFRP tendons, respectively. Moreover, it leads to a decrease in the ultimate deflection by 14.6%, 16.6%, and 16.1% for steel, CFRP, and AFRP tendons, respectively. CFRP tendons show slightly lower ultimate load and higher ultimate deflection than steel ones. Replacing steel tendons with AFRP ones results in decreases in the ultimate load by 7.41% (for 60% prestress level) to 17.6% (for 0% prestress level) and an increase in the ultimate deflection by 7.0%.

## 11.3.2 Stress increase in external tendons

Fig. 11.12A illustrates the effect of tendon type and prestress level on the relation between load and tendon stress increase. Since the prestress loss due to elastic compression is smaller for a lower tendon modulus of elasticity, the effective prestress in AFRP tendons is generally higher than that in CFRP or steel tendons. The tendon stress development consists of two distinct stages separated by the yielding of the steel beam. In the first stage, the stress increases slowly, while the load increases rapidly. In the second stage, a significant tendon stress increment and a limited increase in the load are observed. The stress increases in CFRP and steel tendons are very similar over the entire loading process. The stress increase in GFRP tendons is significantly slower than that in steel tendons due to a significantly lower $E_p$.

Fig. 11.12B shows the variation of the stress increment at ultimate of external steel, CFRP, and AFRP tendons with the prestress level. The ultimate stress increments of CFRP and AFRP tendons slightly decrease, in approximately a linear manner, with an increasing prestress level. For steel tendons, the ultimate stress increment decreases slightly as the prestress level increases up to 40%, and thereafter a high prestress level results in a substantially lower stress increment at ultimate. This can be attributed to the fact that at ultimate, the steel tendons are approximately within

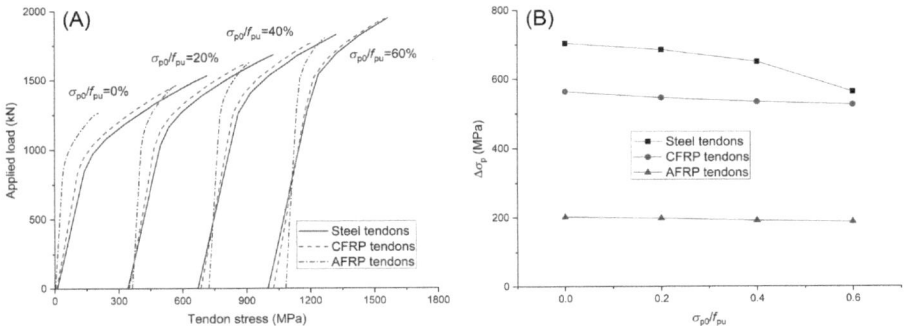

**Figure 11.12** Stress increase in external tendons: (A) load versus tendon stress curves; (B) variation of ultimate tendon stress increment with the prestress level.

the elastic range for a low prestress level, while they exhibit a notable nonlinearity for a prestress level greater than 40%. The ratio of the ultimate tendon increment of AFRP tendons to that of CFRP tendons is 0.36, which is slightly higher than the elastic modulus ratio of 0.33. This can be explained by the fact that at ultimate, AFRP tendons develop higher strain than CFRP tendons. The ratio of the ultimate stress increment of FRP tendons to that of steel tendons is also related to the elastic modulus ratio in the same way if the steel tendons are within their elastic range.

### 11.3.3  Neutral axis depth

Fig. 11.13A shows the effect of tendon type and prestress level on the load versus neutral axis depth curve of the prestressed composite girders. When a prestressed composite girder is subjected to its self-weight load only, the initial neutral axis depth depends on the prestress level. For zero prestress level, the initial neutral axis is located at the centroidal axis of the composite section. For a prestress level of 20%, 40%, or 60%, the initial neutral axis is located above the centroidal axis. As the applied load increases, the neutral axis depth for zero prestress level remains almost unchanged until the yielding of the steel beam and quickly decreases thereafter. For a prestress level of 20%, 40%, or 60%, the neutral axis depth decreases rapidly from a positive value to negative infinity at which the hogging curvature vanishes. At the beginning of sagging curvature, the neutral axis depth jumps suddenly from negative infinity to positive infinity. Thereafter, the neutral axis depth shows a rapid decrease with increasing load until crack development in the concrete slab stabilizes. Then the movement of the neutral axis gradually slows down up to the yielding of the steel beam, followed by a quick decrease in its depth up to ultimate. Before the yielding of the steel beam, changes in the neutral axis depth for different tendon types are similar. After yielding, AFRP tendons result in a much quicker decrease in the neutral axis depth with increasing load compared to CFRP or steel tendons.

Fig. 11.13B shows that at failure, the neutral axis depth increases as the prestress level increases. At specific prestress levels, the neutral axis depths for CFRP and

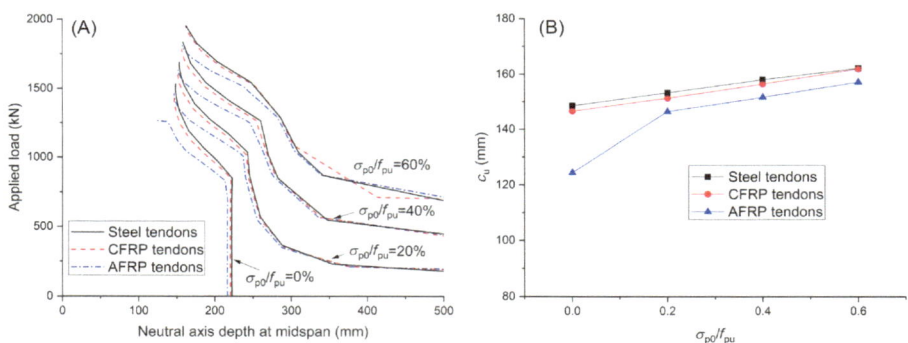

**Figure 11.13** Neutral axis depth characteristics: (A) load versus neutral axis depth curves; (B) variation of neutral axis depth at ultimate with the prestress level.

steel tendons are almost identical. AFRP tendons lead to smaller neutral axis depth than steel ones. The difference is particularly notable at zero prestress level.

### 11.3.4  Stresses and strains in reinforcing steel bars and steel beam

Fig. 11.14 illustrates the development of strains and stresses in reinforcing bars for different tendon types and various prestress levels. Compressive strains or stresses in both the top and bottom bars increase slowly until yielding of the steel beam. Afterward, the strain or stress in the top bars increases quickly, while the compressive strain or stress in the bottom bars gradually vanishes and then the tensile strain or stress occurs. The strain or stress developments in reinforcing bars for different types of tendons differ after the steel beam yields. AFRP tendons lead to a faster increase of bar strain or stress than CFRP or steel tendons. At failure, all the top bars have slightly yielded, while all the bottom bars are far below their yield strength. AFRP tendons lead to a higher ultimate strain or stress in bottom bars than CFRP or steel tendons. The higher the prestress level, the lower the ultimate strain or stress in the bottom bars.

The effect of tendon type and prestress level on the development of strain and stress in the steel beam is illustrated in Fig. 11.15. The strain and stress in the bottom flange increase linearly with increasing load until yielding. Afterward, the increase in strain becomes faster. The strain and stress in the top flange are nearly unchanged until the yielding of the extreme tension face and develop quickly afterward. AFRP tendons lead to a faster increase in postyielding strain in the steel beam than CFRP or steel tendons. The type of external tendons has practically no influence on the bottom flange stress development. At failure, the top flange does not reach its yield strength. AFRP tendons result in higher ultimate stress in the top flange than CFRP or steel tendons. The higher the prestress level, the lower the ultimate stress in the top flange.

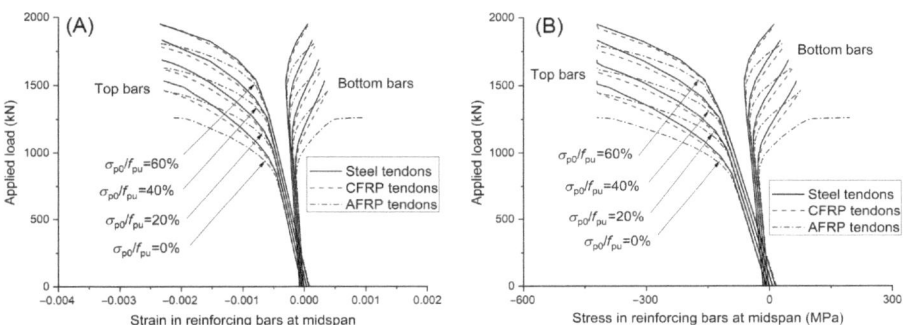

**Figure 11.14** Strain and stress in reinforcing bars: (A) load versus bar strain curves; (B) load versus bar stress curves.

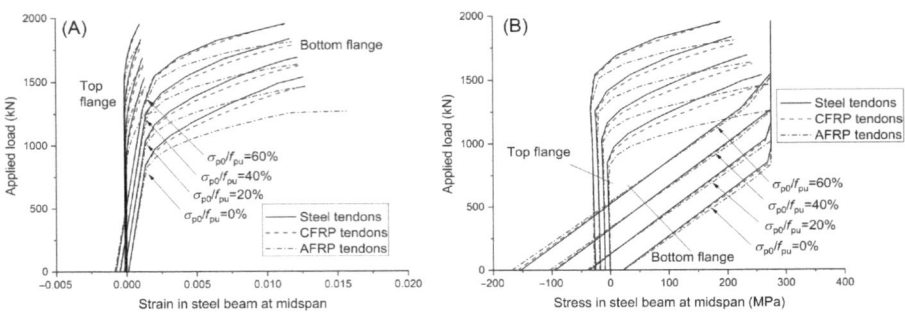

**Figure 11.15** Strain and stress in steel beam: (A) load versus steel strain curves; (B) load versus steel stress curves.

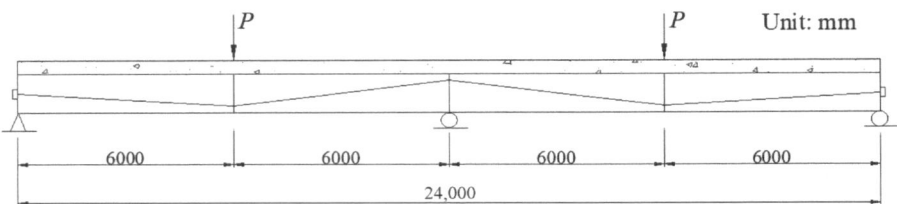

**Figure 11.16** Continuous prestressed composite girders.

## 11.4   Behavior of continuous prestressed composite girders with external FRP tendons

Fig. 11.16 shows two-span continuous prestressed composite girders with a total length of 24.0 m and under center-point loads at each span. The distances between the centroid of external tendons and the steel bottom flange are 295, 545, and 45 mm at outer, inner supports, and midspan, respectively. All the other parameters are identical to those of the simply supported girders described in Section 11.3. The influence of the tendon type and prestress level on the general behavior of continuous prestressed composite girders was found similar to that of simply supported ones. Therefore this section presents only results relevant to the secondary moments and moment redistribution in continuous prestressed composite girders.

### 11.4.1   Secondary moment

The magnitude of secondary moments in statically indeterminate prestressed composite girders could be very high as the tendon line in these girders is commonly located far away from the concordant line. The secondary moment varies significantly with the shift of the tendon line. It has been proved in Chapter 10 that tendon shift with linear transformation has no effects regarding the full-range behavior of

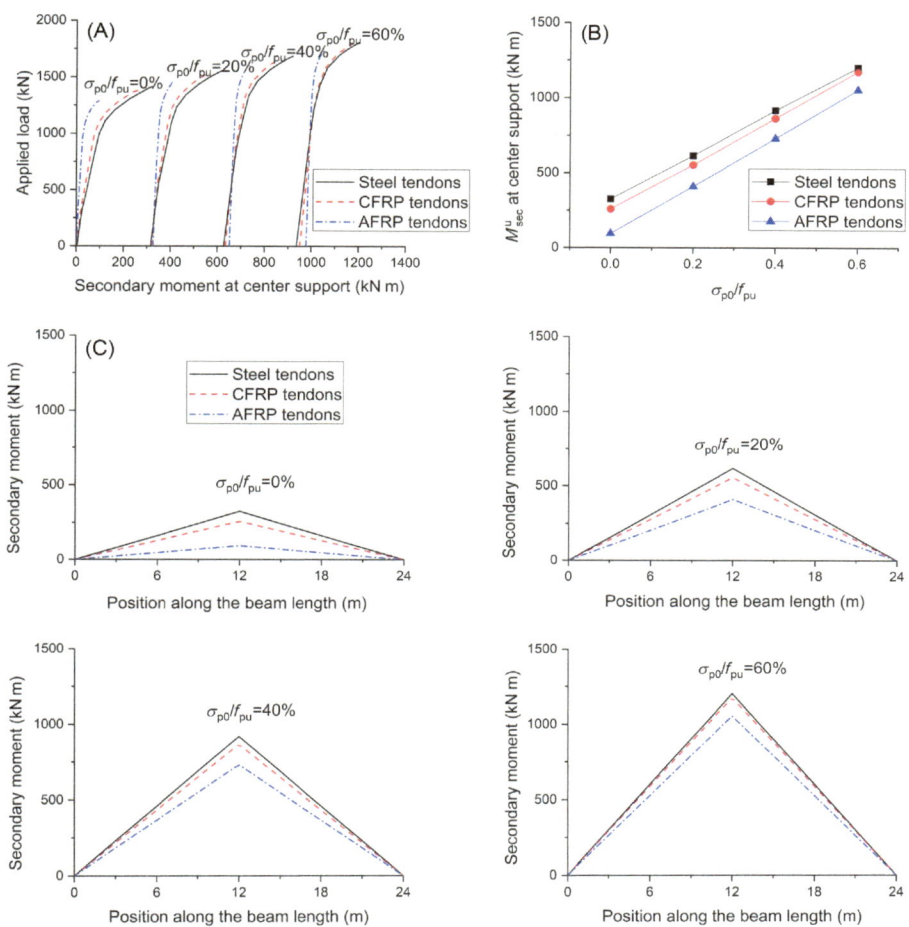

**Figure 11.17** Secondary moment characteristics for different tendon types and prestress levels: (A) load versus secondary moment curves; (B) variation of secondary moment at ultimate with the prestress level; (C) distribution of secondary moments.

continuous prestressed composite girders. By utilizing linear transformation, an approach for quantifying secondary moments was proposed in Chapter 10, that is: locate the position of linear-transformation concordant tendon line; compute bending moments for nonconcordant and concordant tendon lines; calculate the secondary moment by subtracting the moment of the concordant tendon line from the moment of the nonconcordant tendon line.

Fig. 11.17A shows the effect of tendon type and prestress level on the load versus secondary moment curves of the prestressed composite girders. It is noted that at the initial state, significant secondary moments could be existent. For instance, at 60% prestress level, the initial secondary moments at the center support for AFRP,

CFRP, and steel tendons are 977.1, 950.6, and 935.8 kN m, respectively. The secondary moment develops slowly before the yielding of the steel beam, and thereafter, a rapid increase is observed. The secondary moment development for AFRP tendons is slower than that for CFRP or steel tendons due to a slower increase in tendon stress. At failure, the secondary moment and its distribution are illustrated in Fig. 11.17B and C, respectively. An approximate linear relation between the secondary moment and the prestress level can be observed. AFRP tendons lead to smaller secondary moments than CFRP or steel tendons and that difference is particularly notable at a low prestress level.

## 11.4.2  Moment redistribution

Fig. 11.18A-C shows the effect of prestress level on the distribution of ultimate moments for steel, CFRP, and AFRP tendons, respectively. The higher the prestress level, the larger the midspan moment at ultimate. However, the ultimate moments at the center support for different prestress levels are almost identical. The above observation is attributed to the influence of secondary moments. As can be seen in Fig. 11.17, there are positive secondary moments over the span, which counteract the center support moment and enhance the midspan moment. In addition, a higher prestress level produces significantly higher secondary moments, which lead to a larger reduction in the center support moment and a larger increment in the midspan moment. Therefore it is concluded that if the secondary moments are minimized by using a concordant tendon profile, a high prestress level would cause a higher center support moment, but the difference of the midspan moments for different prestress levels would be less pronounced. For a given tendon type, the deviations between actual and elastic moments at the center support for different prestress levels are almost identical, indicating the negligible influence of prestress level on moment redistribution.

Fig. 11.19 shows the load−moment curves for different types of external tendons. The dashed lines represent the development of elastic moments. The results are produced for 60% prestress level and by considering and neglecting secondary moments. These graphs clearly demonstrate how the secondary moments influence the moment development and how the actual moments ($M$) differ relative to the elastic moments ($M_e$) during the loading process. It is seen that the secondary moment substantially reduces the center support moment but increases the midspan moment. The actual moment starts to differ from the elastic moment at the cracking load due to the redistribution of moments. The difference tends to be more pronounced as the load increases.

Fig. 11.20 shows the load versus degree of moment redistribution curves for the prestressed composite girders. A higher prestress level causes a larger cracking load at which moment redistribution initiates. At any inelastic load level, the presence of secondary moments results in an increase in moment redistribution over the center support and this observation becomes more apparent at high load levels. The influence of secondary moments on moment redistribution over the midspan is not as important as that over the center support. The secondary moment may increase or

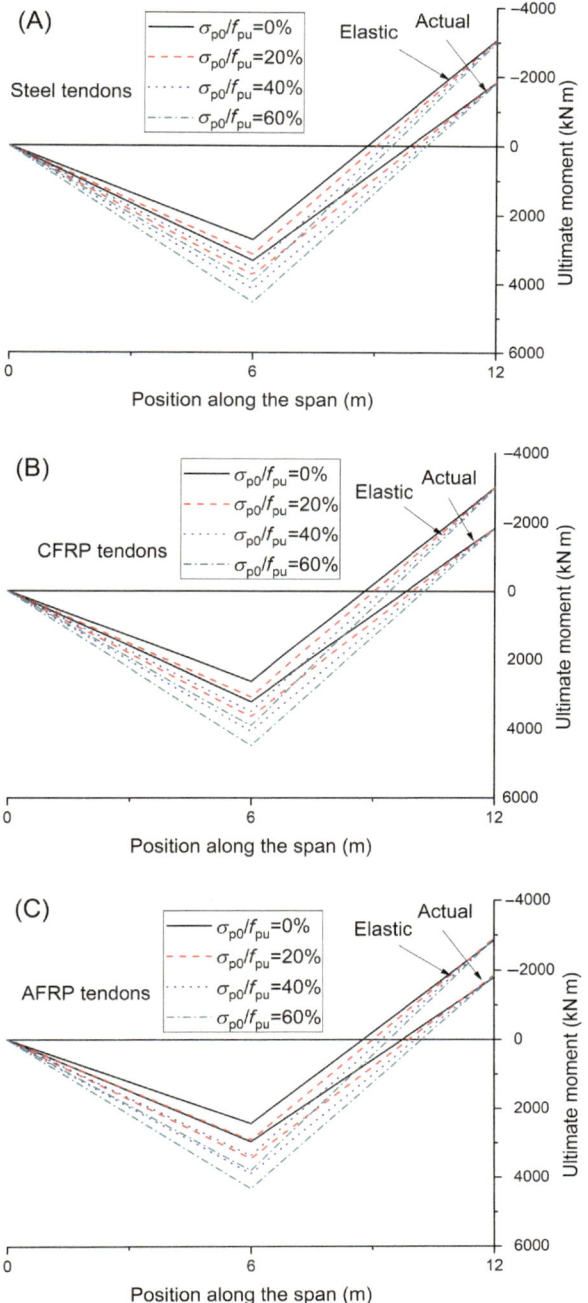

**Figure 11.18** Moment distribution diagrams for various prestress levels: (A) steel tendons; (B) CFRP tendons; (C) AFRP tendons.

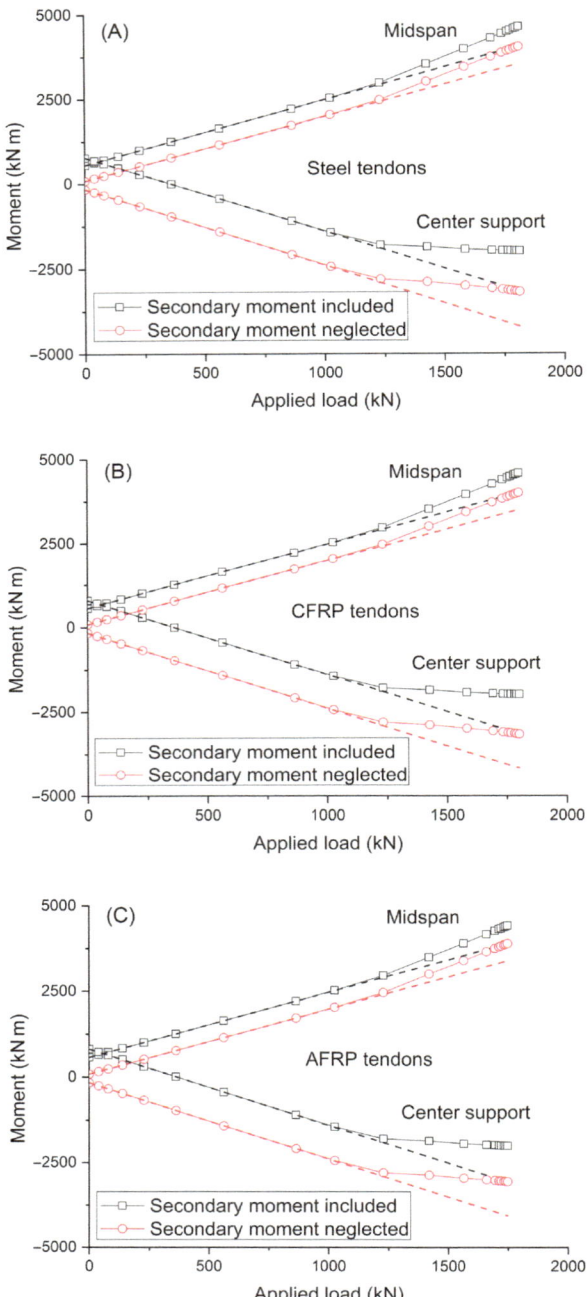

**Figure 11.19** Load−moment curves for the prestress level of 60%: (A) steel tendons; (B) CFRP tendons; (C) AFRP tendons.

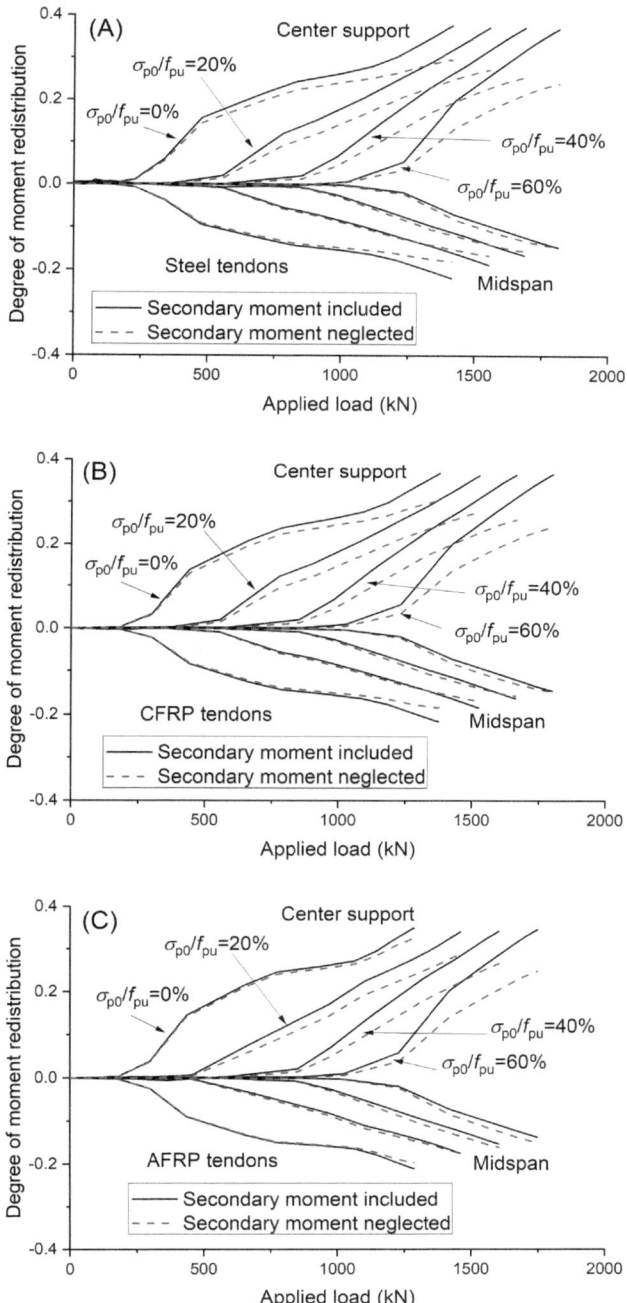

**Figure 11.20** Load versus degree of moment redistribution curves for various prestress levels: (A) steel tendons; (B) CFRP tendons; (C) AFRP tendons.

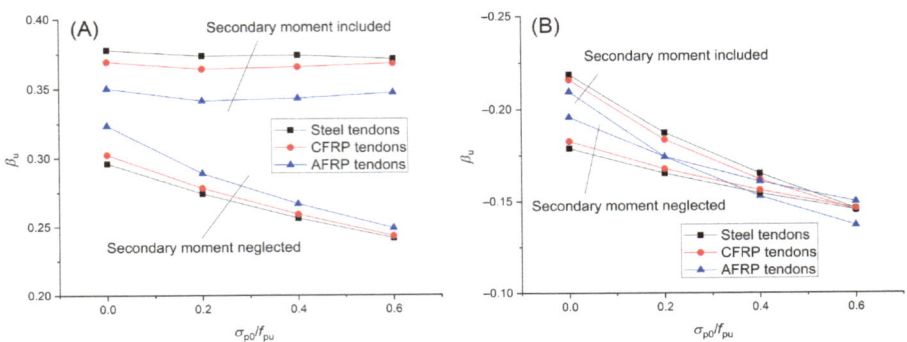

**Figure 11.21** Variation of moment redistribution at ultimate with the prestress level: (A) center support; (B) midspan.

reduce the moment redistribution at the midspan during loading, depending on the load and prestress levels.

Fig. 11.21 illustrates moment redistribution variations with varying prestress levels for different tendon types. When secondary moments are considered, the influence of prestress level on moment redistribution over the center support appears to be insignificant, while a higher prestress level results in a substantially lower redistribution of moments at the midspan. In addition, AFRP and CFRP tendons lead to substantially and slightly lower redistribution values, respectively, when compared to steel tendons. When secondary moments are neglected, the moment redistribution at the critical sections decreases as the prestress level increases. In this case, AFRP tendons result in higher moment redistribution than CFRP or steel tendons. Over the center support, the secondary moment considerably increases the degree of redistribution, especially for high prestress levels. AFRP tendons exhibit a less pronounced influence of secondary moments on moment redistribution compared to CFRP or steel tendons. At the midspan, the presence of secondary moments leads to an increase in the decrease rate of moment redistribution with increasing prestress level. In other words, the prestress level affecting moment redistribution at the midspan is mitigated when neglecting secondary moments.

## 11.5 Conclusions

In steel−concrete composite girders, using CFRP rebars is more effective in reducing the crack width in the concrete slab over the center support than using GFRP or steel rebars. At a given $\rho_r$ level, FRP rebars lead to smaller ultimate deflection despite its lower elastic modulus in comparison with steel rebars. The ultimate load of composite girders with steel rebars is close to that with CFRP rebars while higher than that with GFRP rebars. Both the type and ratio of rebars have practically no impact on the ultimate curvature and neutral axis position over the midspan, and hence, on the ultimate stress and strain in concrete and structural steel in the section. By contrast, those response characteristics at the center support are markedly influenced by the rebars. At a given $\rho_r$ level, CFRP rebars lead to higher center support moments at ultimate due to substantially higher stress developed in the rebars, while GFRP rebars develop smaller ultimate stress, resulting in lower ultimate moments when compared to steel rebars. CFRP rebars lead to much lower moment redistribution at ultimate than steel rebars, while moment redistribution in composite girders with GFRP rebars is comparable to that with steel rebars.

The behavior of prestressed steel—concrete composite girders with external CFRP tendons is similar to that with external steel tendons. It is therefore feasible to replace steel tendons with CFRP ones without compromising the structural performance. Replacing steel tendons with AFRP ones leads to a substantial decrease in ultimate load and an increase in deformation capacity. Moreover, AFRP tendons result in smaller neutral axis depth and higher strains in the steel beam than steel tendons. The increase in ultimate load with increasing prestress level is more effective for AFRP tendons than for CFRP or steel tendons. Therefore using AFRP tendons at a high prestress level while CFRP tendons at a low prestress level may be recommended by considering both the cost and performance factors.

For continuous prestressed steel—concrete composite girders, the prestress level significantly influences the midspan moment but hardly influences the center support moment due to the effect of secondary moments. The presence of secondary moments results in substantial increases in moment redistribution over the center support, while its influence on moment redistribution over the midspan is not important. The effect of secondary moments on moment redistribution for AFRP tendons is substantially less important than that for CFRP or steel tendons. AFRP tendons result in lower moment redistribution than CFRP or steel ones when considering secondary moments. Minimizing the secondary moments leads to the opposite observation. The effect of prestress level on moment redistribution over the center support is not important when considering secondary moments. If secondary moments are minimized, this effect becomes rather important. Moreover, minimizing the secondary moments reduces the prestress level influence on moment redistribution over the midspan.

# References

ACI Committee 440. (2004). *Prestressing concrete structures with FRP tendons*. ACI 440.4R-04, Farmington Hills, MI.

ACI Committee 440. (2015). Guide for the design and construction of structural concrete reinforced with FRP bars. ACI 440.1R-15, Farmington Hills, MI.

Al-Saidy, A. H., Klaiber, F. W., & Wipf, T. J. (2007). Strengthening of steel—concrete composite girders using carbon fiber reinforced polymer plates. *Construction and Building Materials*, *21*(2), 295—302. Available from https://doi.org/10.1016/j.conbuildmat.2005.08.018.

Aly, M.Y.E., & El-Hacha, R. (2009). Strength evaluation of steel-concrete composite girders strengthened with prestressed FRP laminates. In *Second Official Regional Conference of International Institute for FRP in Construction for Asia-Pacific Region (APFIS 2009)*, Seoul, pp. 9-11.

Deng, J., Lee, M. M. K., & Li, S. (2011). Flexural strength of steel-concrete composite beams reinforced with a prestressed CFRP plate. *Construction and Building Materials*, 25(1), 379−384. Available from https://doi.org/10.1016/j.conbuildmat.2010.06.015.

Fam, A., MacDougall, C., & Shaat, A. (2009). Upgrading steel−concrete composite girders and repair of damaged steel beams using bonded CFRP laminates. *Thin-Walled Structures*, 47(10), 1122−1135. Available from https://doi.org/10.1016/j.tws.2008.10.014.

Ghafoori, E., & Motavalli, M. (2015). Innovative CFRP-prestressing system for strengthening metallic structures. *Journal of Composites for Construction*, 19(6), 04015006. Available from https://doi.org/10.1061/(asce)cc.1943-5614.0000559.

Ju, M., Park, Y., & Park, C. (2017). Cracking control comparison in the specifications of serviceability in cracking for FRP reinforced concrete beams. *Composite Structures*, 182, 674−684. Available from https://doi.org/10.1016/j.compstruct.2017.09.016.

Lou, T., & Karavasilis, T. L. (2019). Numerical evaluation of prestressed steel-concrete composite girders with external FRP or steel tendons. *Journal of Constructional Steel Research*, 162, 105698. Available from https://doi.org/10.1016/j.jcsr.2019.105698.

Lou, T., Wu, S., & Chen, B. (2022). Effect of reinforcement on the response of continuous steel-concrete composite beams. *Case Studies in Construction Materials*, 16, e00929. Available from https://doi.org/10.1016/j.cscm.2022.e00929.

Pang, M., Shi, S., Hu, H., & Lou, T. (2021). Flexural behavior of two-span continuous CFRP RC beams. *Materials*, 14(22), 6746. Available from https://doi.org/10.3390/ma14226746.

Santos, P., Laranja, G., França, P. M., & Correia, J. R. (2013). Ductility and moment redistribution capacity of multi-span T-section concrete beams reinforced with GFRP bars. *Construction and Building Materials*, 49, 949−961. Available from https://doi.org/10.1016/j.conbuildmat.2013.01.014.

Schmidt, J. W., Bennitz, A., Täljsten, B., Goltermann, P., & Pedersen, H. (2012). Mechanical anchorage of FRP tendons - A literature review. *Construction and Building Materials*, 32, 110−121. Available from https://doi.org/10.1016/j.conbuildmat.2011.11.049.

Siwowski, T. W., & Siwowska, P. (2018). Experimental study on CFRP-strengthened steel beams. *Composites Part B: Engineering*, 149, 12−21. Available from https://doi.org/10.1016/j.compositesb.2018.04.060.

Tavakkolizadeh, M., & Saadatmanesh, H. (2003). Strengthening of steel-concrete composite girders using carbon fiber reinforced polymers sheets. *Journal of Structural Engineering*, 129(1), 30−40. Available from https://doi.org/10.1061/(asce)0733-9445(2003)129:1(30).

Teng, J. G., Yu, T., & Fernando, D. (2012). Strengthening of steel structures with fiber-reinforced polymer composites. *Journal of Constructional Steel Research*, 78, 131−143. Available from https://doi.org/10.1016/j.jcsr.2012.06.011.

Zhao, X. L., & Zhang, L. (2007). State-of-the-art review on FRP strengthened steel structures. *Engineering Structures*, 29(8), 1808−1823. Available from https://doi.org/10.1016/j.engstruct.2006.10.006.

# Finite element modeling at long-term sustained loads

**12**

## 12.1 Introduction

For a prestressed concrete or steel−concrete composite structure under sustained loads, the section stress and strain are subject to change with time due to time-dependent effects resulting from creep and shrinkage of concrete and relaxation of prestressing tendons (Ghali et al., 2002). If the structure is statically indeterminate, the time-dependent effects will also result in the redistribution of moments in addition to the stress redistribution in cross sections (Kwak & Seo, 2002). The performance of prestressed structures over a long period of time is subject to inevitable losses and needs to be reasonably assessed.

The theoretical simulation of the long-term behavior of prestressed concrete or composite structures is not an easy task. One of the challenges is to model accurately and efficiently the creep effect since creep of concrete is associated with the history of the applied stress. A number of methods have been developed to simulate the creep behavior of concrete members (Bažant, 2001; Bažant & Najjar, 1973). The utilization of the effective modulus is known to be the simplest method for a creep analysis (Lopes & Lopes, 2012). This method may produce satisfactory results when the aging effects are negligible, but it will lead to an overestimation of the creep when the aging effects are present. Most of the available models were based on the age-adjusted effective modulus method (AEMM) by introducing an aging coefficient in the effective modulus (Hosseini & Jefferson, 1998; Rodriguez-Gutierrez & Aristizabal-Ochoa, 2007). The values of the aging coefficient are often taken from tables or charts and, therefore, AEMM is a simplified method rather than a refined method. Another challenge is to model accurately the relaxation of prestressing tendons, taking into account the interaction between different time-dependent effects. Commonly, this interaction is considered approximately by using a relaxation reduction coefficient obtained from tables or charts (Gutiérrez et al., 1996). Refined numerical methods have also been developed to model the concrete creep and/or tendon relaxation (Au & Si, 2011; Hamed & Bradford, 2012).

This chapter presents a time-dependent analysis method of prestressed concrete or steel−concrete composite members at sustained service loads (Lou et al., 2015, 2016), taking into account the creep and shrinkage of concrete and relaxation of prestressing tendons. Several numerical examples are given to illustrate the reliability and applicability of the time-dependent numerical model.

Prestressed Members with External Fiber-Reinforced Polymer (FRP) Tendons. DOI: https://doi.org/10.1016/B978-0-443-23877-2.00012-2

## 12.2 Concrete creep, concrete shrinkage, and tendon relaxation

### 12.2.1 Concrete creep

At service conditions, creep can be considered to be proportional to the applied stress. Therefore the creep strain $\varepsilon_c^{cr}(t)$ under sustained stress is given by

$$\varepsilon_c^{cr}(t) = \sigma_c(\tau)C(t,\tau) \tag{12.1}$$

where $\sigma_c(\tau)$ is the concrete stress at age $\tau$; and $C(t,\tau)$ is the creep function, defined as the creep strain at age $t$ caused by a unit stress applied at age $\tau$.

When the applied stress is subject to a gradual change with time (see Fig. 12.1), which is generally true in practical applications, the creep strain due to the applied stress is calculated by applying the principle of superposition as follows:

$$\varepsilon_c^{cr}(t) = \sigma_c(t_0)C(t,t_0) + \int_{t_0}^{t} C(t,\tau)\frac{\partial \sigma_c(\tau)}{\partial \tau}\,d\tau \tag{12.2}$$

where $t_0$ is the age at which the initial stress is applied. This equation indicates that creep is associated with the history of the applied stress.

As illustrated in Fig. 12.1, time is divided into a number of small intervals to apply an incremental method. Generally, the stresses for each element at various times have to be stored for integrating Eq. (12.2). However, this may have some limitations in application to large structures. To overcome this problem, the following Dirichlet series creep function, initially proposed by Zienkiewicz and Watson (1966), is adopted herein:

$$C(t,\tau) = \sum_{k=1}^{m} \phi_k(\tau)[1 - e^{-r_k(t-\tau)}] \tag{12.3}$$

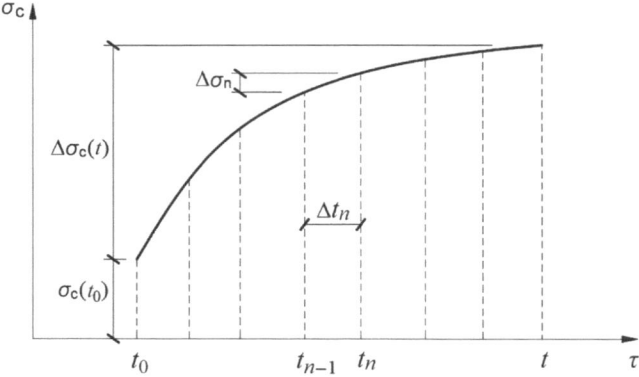

**Figure 12.1** Gradually varying stress with time.

where m, $\phi_k(\tau)$ and $r_k$ are the empirical parameters to be determined by experimental data.

By utilizing the above creep function, the creep strain increment at time interval $\Delta t_n (= t_n - t_{n-1})$, $\Delta \varepsilon_c^{cr}$, is given by (Zhu, 2009)

$$\Delta \varepsilon_c^{cr} = \varepsilon_c^{cr}(t_n) - \varepsilon_c^{cr}(t_{n-1}) = \sum_{k=1}^{m}(1 - e^{-r_k \Delta t_n})\omega_{kn} + C(t_n, t_{n-1/2})\, \Delta \sigma_n \tag{12.4}$$

in which $t_{n-1/2}$ represents the middle time between time $t_{n-1}$ and time $t_n$; $\Delta \sigma_n$ is the stress increment at time interval $\Delta t_n$; $\omega_{kn}$ is obtained from the following recursive formula:

$$\omega_{kn} = \omega_{k(n-1)}e^{-r_k \Delta t_{n-1}} + \Delta \sigma_{n-1}\phi_k(t_{(n-1)-1/2})e^{-r_k \Delta t_{n-1}/2} \tag{12.5a}$$

$$\omega_{k1} = \sigma_c(t_0)\phi_k(t_0) \tag{12.5b}$$

From Eqs. (12.4) and (12.5), it can be seen that, instead of recording the entire stress history, only the value of $\omega_{k(n-1)}$ needs to be stored, thus ensuring an efficient creep analysis.

## 12.2.2 Concrete shrinkage

Shrinkage of concrete is defined as the volume change which is independent of the imposed stress. Therefore the shrinkage strain can be conveniently calculated using the available shrinkage models proposed by the codes or by the investigators. The free shrinkage is commonly expressed as follows (Ghali et al., 2002):

$$\varepsilon^{sh}(t) = \varepsilon_{sh}^{\infty} \beta_{sh}(t) \tag{12.6}$$

where $\varepsilon_{sh}^{\infty}$ is the ultimate shrinkage strain, which is taken, unless otherwise stated, equal to 366 $\mu\varepsilon$; and $\beta_{sh}(t)$ is a function that may be determined by (Ghali et al., 2002)

$$\beta_{sh}(t) = \frac{t}{D + t} \tag{12.7}$$

where $D$ is a coefficient. Based on the experimental results of the shrinkage tests, Fan et al. (2010) recommended that $D = 29.0$ for C20 concrete and 32.5 for C30 concrete. In this chapter, $D = 35.0$ is adopted.

## 12.2.3 Tendon relaxation

The relaxation of stress-relieved steel wires or strands, $\sigma_{pr}$, may be evaluated by the following equation (Magura et al., 1964):

$$\frac{\sigma_{pr}}{\sigma_{p0}} = -\frac{\log(\tau - t_0)}{10}\left(\frac{\sigma_{p0}}{f_{py}} - 0.55\right) \tag{12.8}$$

in which $(\tau - t_0)$ is the time in hours after stressing; $\sigma_{p0}$ is the initial stress immediately after stressing; and $f_{py}$ is the yield stress of prestressing steel. The ratio of the yield stress to the ultimate tensile strength generally varies between 0.8 and 0.9, depending on the type of prestressing steel.

It should be noted that the above relaxation equation is subject to the condition that the tendon length is kept constant and $\sigma_{p0}$ is the only applied stress. In prestressed concrete or composite members, however, the applied stress would be influenced by some causes such as the prestress transfer, the load application, and the interaction between the creep and shrinkage of concrete and the relaxation of prestressing tendons. Therefore the initial stress for computing the stress relaxation at each time interval should be appropriately adjusted according to the change of the tendon stress as a result of these causes. The procedure for computing the actual relaxation of prestressing tendons is illustrated in Fig. 12.2. Denote by $\sigma_{p0(1)}$ the initial prestress at time $t_0$. At time $t_1$, the prestress varies from $\sigma_{p0(1)}$ to $\sigma_{p1}$ due to the tendon relaxation $\Delta\sigma_{pr1}$ and also to other causes. Compute the fictitious initial prestress $\sigma_{p0(2)}$ using Eq. (12.8) such that $\sigma_{p0(2)}$ would be relaxed to $\sigma_{p1}$ from time $t_0$ to time $t_1$. Based on the fictitious initial prestress $\sigma_{p0(2)}$, the tendon relaxation $\Delta\sigma_{pr2}$ between time $t_1$ and time $t_2$ is then calculated utilizing Eq. (12.8). Continue the process and the tendon relaxation $\Delta\sigma_{pr}$ at time interval $\Delta t_n$ is then calculated from

$$\Delta\sigma_{pr} = \sigma_{pr}(t_n) - \sigma_{pr}(t_{n-1}) \tag{12.9}$$

The total relaxation at time $t_n$ is obtained by summing up the relaxation at all time intervals.

## 12.3   Beam element

### 12.3.1  General formulation

The finite element method is formulated based on the classical Euler–Bernoulli theory which is well applicable to slender members where the shear deformation is

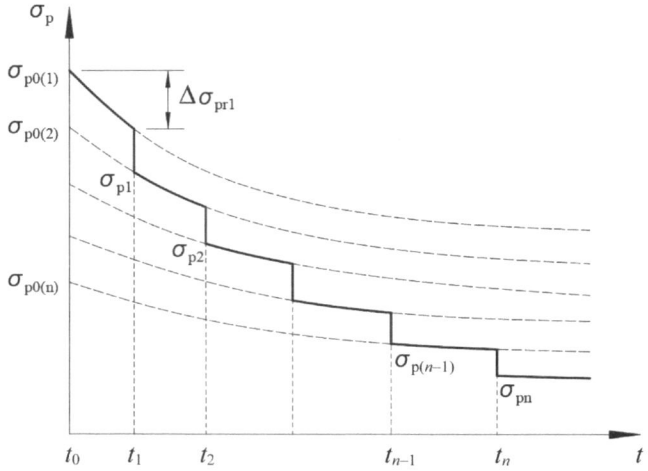

**Figure 12.2** Computation of relaxation of prestressing tendons.

negligible. Consider a plane beam element with two nodes by which the local coordinate system $(x, y)$ is defined, as shown in Fig. 12.3. The origin and direction of the local coordinate system would continuously change with continuous member deformations. Each node has three degrees of freedom, namely, axial displacement $u$, transverse displacement $v$, and rotation $\theta$. Assume that the axis and transverse displacements of any point on the element are a linear function and a cubic polynomial, respectively. Then, these displacements can be described in terms of element nodal displacements as follows:

$$f = Nu^e \tag{12.10}$$

$$f = \{ u \quad v \}^T; \ u^e = \{ u_i \quad v_i \quad \theta_i \quad u_j \quad v_j \quad \theta_j \}^T \tag{12.11}$$

$$N = \begin{bmatrix} N_1 & 0 & 0 & N_4 & 0 & 0 \\ 0 & N_2 & N_3 & 0 & N_5 & N_6 \end{bmatrix} \tag{12.12}$$

in which $u^e$ is the element nodal displacements; $N_1 = 1 - p$; $N_2 = 1 - 3p^2 + 2p^3$; $N_3 = l(p - 2p^2 + p^3)$; $N_4 = p$; $N_5 = 3p^2 - 2p^3$; $N_6 = l(-p^2 + p^3)$; $p = x/l$ where $l$ is the length of the element.

The axial strain $\varepsilon_O$ at any point on the element is given by: $\varepsilon_O = u' + \left[ \left( u' \right)^2 + \left( v' \right)^2 \right] / 2$, where a superimposed prime represents differentiation with respect to $x$. By neglecting the infinitesimal term $(u')^2$, the preceding equation is approximated as

$$\varepsilon_O = u' + \left( v' \right)^2 / 2 \tag{12.13}$$

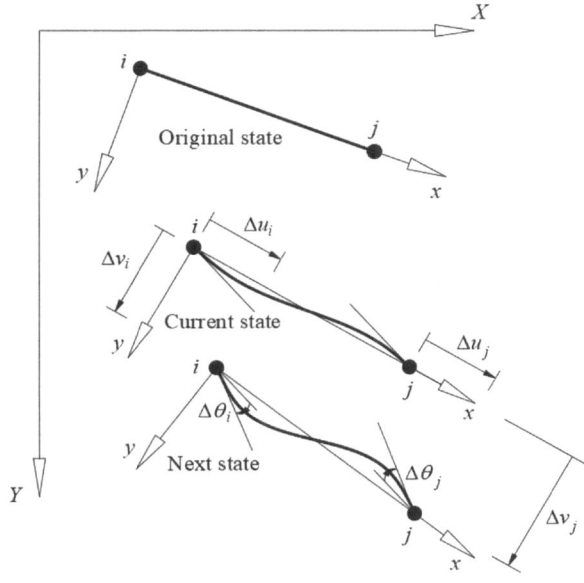

**Figure 12.3** Coordinate system and element at different states.

Assuming that the shear deformation is negligible, that a plane section remains plane after deformations, and that bonded tendons and rebars perfectly bond with the surrounding concrete, the axis strain $\varepsilon$ at any fiber of a cross-section is expressed by

$$\varepsilon = \varepsilon_O - v'' y \tag{12.14}$$

Combining Eqs. (12.10)−(12.14), the incremental strain−displacement relationship at the current state can be determined by

$$\Delta \varepsilon = (\boldsymbol{B} + \Delta \boldsymbol{u}^{\mathrm{eT}} \boldsymbol{J}^T \boldsymbol{J}/2) \Delta \boldsymbol{u}^{\mathrm{e}} \tag{12.15}$$

in which

$$\boldsymbol{B} = \begin{bmatrix} N_1' & -N_2'' y & -N_3'' y & N_4' & -N_5'' y & -N_6'' y \end{bmatrix} \tag{12.16}$$

$$\boldsymbol{J} = \begin{bmatrix} 0 & N_2' & N_3' & 0 & N_5' & N_6' \end{bmatrix} \tag{12.17}$$

The tangent strain−displacement relationship at the next state can be expressed by taking the differential form of Eq. (12.15) as follows:

$$\mathrm{d}\varepsilon = \boldsymbol{B}\mathrm{d}\boldsymbol{u}^{\mathrm{e}} + \Delta \boldsymbol{u}^{\mathrm{eT}} \boldsymbol{J}^T \boldsymbol{J} \mathrm{d}\boldsymbol{u}^{\mathrm{e}}/2 + \mathrm{d}\boldsymbol{u}^{\mathrm{eT}} \boldsymbol{J}^T \boldsymbol{J} \Delta \boldsymbol{u}^{\mathrm{e}}/2 = (\boldsymbol{B} + \Delta \boldsymbol{u}^{\mathrm{eT}} \boldsymbol{J}^T \boldsymbol{J})\mathrm{d}\boldsymbol{u}^{\mathrm{e}} = \mathrm{d}\boldsymbol{u}^{\mathrm{eT}}(\boldsymbol{B}^{\mathrm{T}} + \boldsymbol{J}^T \boldsymbol{J} \Delta \boldsymbol{u}^{\mathrm{e}}) \tag{12.18}$$

The virtual work equations at the next state can be written as

$$\mathrm{d}\boldsymbol{u}^{\mathrm{eT}} \left( \boldsymbol{P}^{\mathrm{e}} + \Delta \boldsymbol{P}^{\mathrm{e}} \right) = \int_V \mathrm{d}\varepsilon (\sigma + \Delta \sigma) \mathrm{d}V \tag{12.19}$$

where $\boldsymbol{P}^{\mathrm{e}}$ is the element equivalent nodal loads; the integration is performed over the entire volume of an element. Substitution of Eq. (12.18) into Eq. (12.19) yields the element equilibrium equations at the next state:

$$\boldsymbol{P}^{\mathrm{e}} + \Delta \boldsymbol{P}^{\mathrm{e}} = \int_V (\boldsymbol{B}^{\mathrm{T}} + \boldsymbol{J}^{\mathrm{T}} \boldsymbol{J} \Delta \boldsymbol{u}^{\mathrm{e}})(\sigma + \Delta \sigma) \mathrm{d}V \tag{12.20}$$

On the other hand, the tangent strain−displacement relationship and the element equilibrium equations at the current state are expressed as follows:

$$\mathrm{d}\varepsilon = \boldsymbol{B}\mathrm{d}\boldsymbol{u}^{\mathrm{e}} \tag{12.21}$$

$$\boldsymbol{P}^{\mathrm{e}} = \int_V \boldsymbol{B}^{\mathrm{T}} \sigma \mathrm{d}V \tag{12.22}$$

By subtracting Eq. (12.22) from Eq. (12.20) and ignoring the higher-order term $\int_V \boldsymbol{J}^T \boldsymbol{J} \Delta \sigma dV \Delta \boldsymbol{u}^e$, the following incremental equilibrium equations for an element at the current state can be established:

$$\Delta \boldsymbol{P}^e = \int_V \boldsymbol{B}^T \Delta \sigma dV + \int_V \sigma \boldsymbol{J}^T \boldsymbol{J} dV \Delta \boldsymbol{u}^e \tag{12.23}$$

### 12.3.2 Method of analysis at immediate loads

The instantaneous tangent stress–strain relationship for materials is given by

$$d\sigma = E d\varepsilon = E \boldsymbol{B} d\boldsymbol{u}^e \tag{12.24}$$

where $E$ represents the tangent modulus of materials. By replacing the incremental operator with the differential operator in Eq. (12.23) and substituting Eq. (12.24) into this equation, the following element tangent equilibrium equations for immediate loads can be established (inclusion of time-dependent effects resulting from concrete creep, shrinkage, and tendon relaxation will be discussed in the next section):

$$d\boldsymbol{P}^e = (\boldsymbol{K}_1^e + \boldsymbol{K}_2^e) d\boldsymbol{u}^e \tag{12.25}$$

$$\boldsymbol{K}_1^e = \int_V \boldsymbol{B}^T E \boldsymbol{B} dV; \quad \boldsymbol{K}_2^e = \int_V \sigma \boldsymbol{J}^T \boldsymbol{J} dV \tag{12.26}$$

where $\boldsymbol{K}_1^e$ is the material stiffness matrix representing the material nonlinearity; and $\boldsymbol{K}_2^e$ is the geometric stiffness matrix representing the geometric nonlinearity or large-displacement effects. The forms of $\boldsymbol{K}_1^e$ and $\boldsymbol{K}_2^e$ are evaluated at the center of the element by employing a layered integration for the cross-section.

### 12.3.3 Method of time-dependent analysis

The total concrete strain at time $t$, $\varepsilon_c(t)$, may be expressed as follows:

$$\varepsilon_c(t) = \varepsilon_c^m(t) + \varepsilon_c^{cr}(t) + \varepsilon_c^{sh}(t) \tag{12.27}$$

where $\varepsilon_c^m(t)$, $\varepsilon_c^{cr}(t)$ and $\varepsilon_c^{sh}(t)$ are the mechanical strain, creep strain, and shrinkage strain, respectively.

The incremental stress–strain relationship for concrete for $\Delta t_n$ is

$$\Delta \sigma_n = E_c \Delta \varepsilon_c^m = E_c (\Delta \varepsilon_c - \Delta \varepsilon_c^{cr} - \Delta \varepsilon_c^{sh}) \tag{12.28}$$

where $E_c$ represents the tangent modulus for concrete; and $\Delta\varepsilon_c^m$ represents the mechanical strain increment. Substituting Eq. (12.4) into Eq. (12.28) and rearranging the equation yields

$$\Delta\sigma_n = \overline{E}_c(\Delta\varepsilon_c - \eta_n - \Delta\varepsilon_c^{sh}) \tag{12.29}$$

where

$$\overline{E}_c = \overline{E}_c(t_{n-1/2}) = \frac{E_c(t_{n-1/2})}{1 + C(t_n, t_{n-1/2})E_c(t_{n-1/2})} \tag{12.30}$$

$$\eta_n = \sum_{k=1}^{m}(1 - e^{-r_k\Delta t_n})\omega_{kn} \tag{12.31}$$

When the relaxation of prestressing tendons is considered, the incremental stress−strain relationship for bonded tendons is expressed by

$$\Delta\sigma_p = E_p\Delta\varepsilon_p + \Delta\sigma_{pr} \tag{12.32}$$

where $\Delta\sigma_p$ and $\Delta\varepsilon_p$ are the tendon stress increment and strain increment, respectively; $\Delta\sigma_{pr}$ is the tendon relaxation at a time interval; and $E_p$ is the tendon tangent modulus.

Substituting the incremental stress−strain equations for $\Delta\sigma$ in Eq. (12.23) and rearranging the resulting equation, the following time-dependent incremental control equations for an element can be established:

$$\Delta P^e + (\Delta P^e)^{cr} + (\Delta P^e)^{sh} + (\Delta P^e)^{pr} = (\overline{K}_1^e + K_2^e)\Delta u^e \tag{12.33}$$

$$(\Delta P^e)^{cr} = \int_{V1} B^T\overline{E}_c\eta_n d(V1) \tag{12.34a}$$

$$(\Delta P^e)^{sh} = \int_{V1} B^T\overline{E}_c d(V1)\Delta\varepsilon_c^{sh} \tag{12.34b}$$

$$(\Delta P^e)^{pr} = -\int_{V2} B^T\Delta\sigma_{pr} d(V2) \tag{12.34c}$$

where $(\Delta P^e)^{cr}$, $(\Delta P^e)^{sh}$, and $(\Delta P^e)^{pr}$ represent the equivalent load increments contributed by concrete creep, concrete shrinkage, and relaxation of bonded tendons, respectively. The integrations in Eq. (12.34a) and (12.34b) are performed over the entire concrete volume ($V1$) of an element, while the integration in Eq. (12.34c) is performed over the entire bonded tendon volume ($V2$) of an element. The material stiffness matrix $\overline{K}_1^e$ is a modification of $K_1^e$ in Eq. (12.26) by replacing the concrete modulus $E_c$ with $\overline{E}_c$, while the geometric stiffness matrix $K_2^e$ in Eq. (12.33) is the same as the one indicated in Eq. (12.26). After assembling the time-dependent equilibrium equations for a structure in the global coordinate system and applying appropriate boundary conditions, the traditional incremental-iterative method is used for the numerical solution of a time-dependent analysis.

### 12.3.4 Contribution of external tendons

The external tendons can be considered as an assemblage of tendon segments each of which spans a beam element, as shown in Fig. 12.4, in which $e_i$ and $e_j$ are the tendon eccentricities at nodes $i$ and $j$, respectively. For external tendon systems, the tendon eccentricities will change with member deformations, except at anchorage and deviator points. At each step, the location of each tendon segment can be determined according to the current coordinates of the anchorage and deviator points. The eccentricities $e_i$ and $e_j$ are then updated in terms of the coordinates of the tendon segment joints ($pi$ and $pj$) and the beam element nodes ($i$ and $j$), thus allowing the second-order effects to be considered in the numerical procedure. Meanwhile, the tendon strain increment is calculated by the elongation of the entire tendon, and then the tendon strain, stress, and force ($N_p$) can be obtained. When the tendon eccentricities and force are determined, the equivalent nodal loads resulting from external prestressing can be easily computed, as can be seen in Fig. 12.4. Details about the numerical treatment of external prestressing tendons have been described in Chapter 2. It should be noted that in the calculation of the tendon stress in members at long-term sustained loads, the prestress loss due to tendon relaxation should be considered.

## 12.4 Numerical examples

### 12.4.1 Bonded and unbonded prestressed concrete beams

Breckenridge and Bugg (1964) performed long-term tests on bonded and unbonded prestressed concrete beams under sustained loading of up to 8 years. The beams had I cross-section and were of 12.2 m span (12.8 m long), as illustrated in

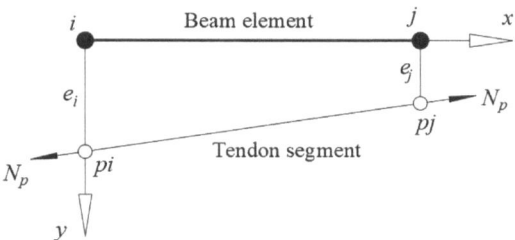

**Figure 12.4** Contribution of external prestressing to beam element.

Fig. 12.5. The bonded or unbonded tendons consisted of two 28.58-mm diameter high-strength steel bars having the modulus of elasticity and yield stress of 169 GPa and 896 MPa, respectively. The reinforcing steel consisted of eight #3 bars (bar area of 71.26 mm$^2$ each) having a modulus of elasticity of 200 GPa. At 8 days after concrete casting, the beams were posttensioned with an initial tendon stress of 683.8 MPa. At 22 days after the casting of concrete, loads $P$ were applied. Different levels of sustained loads were used in the tests. In this analysis, two service load conditions are considered, namely, $P = 0$ and 67.6 kN. The concrete modulus of elasticity at loading was 24.1 GPa.

Zhu (2009) proposed the following creep data for the preliminary design of large concrete dams: $m = 2$, $\phi_1(\tau) = \beta_1(1 + 9.2\tau^{-0.45})/E_0$, $r_1 = 0.3$, $\beta_1 = 0.23$, $\phi_2(\tau) = \beta_2(1 + 1.7\tau^{-0.45})/E_0$, $r_2 = 0.005$, $\beta_2 = 0.52$. $E_0$ is associated with the time-dependent modulus of elasticity, that is, $E_c(\tau) = E_0(1 - e^{-0.4\tau^{0.34}})$. In the present analysis, the creep data proposed by Zhu (2009) are modified, that is, $r_1 = 0.1$, $\beta_1 = 0.46$, and $\beta_2 = 1.04$, to capture the time-dependent behavior of the prestressed concrete beam specimens tested by Breckenridge and Bugg (1964). Using these creep data, the beam element discussed in Section 12.3 is used to numerically predict the long-term behavior of the prestressed concrete beam specimens. In the analysis, the self-weight of the specimens is converted into uniform load, and the span of the unbonded prestressed concrete specimens is divided into 20 beam elements.

Fig. 12.6 compares the predicted time-dependent deflections (using the original Zhu data and the modified creep data) with the experimental ones of two unbonded prestressed concrete beams for a period of 7 years. It is seen that when the creep data proposed by Zhu (2009) are used, the analysis considerably underestimates the long-term deflections. On the other hand, when the modified Zhu data are used in the analysis, the calculated and experimental values are shown to be in satisfactory agreement. Therefore the modified concrete creep data is used in the following time-dependent analysis.

To examine the influence of element mesh on the numerical results, three different element layouts are used in the time-dependent analysis of two bonded prestressed concrete beam specimens, namely, 12, 20, and 40 elements for the span. A comparison between numerical and test results in relation to the variation of

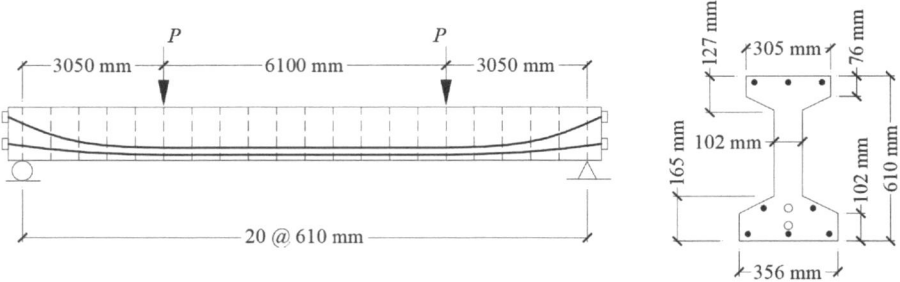

**Figure 12.5** Details of long-term tests for prestressed concrete beams.

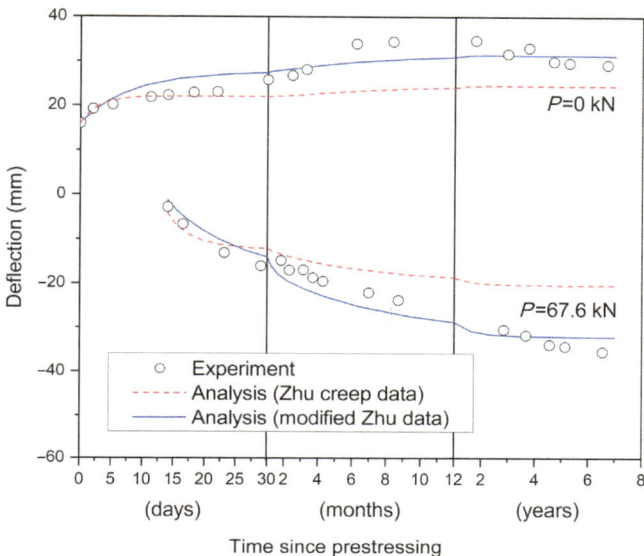

**Figure 12.6** Comparison of predicted long-term deflections with experimental data for unbonded prestressed concrete beam specimens.

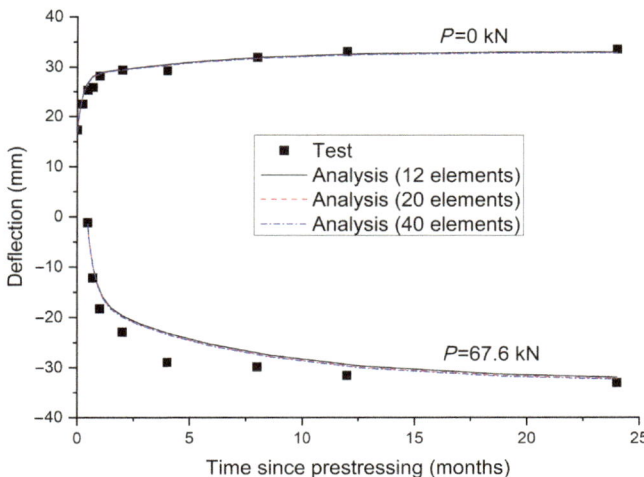

**Figure 12.7** Comparison between predicted deflections and test data for bonded prestressed concrete beam specimens.

midspan deflections with time up to 24 months is shown in Fig. 12.7. It is observed that the results of time-dependent analysis are insensitive to the element mesh. The discrepancy between the time-dependent deflections generated by using different element layouts is practically negligible. Therefore a mesh by adopting an element length of one to two times the cross-sectional depth would be sufficient for a time-dependent analysis. It is also seen that for the specimen without external loads, the proposed analysis reproduces the long-term camber with remarkable accuracy. For the specimen under sustained service loads, although the analysis leads to a bit underestimation of the long-term downward deflection, a satisfactory agreement between the numerical and experimental data can also be observed.

## 12.4.2 Prestressed concrete columns

Lin and Lakhwara (1966) tested two prestressed concrete slender columns up to failure to investigate their elastic and plastic behavior. These two columns are designated herein as PC-I and PC-II. The columns were identical in all aspects except for the eccentricity $e$ of the applied axial load: $e = 76.2$ mm for PC-I and 50.8 mm for PC-II. The columns were of a $152.4 \times 152.4$ mm square cross-section and were 3073.4 mm long, as shown in Fig. 12.8A. Each column was posttensioned with a 15.875 mm diameter high-strength Grade 160 rod having the ultimate tensile strength, yield stress, and elastic modulus of 1103.2 MPa, 896.3 MPa, and 165 GPa, respectively. The effective prestress was 786.6 MPa. The nonprestressed steel consisted of four 12.7 mm bars

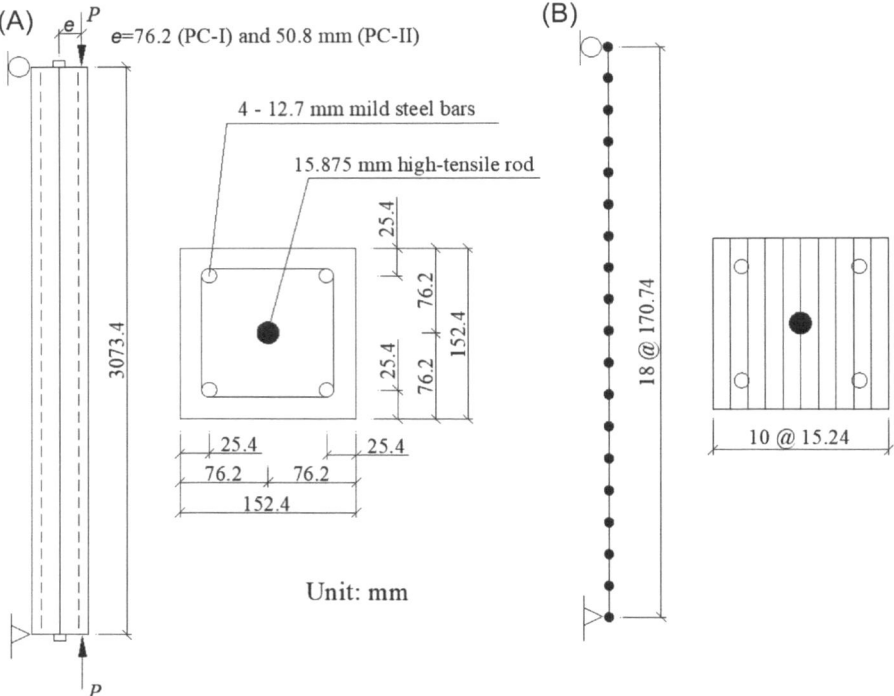

**Figure 12.8** Prestressed concrete test columns and finite element model: (A) column details; (B) finite element model.

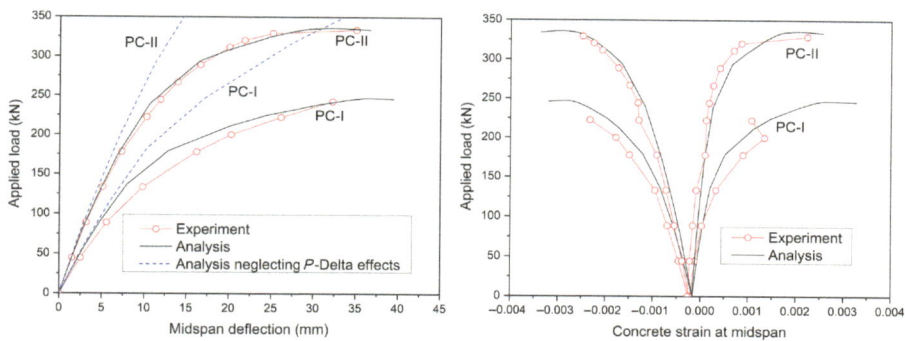

**Figure 12.9** Comparison between computational and experimental results.

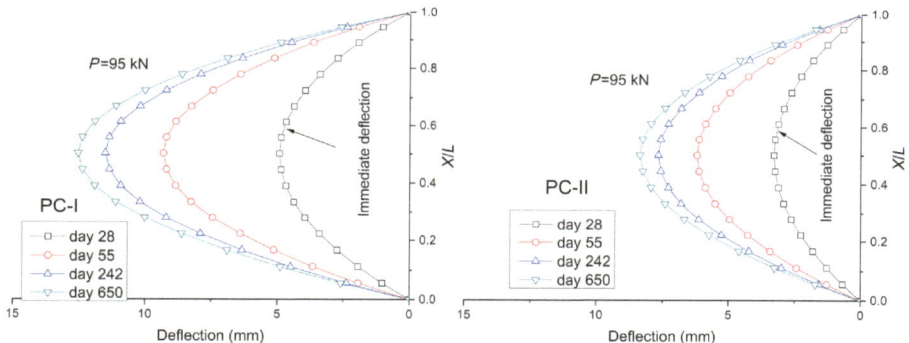

**Figure 12.10** Deflection curves at different times for the prestressed concrete slender columns.

having a yield strength of 351 MPa. The elastic modulus of nonprestressed steel is taken as 200 GPa. The concrete cylinder compressive strength was 42.1 MPa.

The finite element model of the columns is shown in Fig. 12.8B. Each column is divided into 18 elements having the same length of 170.74 mm. Each element is subdivided into 10 concrete layers, 2 nonprestressed steel layers each of which represents two bars at one side, and 1 tendon layer which represents the Grade 160 rod. Fig. 12.9 shows a comparison between the computational and experimental results regarding the load−deflection response and load versus concrete strain response. In general, a fairly good agreement between the model predictions and experimental data can be observed. The results of the analysis neglecting the *P*-Delta effects confirm the significance of the *P*-Delta effects in slender columns.

In the time-dependent modeling, it is assumed that the tendon was tensioned to the initial prestressing force and then anchored to the column at the concrete age of 28 days and that the external load was also applied at day 28. At an eccentric load of 95 kN, the lateral deflection curves at different times for the prestressed concrete columns are shown in Fig. 12.10, where *X/L* represents the ratio of the distance from a support to the span length. At day 28, there are initial lateral deflections produced by the eccentric load. Due to time-dependent effects, the deflections at days 55, 242, and 650 are increased to approximately 1.9, 2.3, and 2.5 times, respectively. As can be seen in Fig. 12.10, the long-term response characteristics for slender prestressed concrete columns under sustained eccentric loads are quite important, partly attributed to the interaction between *P*-Delta and time-dependent effects. To examine the impact of *P*-Delta effects on long-term behavior, a comparison between the midspan deflections obtained by time-dependent analyses considering and neglecting the *P*-Delta effects is shown in Fig. 12.11. It is seen that, when the *P*-Delta effects are ignored, the long-term deflection is obviously reduced. At day 650, neglect of *P*-Delta effects results in decreasing the time-dependent deflection by 15.1% for both columns.

## 12.4.3   Steel−concrete composite beams

Two steel−concrete composite beams, designated as B1 and B2 (Bradford & Gilbert, 1991), are used herein to calibrate the proposed time-dependent numerical model. Beam B1 was under a sustained uniform load of 7.52 kN/m in addition to its self-weight load, while B2 was only under its self-weight load. Both specimens were simply supported over a span of 5.9 m. The slab was 1000 × 70 mm, and the steel beam was made of 200 mm UB25.4. The dimension details are given in Fig. 12.12. The short-term concrete modulus of elasticity was 25.1 GPa. The free shrinkage strain at day 220

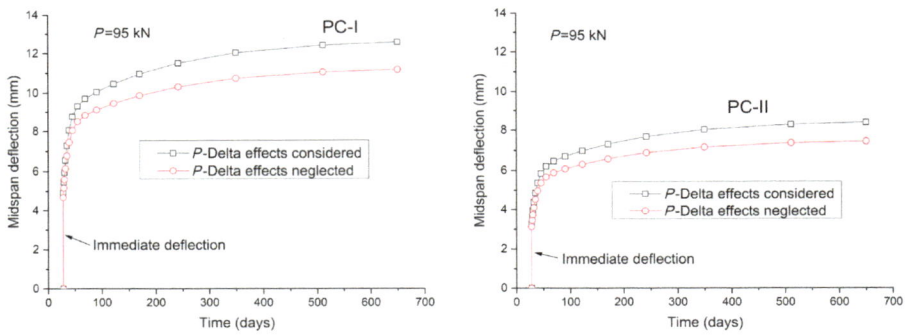

**Figure 12.11** Comparison between time-dependent deflections obtained by analyses considering and neglecting *P*-Delta effects.

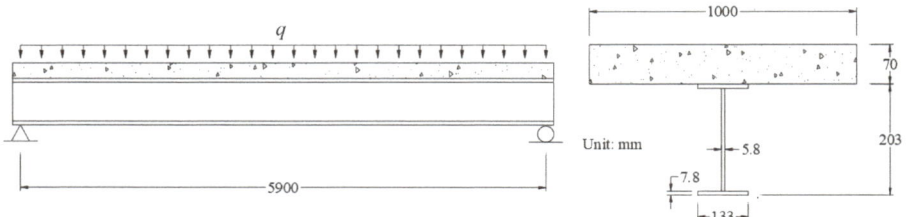

**Figure 12.12** Details of long-term tests for steel−concrete composite beam specimens.

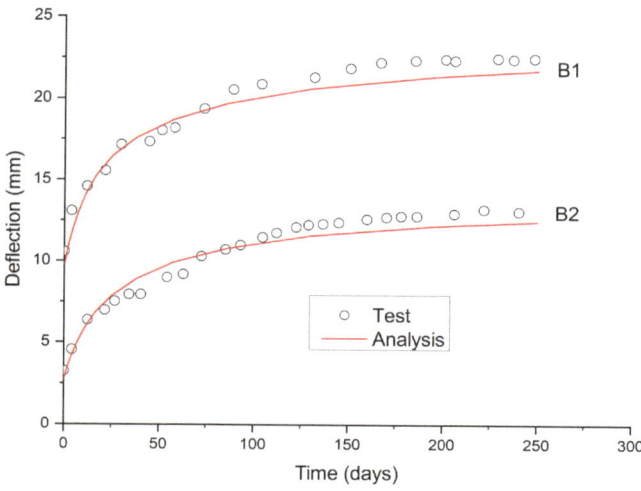

**Figure 12.13** Comparison of long-term deflections obtained by numerical analysis with experimental data for composite beams.

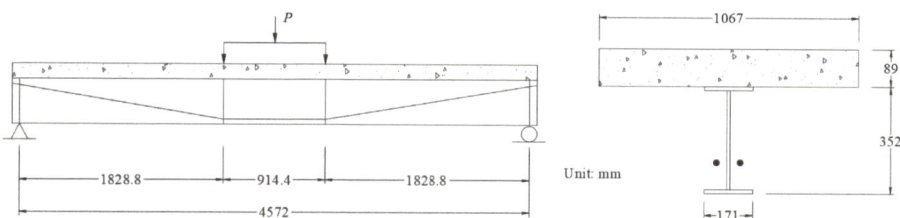

**Figure 12.14** Details of externally prestressed steel—concrete composite test beam.

was 410 $\mu\varepsilon$. The steel modulus of elasticity was 200 GPa. Fig. 12.13 shows a comparison of the predicted long-term deflections against the experimental data for the two composite beam specimens. Fairly good agreement between the numerical and experimental results can be observed, indicating that the proposed model can capture reasonably the time-dependent behavior of composite beams.

An externally prestressed steel—concrete composite beam tested by Ayyub et al. (1990) is also used to demonstrate the applicability of the time-dependent method. Details of the prestressed composite beam are shown in Fig. 12.14. The steel tendons were 32 mm below the top flange of the steel beam at the end anchorage points and 30 mm above the bottom flange at the deviator points. The tendon area and initial prestress were 279 mm$^2$ and 957 MPa, respectively. The thicknesses of the top flange, bottom flange, and web of the steel beam were 10, 10, and 7 mm, respectively. The material properties are as follows: $E_c = 36.0$ GPa; $E_p = 195$ GPa; $f_{py} = 1620$ MPa; $E_s = 200$ GPa. Assume external cables were stressed to the initial prestress and assembled to the composite beam at 28 days after the concrete cast.

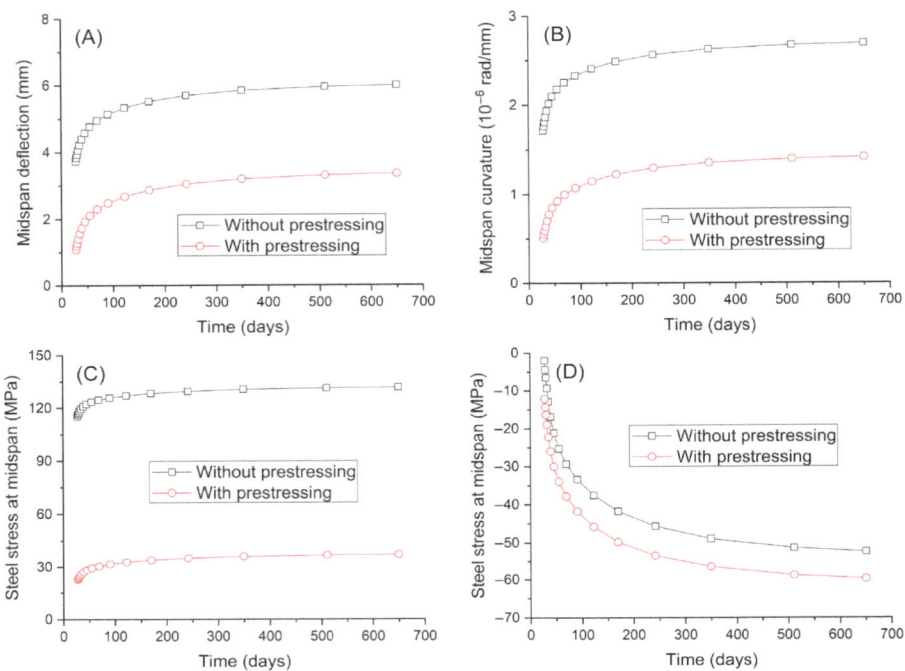

**Figure 12.15** Time-dependent behavior of composite beams with and without prestressing:
(A) deflection; (B) curvature; (C) steel stress in the bottom flange; (D) steel stress in the top flange.

The external loads were also assumed to be applied at day 28. The composite beam
without external prestressing is also analyzed for comparison.

Fig. 12.15 shows the variation of deflection, curvature, and steel stresses at the bottom
and top flanges with time for the composite beams under external loads of 120 kN. It is
seen that although the total (immediate plus time-dependent) deformation or steel stress
of the beam with external prestressing is significantly different from that of the beam
without external prestressing, the time-dependent response characteristics appear to be
almost identical for the beams with and without external prestressing. From day 28 to
day 650, the variations of midspan deflection, curvature, and steel stresses in the bottom
and top flanges due to time-dependent effects for the composite beam without external
prestressing are 2.26 mm, 0.98 rad/km, 16.23 MPa, and $-50.41$ MPa, respectively.
When strengthening of the composite beam with external prestressing, the corresponding
values are 2.27 mm, 0.91 rad/km, 14.26 MPa, and $-47.62$ MPa, respectively.

## 12.5  Conclusions

This chapter presents a numerical model for predicting the long-term behavior of
prestressed concrete and steel−concrete composite members at service load

conditions, taking into account the creep and shrinkage of concrete and the relaxation of prestressing tendons. The time-dependent finite element method is developed by applying the Euler—Bernoulli beam theory in combination with a linear creep law and a rational tendon relaxation model considering its interaction with other time-dependent effects. The equivalent nodal loads in the time-dependent control equations are contributed by concrete creep, concrete shrinkage, and tendon relaxation (if bonded tendons are used) in addition to the contribution made by the applied loads and external prestressing (if external tendons are used). The proposed model is able to simulate effectively the time-dependent effects in prestressed concrete and steel—concrete composite members, taking into consideration the interaction between these effects.

Some numerical examples are presented to illustrate the reliability and applicability of the proposed model. The numerical predictions using the modified Zhu creep data showed good agreement with the experimental results. It should be noted that the results of the proposed time-dependent analysis are highly dependent on the material models. The empirical parameters used in these models are of critical importance to the accuracy of the analysis. Extensive experimental and numerical work should be carried out to calibrate the reliability of these parameters.

In the following chapters (Chapters 13 and 14), the time-dependent numerical method presented in this chapter is applied to modeling the long-term behavior of prestressed concrete members with FRP or steel tendons and steel—concrete composite girders.

# References

Au, F. T. K., & Si, X. T. (2011). Accurate time-dependent analysis of concrete bridges considering concrete creep, concrete shrinkage and cable relaxation. *Engineering Structures*, *33* (1), 118—126. Available from https://doi.org/10.1016/j.engstruct.2010.09.024.

Ayyub, B. M., Sohn, Y. G., & Saadatmanesh, H. (1990). Prestressed composite girders under positive moment. *Journal of Structural Engineering*, *116*(11), 2931—2951. Available from https://doi.org/10.1061/(asce)0733-9445(1990)116:11(2931).

Bažant, Z. P. (2001). Prediction of concrete creep and shrinkage: past, present and future. *Nuclear Engineering and Design*, *203*(1), 27—38. Available from https://doi.org/10.1016/s0029-5493(00)00299-5.

Bažant, Z. P., & Najjar, L. J. (1973). Comparison of approximate linear methods for concrete creep. *Journal of the Structural Division*, *99*(9), 1851—1874. Available from https://doi.org/10.1061/jsdeag.0003602.

Bradford, M. A., & Gilbert, R. I. (1991). Time-dependent behaviour of simply-supported steel-concrete composite beams. *Magazine of Concrete Research*, *43*(157), 265—274. Available from https://doi.org/10.1680/macr.1991.43.157.265.

Breckenridge, R. A., & Bugg, S. L. (1964). Effects of long-time loads on prestressed concrete beams. *PCI Journal*, *9*(6), 75—89. Available from https://doi.org/10.15554/pcij.12011964.75.89.

Fan, J., Nie, J., Li, Q., & Wang, H. (2010). Long-term behavior of composite beams under positive and negative bending. I: Experimental study. *Journal of Structural Engineering*, *136*(7), 849—857. Available from https://doi.org/10.1061/(asce)st.1943-541x.0000175.

Ghali, A., Favre, R., & Elbadry, M. (2002). *Concrete structures: stresses and deformation* (3rd). London and New York: Spon Press.

Gutiérrez, S. E., Cudmani, R. O., & Danesi, R. F. (1996). Time-dependent analysis of reinforced and prestressed concrete members. *ACI Structural Journal*, *93*(4), 420–427.

Hamed, E., & Bradford, M. A. (2012). Flexural time-dependent cracking and post-cracking behaviour of FRP strengthened concrete beams. *International Journal of Solids and Structures*, *49* (13), 1595–1607. Available from https://doi.org/10.1016/j.ijsolstr.2012.03.001.

Hosseini, M., & Jefferson, A. D. (1998). Time-dependent behaviour of widened reinforced concrete under-bridge. *Materials and Structures*, *31*(10), 714–719. Available from https://doi.org/10.1007/bf02480449.

Kwak, H. G., & Seo, Y. J. (2002). Numerical analysis of time-dependent behavior of pre-cast pre-stressed concrete girder bridges. *Construction and Building Materials*, *16*(1), 49–63. Available from https://doi.org/10.1016/S0950-0618(01)00027-7.

Lin, T. Y., & Lakhwara, T. R. (1966). Ultimate strength of eccentrically loaded partially prestressed columns. *PCI Journal*, *11*(3), 37–49. Available from https://doi.org/10.15554/pcij.06011966.37.49.

Lopes, A. V., & Lopes, S. M. R. (2012). Importance of a rigorous evaluation of the cracking moment in RC beams and slabs. *Computers and Concrete*, *9*(4), 275–291. Available from https://doi.org/10.12989/cac.2012.9.4.275.

Lou, T., Lopes, S. M. R., & Lopes, A. V. (2015). FE analysis of short- and long-term behavior of simply supported slender prestressed concrete columns under eccentric end axial loads causing uniaxial bending. *Engineering Structures*, *85*, 52–62. Available from https://doi.org/10.1016/j.engstruct.2014.12.023.

Lou, T., Lopes, S. M. R., & Lopes, A. V. (2016). Numerical modeling of externally prestressed steel-concrete composite beams. *Journal of Constructional Steel Research*, *121*, 229–236. Available from https://doi.org/10.1016/j.jcsr.2016.02.008.

Magura, D. D., Sozen, M. A., & Siess, C. P. (1964). A study of stress relaxation in prestressing reinforcement. *PCI Journal*, *9*(2), 13–57. Available from https://doi.org/10.15554/pcij.04011964.13.57.

Rodriguez-Gutierrez, J. A., & Aristizabal-Ochoa, J. D. (2007). Short- and long-term deflections in reinforced, prestressed, and composite concrete beams. *Journal of Structural Engineering*, *133*(4), 495–506. Available from https://doi.org/10.1061/(ASCE)0733-9445(2007)133:4(495).

Zhu, B. F. (2009). *The finite element method theory and application*. Beijing: China Water Power Press.

Zienkiewicz, O. C., & Watson, M. (1966). Some creep effects in stress analysis with particular reference to concrete pressure vessels. *Nuclear Engineering and Design*, *4*(4), 406–412. Available from https://doi.org/10.1016/0029-5493(66)90069-0.

# Long-term behavior of prestressed concrete members with FRP/steel tendons

## 13.1 Introduction

The mechanical properties of prestressed concrete members degrade with time as a consequence of time-dependent effects, especially when fiber-reinforced polymer (FRP) tendons are used since their relaxation is rather pronounced (Saadatmanesh & Tannous, 1999a, 1999b). Therefore a rigorous time-dependent evaluation of prestressed concrete members with FRP tendons would be of practical importance. While a great number of works have been conducted to evaluate the long-term behavior of prestressed concrete members with steel tendons (Au & Si, 2011; Huang et al., 2018; Rodriguez-Gutierrez & Aristizabal-Ochoa, 2007; Xue et al., 2016), there are relatively few studies on the time-dependent assessment of prestressed concrete members with FRP tendons. Pisani (2000a, 2000b) proposed both general and simplified approaches for the time-dependent analysis of concrete beams with bonded aramid FRP (AFRP) tendons. Youakim and Karbhari (2007) performed a comparative study on the time-dependent behavior of prestressed concrete beams using either FRP or steel tendons. They concluded that the long-term prestress loss in FRP tendons is much lower than that in steel tendons due to a lower modulus of elasticity of FRP composites.

The deflection control at long-term sustained loads is an important task for designers to meet the serviceability requirement. It is necessary to develop a simplified equation for calculating the long-term deflection of prestressed concrete members with FRP tendons for design purposes. ACI 318-19 (ACI Committee 318, 2019) recommends an equation that adopts the compressive reinforcement ratio as a key parameter for predicting the long-term deflection of conventional concrete members. This equation, however, may not be applied to prestressed concrete members with FRP or steel tendons as it neglects the prestress-related parameters that may have a notable impact on the long-term deflection.

The second-order effects are known to be the main difference that distinguishes an external tendon member from an internal unbonded tendon one and shall be minimized by attaching external tendons to the concrete member to maintain desired tendon eccentricities. When an externally prestressed concrete member is subjected to long-term sustained loads, the second-order effects caused by time-dependent effects can lead to a change in the structural responses including the material stresses, which in turn would affect the time-dependent effects, particularly the creep effects since creep of concrete is highly dependent on the imposed stress

Prestressed Members with External Fiber-Reinforced Polymer (FRP) Tendons. DOI: https://doi.org/10.1016/B978-0-443-23877-2.00013-4

(Bažant, 2001). The interaction between time-dependent and second-order effects should be fully taken into account in the time-dependent design of externally prestressed concrete members.

This chapter presents a comprehensive assessment on the time-dependent performance of concrete members prestressed with FRP or steel tendons (Lou et al., 2016; Lou & Karavasilis, 2018). Based on the results of the numerical assessment, a simplified equation for calculating the long-term deflection of prestressed concrete members with FRP or steel tendons is proposed for practical purposes. In addition, the time-dependent second-order effects of externally prestressed concrete members are discussed (Lou et al., 2015).

## 13.2 Relaxation models for prestressing steel and FRP tendons

Creep of prestressing tendons is usually evaluated by relaxation tests. When a tendon is stretched and a constant strain is maintained, the tendon stress is subject to loss due to creep effects, referred to as intrinsic relaxation $\sigma_{pr}$. An equation that has been widely used for the intrinsic relaxation of prestressing steel is as follows (Magura et al., 1964):

$$\frac{\sigma_{pr}}{\sigma_{p0}} = -\frac{\log(\tau - t_0)}{k_r}\left(\frac{\sigma_{p0}}{f_{py}} - 0.55\right) \tag{13.1}$$

in which $(\tau - t_0)$ is the relaxation time in hours; $\sigma_{p0}$ is the initial tendon stress at which the relaxation starts; $f_{py}$ is the yield stress of prestressing steel; and $k_r$ is a coefficient, equal to 10 for normal-relaxation steel tendons and 30 for low-relaxation steel tendons.

On the other hand, the relaxation of FRP tendons is very different from that of steel tendons. Saadatmanesh and Tannous (1999a, 1999b) conducted a set of experimental studies to explore the relaxation behavior of AFRP and carbon FRP (CFRP) tendons. Test results of 12 specimens in the air at temperatures of $-30$, 25, and 60°C as well as 24 specimens in three different solutions (i.e., alkaline, acidic, and salt) at temperatures of 25°C and 60°C were reported for each type of tendons. Based on their test data, Saadatmanesh and Tannous (1999a, 1999b) proposed an equation for predicting the intrinsic relaxation of FRP tendons.

$$\frac{\sigma_{pr}}{\sigma_{p0}} = -\frac{\lambda - [a - b\log(\tau - t_0)]}{\lambda} \tag{13.2}$$

in which $\lambda = \sigma_{p0}/f_{pu}$ where $f_{pu}$ is the ultimate tensile strength of prestressing tendons; $a$ and $b$ are constants determined from test data by the regression analysis. The values of $a$ and $b$ depend on the initial prestress level, temperature, and the solution type. For prestressing AFRP in the air at a temperature of 25°C, the values of $a$ and $b$ are 0.37 and 0.0058, respectively, for the initial prestress level of 40%, and are 0.5684 and 0.0145, respectively, for the initial prestress level of 60%. For (Leadline) CFRP tendons in the air at a temperature of 25°C, $a$ is equal to 0.3846 and $b$ is equal to 0.0046 at 40% initial prestress level, while $a$ is equal to 0.5546 and $b$ is equal to 0.0063 at 60% initial prestress level (Saadatmanesh & Tannous, 1999a, 1999b).

In prestressed concrete members, creep and shrinkage of concrete would interact with the relaxation of prestressing tendons and this interaction would lead to an additional stress loss in tendons. Therefore when computing tendon relaxation of prestressed concrete members, the value of $\sigma_{p0}$ in Eqs. (13.1) and (13.2) should be appropriately modified at each time interval in accordance with the change in tendon stress due to concrete creep and shrinkage as well as some other causes such as prestress transfer and load application. The procedure is explained in Chapter 12.

## 13.3    Prestressed concrete members with bonded AFRP/steel tendons

### 13.3.1    Beam details

A reference beam is designed here as shown in Fig. 13.1A to evaluate the long-term behavior of concrete members prestressed with bonded AFRP tendons. The beam is simply supported over 12 m and has a rectangular cross-section with 300 mm in width and 600 mm in depth. The tendons are assumed to be a parabolic profile, with eccentricities at supports and midspan of 0 and 200 mm, respectively. The tendon area $A_p$ is 10 cm$^2$. The main investigation variables include the type of tendons (AFRP and steel tendons), the initial prestress level (929.6 and

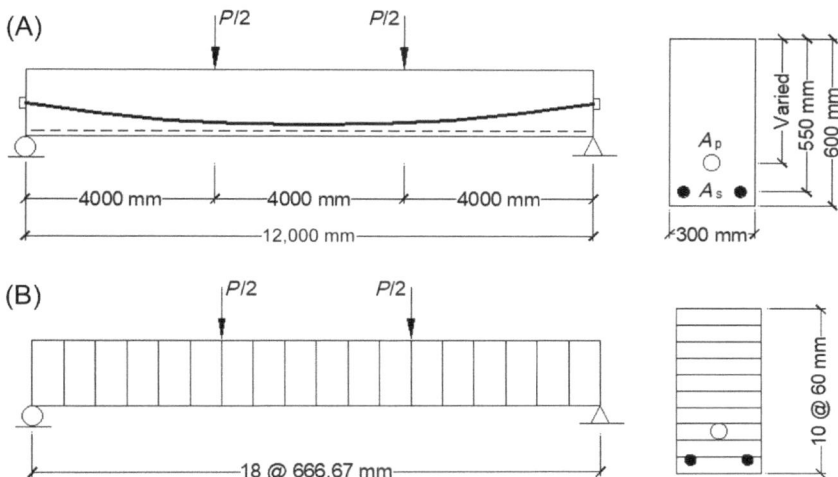

**Figure 13.1** Details of bonded prestressed concrete beams and the finite element model: (A) beam details; (B) finite element model.

1394.4 MPa), the load level (0 and 100 kN) and the amount of bottom nonprestressed steel ($A_s = 0$, 6, and 12 cm$^2$).

The material properties of AFRP tendons are the same as those of specimens
used by Saadatmanesh and Tannous (1999a), that is, the tensile strength, elastic
modulus, and ultimate tensile strain of AFRP tendons are 2324 MPa, 120.7 GPa,
and 1.93%, respectively. The steel tendons are assumed to be low-relaxation strands
with a yield stress and an elastic modulus of 1674 MPa and 195 GPa, respectively.
Two levels of initial prestress are considered, namely, 929.6 and 1394.4 MPa
(i.e., 40% and 60% of the tensile strength of AFRP tendons as used in the tests by
Saadatmanesh and Tannous (1999a). The elastic modulus of nonprestressed steel is
200 GPa. The concrete cylinder compressive strength and elastic modulus at the
age of 28 days are 35 MPa and 34 GPa, respectively. Both prestress transfer and
load application are assumed to take place on day 28.

In the finite element idealization, the beam is divided into 18 beam elements and
each element is subdivided into 10 concrete layers, 1 tendon layer, and 1 nonprestressed steel layer (if any), as shown in Fig. 13.1B. In the analysis, the self-weight
of the member is converted into the uniformly distributed load.

### 13.3.2   Effect of using AFRP tendons instead of steel tendons

Figs. 13.2 and 13.3 show the variation of tendon stress and deflection at midspan
with time for AFRP and steel prestressed concrete members without any external

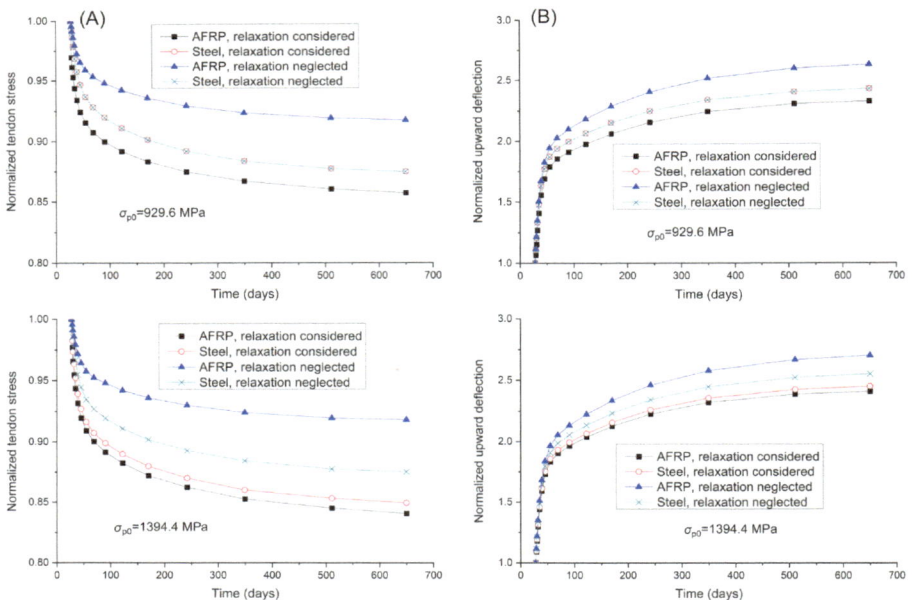

**Figure 13.2** Time-dependent performance of AFRP and steel prestressed concrete beams
without external loads: (A) normalized tendon stress; (B) normalized deflection.

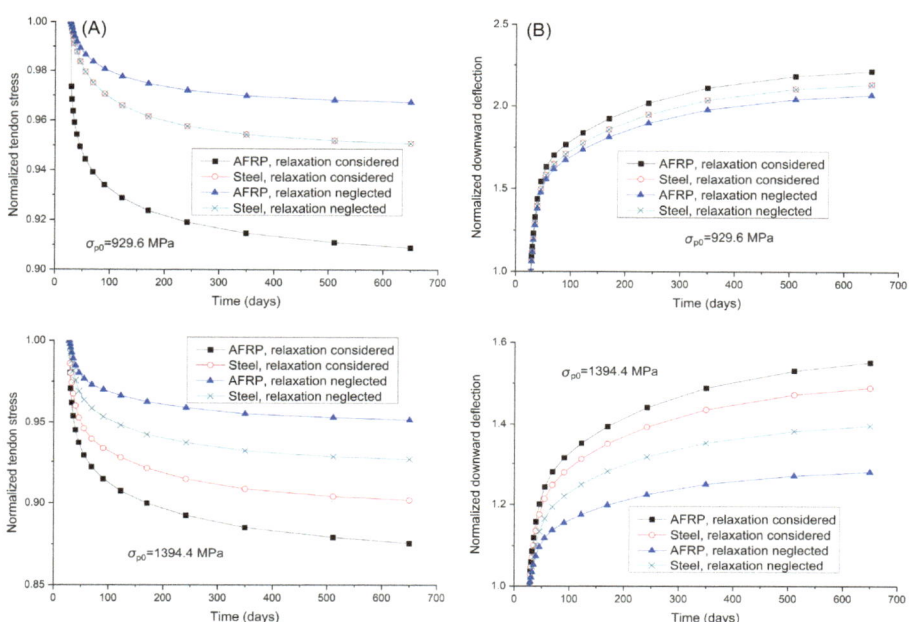

**Figure 13.3** Time-dependent performance of AFRP and steel prestressed concrete beams under external loads of 100 kN: (A) normalized tendon stress; (B) normalized deflection.

loads and with external loads of 100 kN, respectively. The results were produced by considering fully prestressed concrete members (i.e., $A_s = 0$ cm$^2$) and are displayed by using normalized values, that is, long-term response divided by immediate response. To demonstrate the relaxation effect of AFRP and steel tendons, the results of the analysis both considering and neglecting tendon relaxation are illustrated. At an initial prestress of 929.6 MPa, the stress in steel tendons immediately after prestress transfer is 881.3 MPa (this value is used at zero loads as initial stress at which relaxation starts) and, when external loads of 100 kN are applied, the tendon stress goes to 920.8 MPa (this value is used at service loads as initial stress at which relaxation starts). According to Eq. (13.1), the relaxation for steel tendons is negligible. Therefore at this initial prestress level, the time-dependent responses of the steel prestressed concrete member obtained from the analysis considering and neglecting tendon relaxation are identical, as shown in the graphs of the figures.

When the tendon relaxation is neglected, the prestress loss in AFRP tendons due to concrete creep and shrinkage is much smaller than that in steel tendons as a result of a lower modulus of elasticity. Therefore after a period of time, AFRP tendons register a significantly higher stress than steel tendons, as can be seen in Figs. 13.2A and 13.3A. Since prestressing leads to an upward deflection, the AFRP prestressed concrete member exhibits an obviously higher camber at zero external loads but a lower downward deflection at external loads of 100 kN when compared to the steel prestressed concrete member, as can be seen in Figs. 13.2B and 13.3B.

At a given level of initial prestress, the stress relaxation of AFRP tendons is significantly more pronounced than that of steel tendons. As a result, the prestress loss in AFRP tendons due to concrete creep, shrinkage, and tendon relaxation turns out to be higher than that in steel tendons although the prestress loss contributed by concrete creep and shrinkage is significantly lower in AFRP tendons than in steel tendons. Therefore when tendon relaxation is considered, AFRP tendons register a smaller long-term stress than steel tendons as illustrated in Figs. 13.2A and 13.3A. This phenomenon is particularly apparent for a lower initial prestress and when external loads are applied. Consequently, the AFRP prestressed concrete member exhibits a lower camber at zero loads but a higher downward deflection at applied loads of 100 kN compared to the steel one, as shown in Figs. 13.2B and 13.3B. At day 650, a list of the normalized tendon stresses and deflections for AFRP and steel prestressed concrete members at different load and initial prestress levels is given in Table 13.1.

### 13.3.3   Effect of bottom nonprestressed steel

In a prestressed concrete member with FRP tendons, a certain amount of nonprestressed steel is commonly provided to improve the ductile capacity, crack pattern, and failure mode of the member. Three different amounts of bottom nonprestressed steel are considered here, namely, $A_s = 0$, 6, and 12 cm$^2$.

Fig. 13.4 shows some typical aspects (i.e., midspan deflection, axial shortening, concrete strain in extreme compressive fiber and prestress loss at midspan) of time-dependent behavior of the fully ($A_s = 0$ cm$^2$) and partially ($A_s = 6$ and 12 cm$^2$) prestressed concrete members with AFRP tendons and under zero applied loads, while the results for external loads of 100 kN are presented in Fig. 13.5. The results were produced by considering an initial prestress of 929.6 MPa. It is seen that the effect of bottom nonprestressed steel on the long-term behavior is dependent on the load level. At zero applied loads, the bottom nonprestressed steel is subjected to compression due to prestressing effects. In this circumstance, the long-term behavior of the member is heavily influenced by the amount of bottom nonprestressed steel.

**Table 13.1** Normalized tendon stresses and deflections for AFRP and steel prestressed concrete beams at day 650.

| Load (kN) | Initial prestress (MPa) | Normalized tendon stress | | | Normalized deflection | | |
|---|---|---|---|---|---|---|---|
| | | **AFRP** | **Steel** | **Ratio** | **AFRP** | **Steel** | **Ratio** |
| 0 | 929.6 | 0.857 | 0.875 | 0.979 | 2.325 | 2.426 | 0.958 |
| | 1394.4 | 0.841 | 0.849 | 0.991 | 2.409 | 2.450 | 0.983 |
| 100 | 929.6 | 0.909 | 0.951 | 0.956 | 2.216 | 2.135 | 1.038 |
| | 1394.4 | 0.876 | 0.902 | 0.971 | 1.552 | 1.489 | 1.042 |

*Note*: Ratio, ratio of the value for AFRP tendons to that for steel tendons.

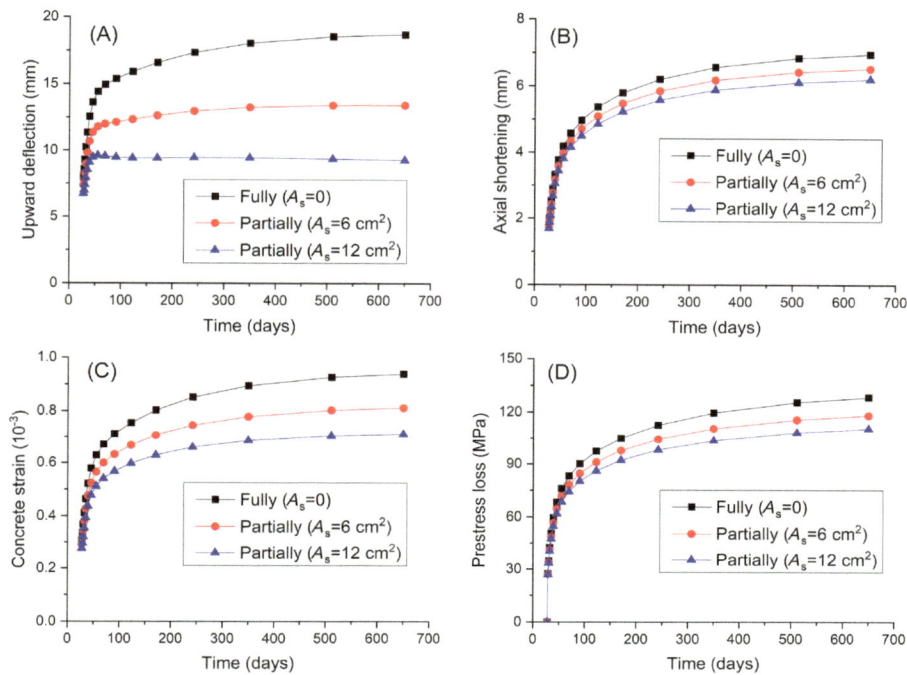

**Figure 13.4** Time-dependent performance of fully and partially AFRP prestressed concrete beams without external loads: (A) upward deflection; (B) axial shortening; (C) concrete strain in extreme compressive fiber; (D) prestress loss.

The partially prestressed concrete members exhibit significantly lower time-dependent deflection and concrete strain as well as obviously smaller time-dependent axial shortening and prestress loss, when compared to the fully prestressed concrete member. At day 650, the time-dependent deflection, axial shortening, concrete strain, and prestress loss of the partially prestressed concrete member with $A_s$ of 12 cm$^2$ are 75.9%, 12.5%, 30.7%, and 13.8% lower than those of the fully prestressed concrete member, respectively.

On the other hand, at applied loads of 100 kN, the bottom nonprestressed steel is subjected to tension. In this case, the discrepancy between long-term response characteristics for fully and partially prestressed concrete members is not so important. The concrete strains due to time-dependent effects for fully and partially prestressed concrete members are almost identical. The partially prestressed concrete members display higher time-dependent deflection but slightly lower axial shortening and prestress loss than the fully prestressed concrete members. At day 650, increasing the amount of nonprestressed steel from 0 to 12 cm$^2$ leads to an increase in time-dependent deflection by 10.2% and to decreases in axial shortening and prestress loss by 3.56% and 3.69%, respectively.

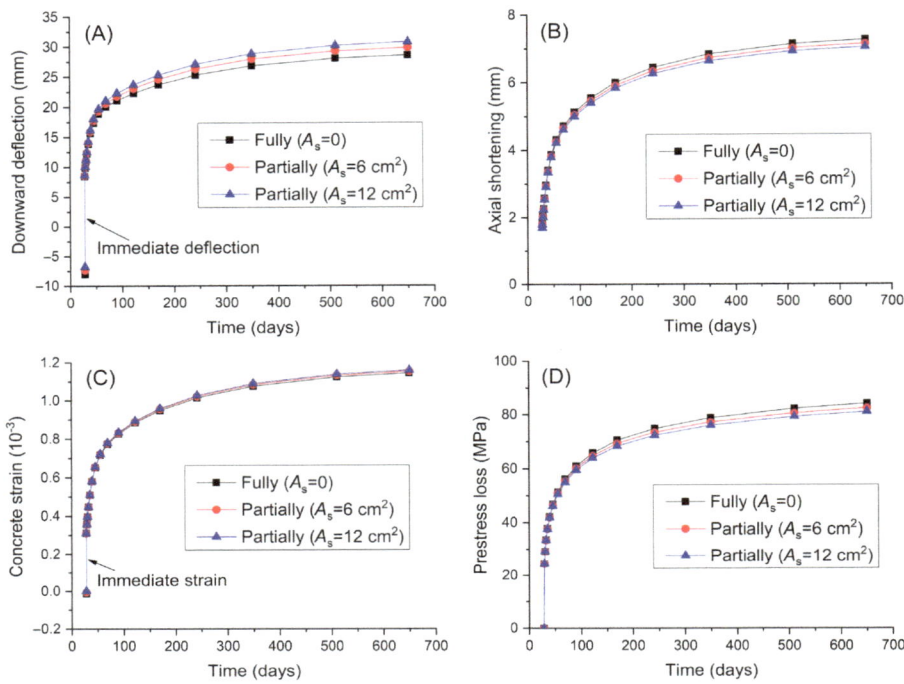

**Figure 13.5** Time-dependent performance of fully and partially AFRP prestressed concrete beams under external loads of 100 kN: (A) downward deflection; (B) axial shortening; (C) concrete strain in extreme compressive fiber; (D) prestress loss.

## 13.4  Prestressed concrete members with unbonded CFRP/steel tendons

### 13.4.1  Beam details

Numerical work is conducted to assess the long-term behavior of concrete beams internally prestressed with unbonded CFRP tendons. The beams have a rectangular cross-section ($300 \times 600$ mm) and are simply supported over a span of 12 m, as shown in Fig. 13.6A. The profile of unbonded tendons is assumed to be parabolic. The tendon eccentricities over the end support ($e_0$) and midspan ($e_1$) are equal to 0 mm and 200 mm, respectively. The area of the prestressing tendons, $A_p$, is equal to 10 cm$^2$, of the tensile reinforcing steel, $A_s$, is equal to 7.2 cm$^2$, and of the compressive reinforcing steel, $A_s'$, is equal to 3.6 cm$^2$. Two load conditions are considered, that is, self-weight and $P = 0$ kN (i.e., members without external loads) or self-weight and $P = 100$ kN (i.e., members with service external loads), where $P$ are the third point loads acting on the members.

The unbonded tendons are made of either CFRP composites or low-relaxation prestressing steel to allow a meaningful comparison to be made. CFRP tendons are assumed to be the same as the Leadline tendon specimens tested by Saadatmanesh and Tannous (1999b).

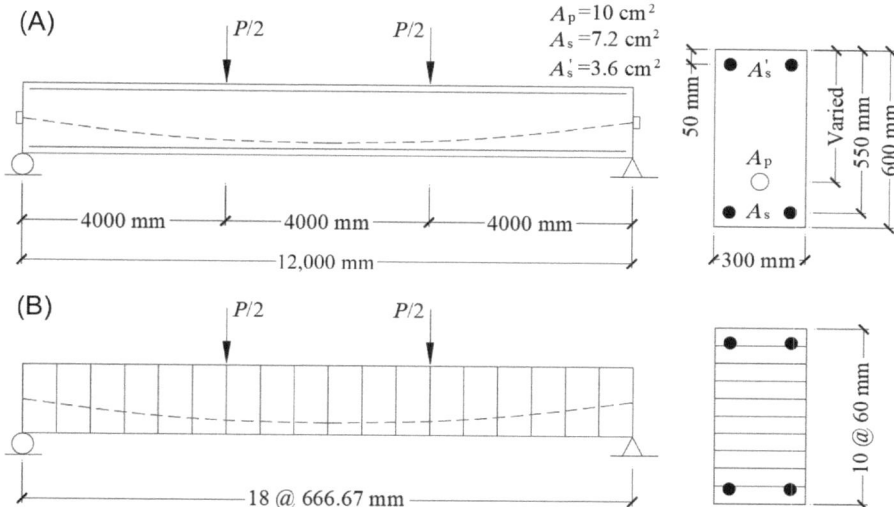

**Figure 13.6** Details of unbonded prestressed concrete beams and the finite element model: (A) beam details; (B) finite element model.

**Table 13.2** Mechanical properties of CFRP and steel tendons.

| Tendon | Tensile strength (MPa) | Yield stress (MPa) | Elastic modulus (GPa) | Rupture strain (%) |
|--------|------------------------|--------------------|-----------------------|--------------------|
| CFRP | 1999.2 | – | 149.6 | 1.34 |
| Steel | 1860 | 1674 | 195 | >3.5 |

The mechanical properties of CFRP and steel tendons are given in Table 13.2. Two initial prestress values are considered, that is, 800 MPa and 1200 MPa. The yield strength and elastic modulus of the reinforcing steel are equal to 450 MPa and 200 GPa, respectively. The cylinder compressive strength and elastic modulus of concrete at day 28 (assumed to be the day of prestress transfer and loading) are equal to 35 MPa and 34 GPa, respectively. The finite element model of the member is illustrated in Fig. 13.6B.

## 13.4.2 Long-term behavior due to concrete creep and concrete shrinkage

A numerical analysis neglecting the relaxation of prestressing tendons is conducted first to isolate the contribution of concrete creep and shrinkage to the time-dependent behavior. Fig. 13.7 shows the long-term prestress loss in CFRP and steel tendons caused by concrete creep and shrinkage for unbonded members with different initial prestress levels and load conditions. Fig. 13.7A and B shows that the prestress loss initially increases quickly and then slows down with time. For a given tendon type and load condition, a higher initial prestress leads to a significantly higher prestress loss. For a given tendon type and initial prestress, the member with service external loads experiences a lower prestress loss than that without external loads. Because the elastic modulus of CFRP tendons (149.6 GPa) is lower than that of steel ones (195 GPa), CFRP tendons exhibit considerably lower prestress loss than steel ones when tendon relaxation is ignored in the numerical analysis. The ratio of the long-term prestress loss in CFRP tendons to that in steel ones ($\sigma_{loss}^{CFRP}/\sigma_{loss}^{steel}$) is around 0.8 over time, independently of the initial prestress and load condition, as displayed in Fig. 13.7C.

The long-term deflection and stress in bottom reinforcing steel at midspan caused by concrete creep and shrinkage are shown in Fig. 13.8. It is seen that at zero external loads, as a result of the lower prestress loss, CFRP tendons result in higher long-term camber and higher compressive stress in the bottom reinforcing steel compared to steel ones. Under service loading ($P = 100$ kN), CFRP tendons lead to lower downward deflection than steel ones, particularly in the case of an initial prestress of 800 MPa. Just after the application of external loads, the bottom reinforcing steel is in tension for an initial prestress of 800 MPa, while it is in compression for an initial prestress of 1200 MPa. Due to the effects of concrete creep and shrinkage, the tensile stress (if any) gradually disappears and compressive stress

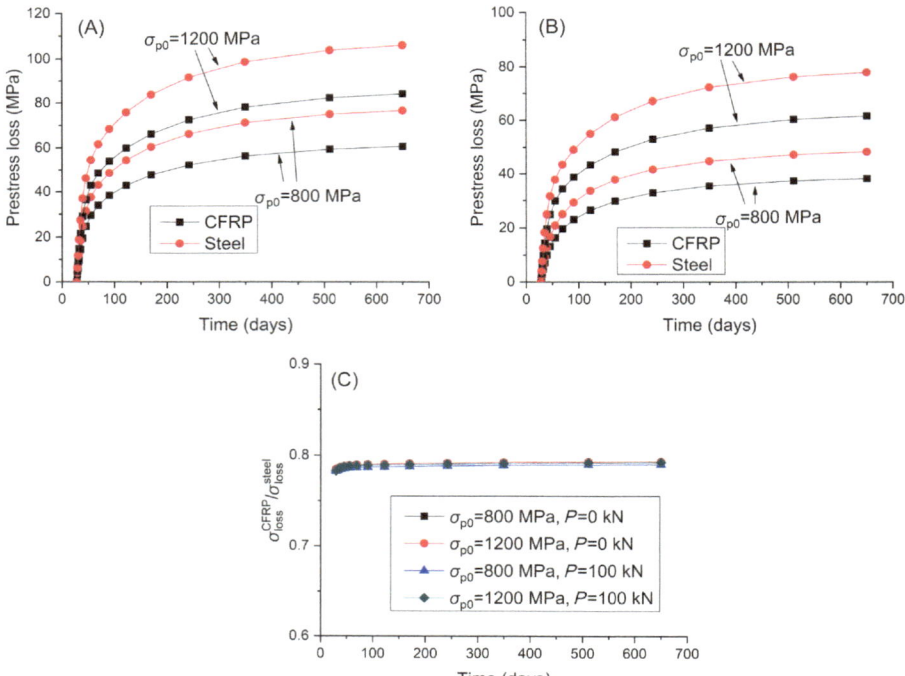

**Figure 13.7** Long-term prestress loss in unbonded CFRP and steel tendons caused by concrete creep and shrinkage: (A) beams without external loads ($P = 0$ kN); (B) beams with service external loads ($P = 100$ kN); (C) ratio of prestress loss in CFRP tendons to that in steel ones.

develops. The long-term compressive stress in the bottom reinforcing steel of members with CFRP tendons is higher than that of members with steel tendons. The latter observation is more pronounced for the initial prestress of 1200 MPa.

### 13.4.3  Long-term behavior due to concrete creep, concrete shrinkage, and tendon relaxation

Stress relaxation of CFRP tendons is significantly higher than that of steel ones. Fig. 13.9 shows the long-term prestress loss in CFRP and steel tendons caused by concrete creep, concrete shrinkage, and tendon relaxation for members with different initial prestress levels and load conditions. In particular, as can be observed in Fig. 13.9A and B, CFRP tendons exhibit much higher prestress loss due to concrete creep, shrinkage, and tendon relaxation despite the smaller modulus of elasticity of CFRP composites when compared to steel tendons. It is seen in Fig. 13.9C that the $\sigma_{\text{loss}}^{\text{CFRP}}/\sigma_{\text{loss}}^{\text{steel}}$ ratio decreases quickly as time passes and tends to stabilize several months later. The initial prestress does not affect the $\sigma_{\text{loss}}^{\text{CFRP}}/\sigma_{\text{loss}}^{\text{steel}}$ ratio in the case of without external loads, while in the case of external loads of 100 kN, a higher prestress level results in lower $\sigma_{\text{loss}}^{\text{CFRP}}/\sigma_{\text{loss}}^{\text{steel}}$ ratio. At a given prestress level, unbonded prestressed concrete members without external loads exhibit a substantially lower $\sigma_{\text{loss}}^{\text{CFRP}}/\sigma_{\text{loss}}^{\text{steel}}$ ratio compared to those with external loads, especially for the initial prestress of 800 MPa.

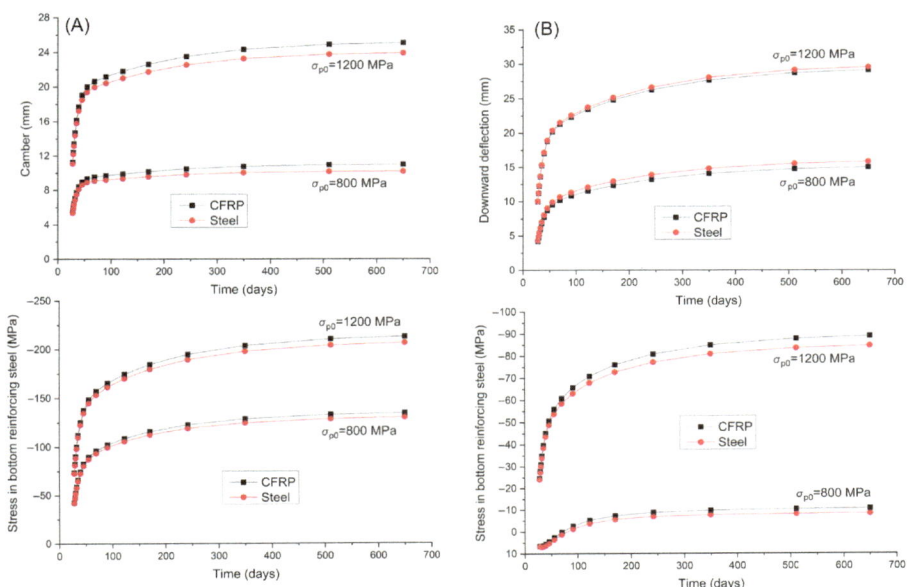

**Figure 13.8** Long-term deflection and stress in bottom reinforcing steel caused by concrete creep and shrinkage: (A) beams without external loads ($P = 0$ kN); (B) beams with service external loads ($P = 100$ kN).

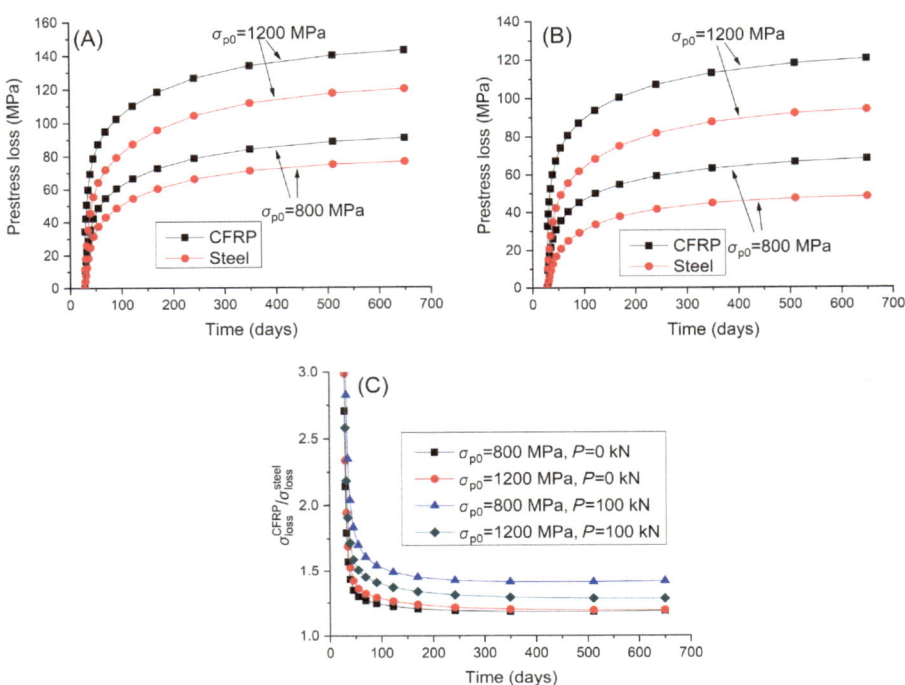

**Figure 13.9** Long-term prestress loss in unbonded CFRP and steel tendons caused by concrete creep, shrinkage, and tendon relaxation: (A) beams without external loads ($P = 0$ kN); (B) beams with service external loads ($P = 100$ kN); (C) ratio of prestress loss in CFRP tendons to that in steel ones.

Fig. 13.10 shows the long-term deflection and stress in bottom reinforcing steel at midspan caused by concrete creep, shrinkage, and tendon relaxation for members with different initial prestress levels and load conditions. After a long period of time, due to higher prestress loss, CFRP tendons exhibit a lower camber (for $P = 0$ kN) and a higher downward deflection (for $P = 100$ kN) compared to steel ones. The development of long-term compressive stress in bottom reinforcing steel of members with CFRP tendons is slower than that of members with steel tendons.

### 13.4.4 Influence of compressive reinforcing steel on long-term behavior

In Section 13.3.3, the long-term behavior of simply supported bonded AFRP prestressed concrete members having various tensile reinforcing bars has been analyzed. The results showed that the tensile reinforcing bars affect heavily the long-term behavior under zero external loads but their effect appears to be less pronounced at a certain level of sustained external loads. The latter conclusion holds true for the prestressed concrete members with unbonded FRP tendons.

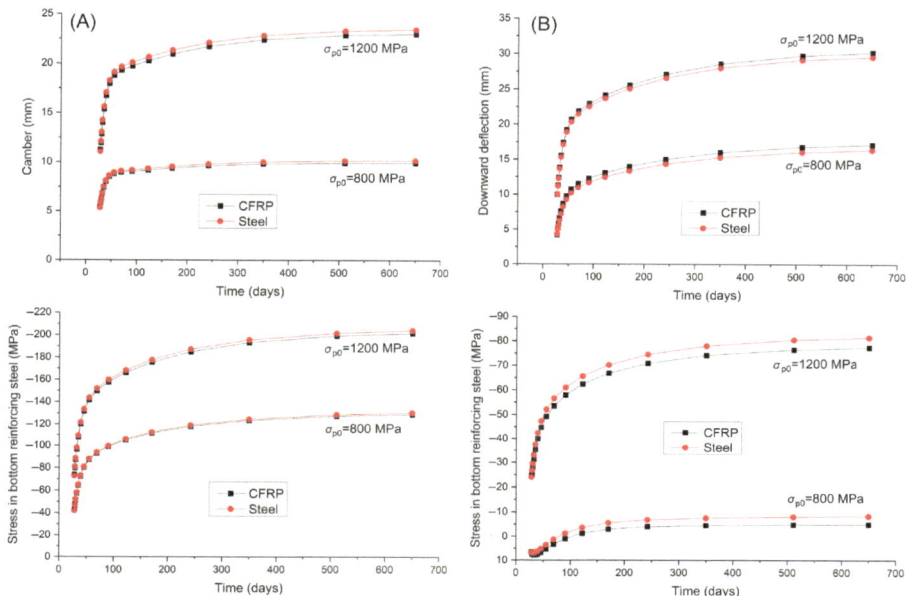

**Figure 13.10** Long-term deflection and stress in bottom reinforcing steel caused by concrete creep, shrinkage, and tendon relaxation: (A) beams without external loads ($P = 0$ kN); (B) beams with service external loads ($P = 100$ kN).

In this section, the effect of the compressive reinforcing steel ratio ($\rho'_s$) on the time-dependent behavior of simply supported prestressed concrete members with internal unbonded CFRP tendons is explored. Fig. 13.11 shows the effect of compressive reinforcing steel ratio on the long-term deformations (i.e., midspan deflection, axial shortening, concrete strains in the bottom and top fibers of the midspan section) for members without external loads, while the results produced for the case that $P = 100$ kN are illustrated in Fig. 13.12. Fig. 13.11 shows that for the case without external loads, the compressive reinforcing steel influences notably the long-term camber and concrete strain of the top fiber, whereas its influence on the long-term axial shortening and concrete strain in the bottom fiber is negligible. As the compressive reinforcing steel ratio increases, the long-term camber increases while the long-term concrete compressive strain in the top fiber decreases substantially. Moreover, increasing the quantity of compressive reinforcing steel leads to a slight decrease in long-term axial shortening and an increase in long-term concrete compressive strain in the bottom fiber.

It has been stated that the presence of reinforcing steel would reduce substantially the long-term camber of prestressed concrete members (Ghali et al., 2002). This statement, however, is not fully correct. When the tensile reinforcing steel is provided or its quantity is increased in prestressed concrete members, the long-term camber is significantly reduced, as confirmed by the results presented in Section 13.3.3. On the other hand, the presence or an increasing amount of compressive reinforcing steel leads to an increase in the long-term camber as mentioned above.

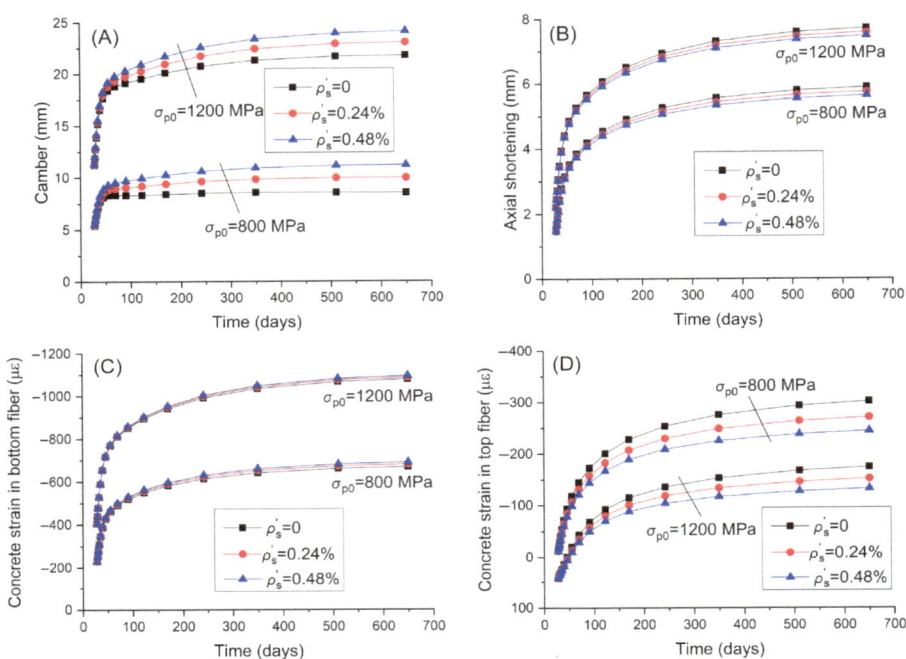

**Figure 13.11** Effect of $\rho'_s$ on the long-term performance of unbonded CFRP tendon beams without external loads ($P = 0$ kN): (A) camber; (B) axial shortening; (C) concrete strain in bottom fiber; (D) concrete strain in top fiber.

Fig. 13.12 shows that at sustained service loads ($P = 100$ kN), the quantity of compressive reinforcing steel has significant influences on the time-dependent performance. The long-term downward deflection, axial shortening, and concrete compressive strain in the top fiber are reduced substantially as the quantity of compressive reinforcing steel increases. At $P = 100$ kN, the concrete strain in the bottom fiber is in tension for the initial prestress of 800 MPa, while for the initial prestress of 1200 MPa is in compression. Increasing the quantity of compressive reinforcing steel leads to a considerable decrease in long-term concrete tensile strain or an increase in long-term concrete compressive strain in the bottom fiber.

## 13.5 Proposed equation for calculating the long-term deflection

The effectiveness of adding compressive reinforcing steel in reducing the long-term deflection of concrete members is taken into account in ACI 318-19

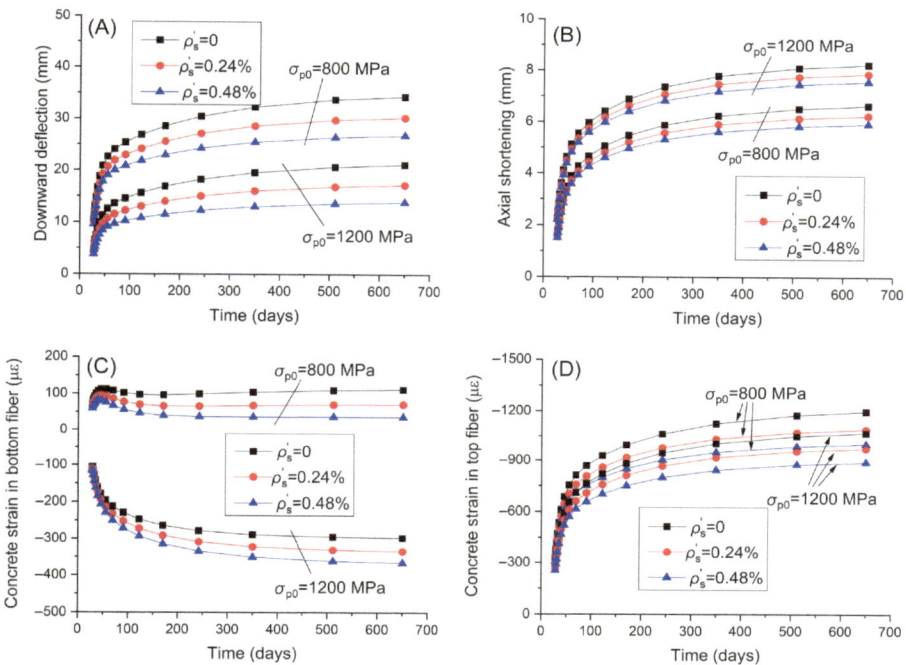

**Figure 13.12** Effect of $\rho'_s$ on the long-term performance of unbonded CFRP tendon beams with service external loads ($P = 100$ kN): (A) deflection; (B) axial shortening; (C) concrete strain in bottom fiber; (D) concrete strain in top fiber.

(ACI Committee 318, 2019) through a reduction factor $\alpha$, that is:

$$\alpha = \frac{1}{1 + 50\rho'_s} \tag{13.3}$$

At service loads of 100 kN, a comparison between the $\alpha$ values predicted by ACI 318-19 and numerical simulations for unbonded CFRP tendon members (see Fig. 13.6) subjected to sustained loading of 12 months is illustrated in Fig. 13.13. Initial prestress values of 800 MPa (i.e., 40% of the ultimate tensile strength) and 1200 MPa (i.e., 60% of the ultimate tensile strength) are used for the investigation. The results of numerical simulations show that as the initial prestress level increases, the influence of the compressive reinforcing steel increases. It is also apparent that ACI 318-19 underestimates the influence of compressive reinforcing steel on the long-term deflection of prestressed concrete members. This underestimation is particularly notable at a high initial prestress level.

To extend Eq. (13.3) to the case of prestressed concrete members with FRP or steel tendons, the following modification is proposed:

$$\alpha = \frac{1}{1 + 50\eta_e\eta_p\rho'_s} \tag{13.4}$$

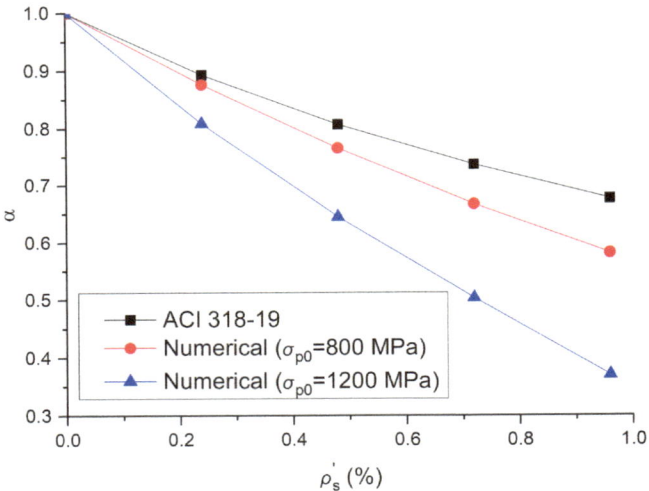

**Figure 13.13** Variation of $\alpha$ with $\rho'_s$ according to ACI 318-19 and numerical simulations for 12-month loading.

where

$$\eta_e = \frac{E_s}{E_p} \tag{13.5}$$

and

$$\eta_p = \begin{cases} 1.0 & \text{for } \dfrac{\sigma_{p0}}{\sigma_{pu}} \leq 0.4 \\[3mm] 2.5\dfrac{\sigma_{p0}}{\sigma_{pu}} & \text{for } \dfrac{\sigma_{p0}}{\sigma_{pu}} > 0.4 \end{cases} \tag{13.6}$$

In the above equations, $\eta_p$ represents the effect of the initial prestress level (ratio of initial prestress to ultimate strength of the tendon material); and $\eta_e$ represents the effect of the tendon modulus of elasticity. For steel tendons, their elastic modulus can be considered equal to that of reinforcing steel, and therefore, $\eta_e$ in that case is equal to 1.0. For FRP tendons, the value of $\eta_e$ depends on the modulus of elasticity of the FRP material.

Fig. 13.14 illustrates a comparison between the $\alpha$ values predicted by ACI 318-19, the proposed Eq. (13.4), and numerical simulations for different periods of sustained loading ($P = 100$ kN) and initial prestress levels in unbonded CFRP tendons. Given an initial prestress level, the difference between the results of numerical simulations for different loading periods is not large, indicating that the reduction factor $\alpha$ is almost time-independent. It is also seen that the results predicted by the proposed equation are in good agreement with those from numerical simulations. Therefore the proposed modification of the ACI 318-19 equation by introducing the parameters $\eta_e$ and $\eta_p$ appears to be reasonable for calculating the reduction factor $\alpha$.

An empirical equation is recommended in ACI 318-19 to calculate the long-term deflection of concrete members:

$$\lambda_\Delta = \frac{\xi}{1 + 50\rho'_s} \tag{13.7}$$

where $\lambda_\Delta$ is the ratio of the time-dependent deflection to the immediate deflection, while is a time-dependent factor, which $\xi$ is equal to 1.0, 1.2, 1.4, and 2.0 for loading periods of 3 months, 6 months, 12 months, and 5 years, respectively.

At service external loads of 100 kN, a comparison between the results predicted by ACI 318-19 and numerical simulations in terms of the variation of $\lambda_\Delta$ with $\rho'_s$ for different loading periods and initial prestress levels is illustrated in Figs. 13.15 and 13.16. Fig. 13.15 shows that for initial prestress of 800 MPa and sustained loading for a period equal to or less than 12 months, ACI 318-19 predicts well the time-dependent deflection of prestressed members without compressive reinforcing steel. However, ACI 318-19 generally overestimates the time-dependent deflection for prestressed members with compressive reinforcing steel. For a 5-year duration of loading, ACI 318-19 leads to a considerable overestimation, indicating that the $\xi$ value of 2.0 specified is rather conservative. Fig. 13.16 shows that for an initial prestress of 1200 MPa, ACI 318-19 substantially overestimates the time-dependent deflection regardless of the duration of loading. Therefore an extension of Eq. (13.7) to the case of prestressed concrete members is deemed necessary. The latter is achieved by using the modified reduction factor in Eq. (13.4) and by replacing $\xi$ with $\xi_p$, that is:

$$\lambda_\Delta = \frac{\xi_p}{1 + 50\eta_e\eta_p\rho'_s} \tag{13.8}$$

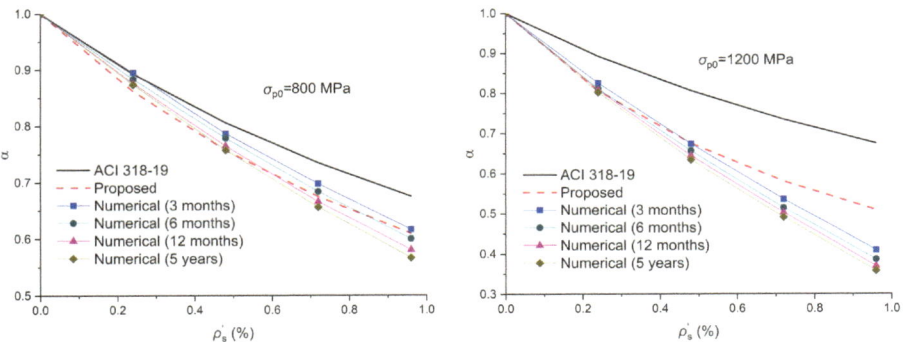

**Figure 13.14** Variation of $\alpha$ with $\rho'_s$ according to ACI 318-19, the proposed equation and numerical simulations for different durations of loading and levels of initial prestress.

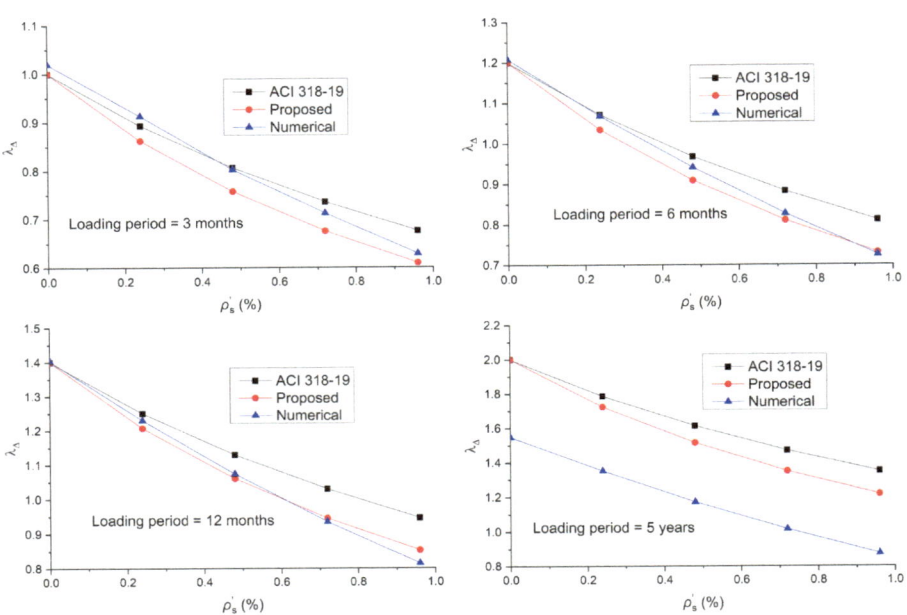

**Figure 13.15** Variation of $\lambda_\Delta$ with $\rho'_s$ according to ACI 318-19, the proposed equation and numerical simulations for initial prestress of 800 MPa.

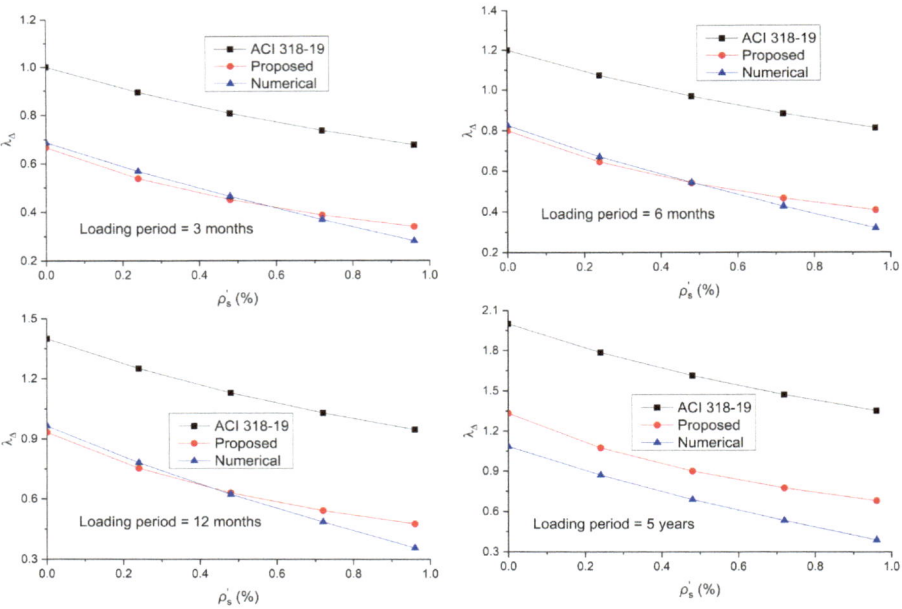

**Figure 13.16** Variation of $\lambda_\Delta$ with $\rho'_s$ according to ACI 318-19, the proposed equation and numerical simulations for initial prestress of 1200 MPa.

where

$$\xi_p = \frac{\xi}{\eta_p} \tag{13.9}$$

A comparison of the predictions obtained from ACI 318-19 (Eq. 13.7), the proposed (Eq. 13.8), and numerical simulations is presented in Figs. 13.15 and 13.16. Moreover, the correlation of the predictions of the simplified equations (ACI 318-19 and proposed) and the results from numerical simulations for $\lambda_\Delta$ is presented in Fig. 13.17. A total of 40 prestressed concrete members with unbonded CFRP tendons are used to produce this correlation. It can be seen in Figs. 13.15–13.17 that the predictions of the proposed equation and the results of numerical simulations are generally in good agreement. Moreover, the proposed equation seems to be far more accurate than the ACI 318-19 equation in the prediction of the time-dependent deflection of prestressed concrete members with unbonded tendons.

## 13.6    Time-dependent second-order effects of externally prestressed concrete members

Simply supported rectangular concrete beams prestressed with straight external tendons, as illustrated Fig. 13.18, are used here for the evaluation of time-dependent second-order effects. The initial tendon depth is 850 mm all along the span. The main variable is the number of deviators. The tendons are normal-relaxation

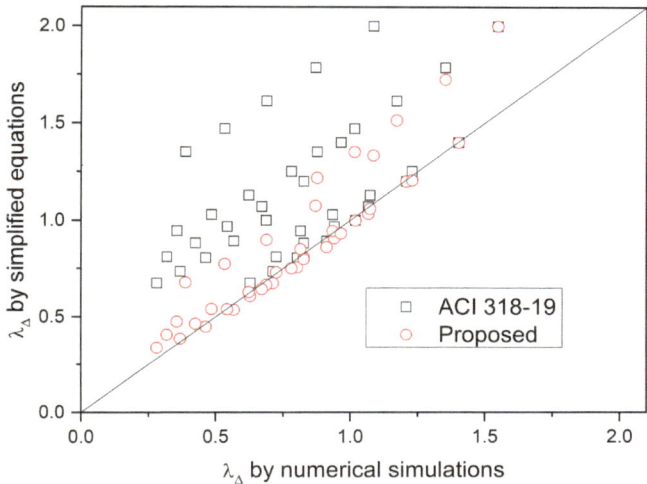

**Figure 13.17** Correlation of $\lambda_\Delta$ values predicted by simplified equations and numerical simulations.

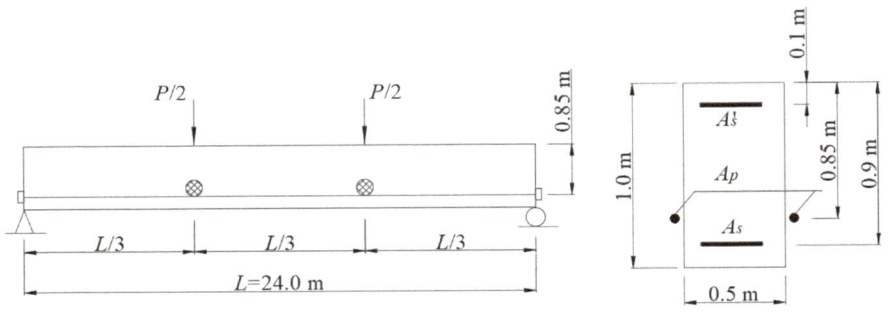

$A_p$=3000 mm$^2$   $A_s$=2000 mm$^2$   $A_s'$=1000 mm$^2$

**Figure 13.18** Details of externally prestressed concrete beams.

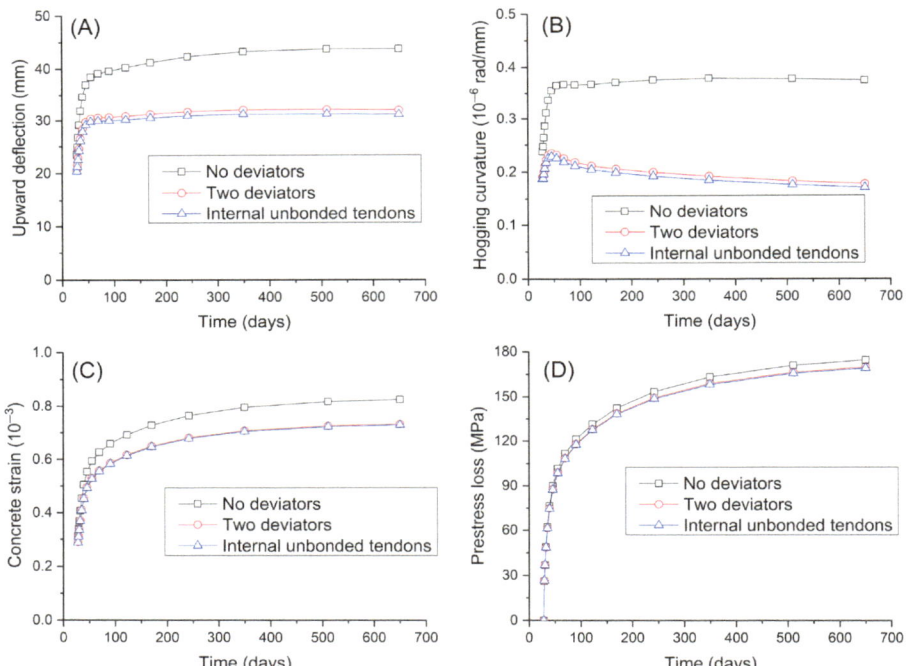

**Figure 13.19** Effect of deviators on the time-dependent behavior at zero loads: (A) midspan deflection; (B) midspan curvature; (C) total concrete strain (absolute value) at the bottom fiber of the midspan section; (D) prestress loss.

prestressing steel with elastic modulus and yield stress of 200 GPa and 1581 MPa, respectively. The tendon area and initial prestress are 3000 mm$^2$ and 1200 MPa, respectively. The areas of bottom and top nonprestressed steel are 2000 and 1000 mm$^2$, respectively. The elastic modulus of nonprestressed steel is 200 GPa.

The concrete cylinder compressive strength and modulus of elasticity at the age of 28 days are 35 MPa and 34 GPa, respectively. Both the prestress transfer and the application of loads $P$ are assumed to take place at a concrete age of 28 days. The beam is divided into 18 equal beam elements, and each element is subdivided into 10 concrete layers and two steel layers (for top and bottom nonprestressed steel); the external tendon is also idealized as 18 segments corresponding to the beam elements.

The time-dependent second-order effects of externally prestressed concrete members are evaluated using three different deviator configurations, namely, no deviators, two deviators at third points, and deviators all along the span (17 deviators) for the simulation of internal unbonded tendons. Figs. 13.19 and 13.20 show the effect of deviators on the typical time-dependent response characteristics for the members subjected to loads of zero and 200 kN, respectively. It is seen that the long-term responses for the external tendon member with two deviators are very similar to those for the internal unbonded tendon member, indicating the second-order effects are negligible when providing two deviators at third points. Therefore the second-order effects can be evaluated by comparing the time-dependent responses for the external tendon members without deviators and with two deviators. It is seen that the member without deviators exhibits significantly higher time-dependent deformations than the member with deviators. At zero loads, the camber and hogging curvature, at day 650, caused by time-dependent effects for the member without deviators are 20.2 mm and $0.14 \times 10^{-6}$ rad/mm, respectively. By the provision of two deviators at third points, the time-dependent camber is reduced by 44.1% to 11.3 mm and the time-dependent hogging curvature is reduced significantly by 109.5% to a negative value of $-0.013 \times 10^{-6}$ rad/mm. At loads of 200 kN, the downward deflection and sagging curvature, at day 650, caused by time-dependent effects for the member without deviators are 55.8 mm and $1.14 \times 10^{-6}$ rad/mm, respectively. By the provision of two deviators at third points, the time-dependent deflection is reduced by 26.8% to 40.9 mm and the time-dependent curvature is reduced by 21.8% to $0.89 \times 10^{-6}$ rad/mm. The second-order effects also have an important influence on the concrete strain, but the influence is not as much as that on the deformations. At day 650, the concrete strain at the maximum compressive fiber of the midspan section caused by time-dependent effects for the members without deviators under loads of zero and 200 kN are $0.51 \times 10^{-3}$ and $1.05 \times 10^{-3}$, respectively. By the provision of two deviators at third points, the concrete strains are reduced by 13.7% to $0.44 \times 10^{-3}$ for zero loads and by 12.3% to $0.92 \times 10^{-3}$ for loads of 200 kN. From Figs. 13.19D and 13.20D, it is seen that the influence of second-order effects on the long-term prestress loss is insignificant. At day 650, the prestress loss caused by time-dependent effects for the member without deviators under loads of zero and 200 kN is 174.4 and 143.4 MPa, respectively. By the provision of two deviators at third points, the prestress loss is reduced by 2.7% to 169.6 MPa for zero loads while it is increased by 3.8% to 148.9 MPa for loads of 200 kN.

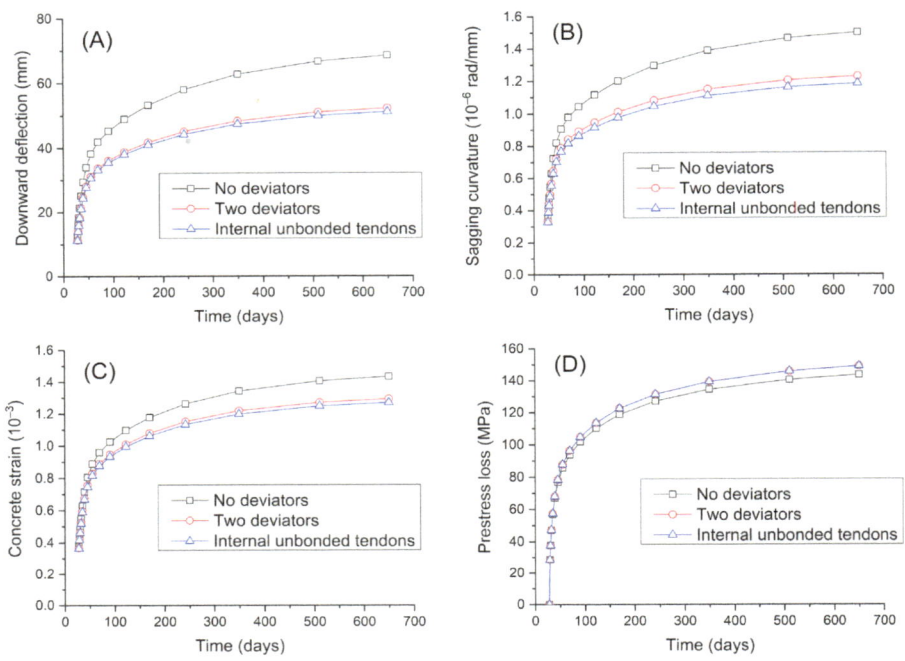

**Figure 13.20** Effect of deviators on the time-dependent behavior at loads of 200 kN: (A) midspan deflection; (B) midspan curvature; (C) total concrete strain (absolute value) at the top fiber of the midspan section; (D) prestress loss.

## 13.7  Conclusions

The long-term performance of prestressed concrete members with FRP or steel tendons is examined by using a time-dependent finite element model presented in Chapter 12. Time-dependent assessments are performed to reveal the effect of using FRP tendons instead of low-relaxation steel ones, the magnitude of the initial prestress, the loading conditions, and the internal reinforcing steel. Moreover, the empirical equation recommended in ACI 318-19 for calculating the long-term deflection is evaluated, and a modified ACI 318-19 equation is proposed. In addition, the time-dependent second-order effects of externally prestressed concrete members are assessed.

Due to a lower modulus of elasticity, FRP tendons exhibit a lower long-term prestress loss caused by concrete creep and shrinkage only (i.e., without considering tendon relaxation) when compared to steel tendons. However, when tendon relaxation is included, FRP tendons exhibit higher long-term prestress loss than steel tendons, especially for members under sustained external loads. The results also demonstrate that when only concrete creep and shrinkage are considered, FRP tendons induce a higher long-term prestress camber but a lower long-term load deflection, when compared to steel tendons. Opposite trends are seen when tendon relaxation is considered in numerical analysis.

The influence of bottom nonprestressed steel on the long-term displacement depends on the load level, that is, its influence is notable at zero loads but appears to be not important at sustained service loads of 100 kN. In the case of members without external loads, increasing the top nonprestressed steel area leads to an increase in long-term camber, a slight decrease in long-term axial shortening, a slight increase in long-term concrete compressive strain in the top fiber, and a decrease in long-term concrete compressive strain in the bottom fiber. On the other hand, under sustained service loads of 100 kN, increasing the top nonprestressed steel reduces substantially the long-term deflection, axial shortening, and concrete compressive strain in the top fiber. A larger top nonprestressed steel area leads to a lower long-term tensile strain (when a lower initial prestress is used) or a higher long-term compressive strain (when a higher initial prestress is used) of concrete in the bottom fiber.

ACI 318-19 underestimates the effect of compressive reinforcing steel on the long-term deflection of prestressed concrete members with FRP tendons, especially at a high initial prestress level. ACI 318-19 substantially overestimates the time-dependent deflection for the initial prestress level of 60%. A modification of the ACI 318-19 equation is proposed to calculate the long-term deflection of prestressed concrete members with FRP or steel tendons. By introducing the prestress-related parameters $\eta_e$, $\eta_p$, and $\xi_p$, the proposed equation is shown to predict well the time-dependent deflection of prestressed concrete members with FRP tendons.

The time-dependent second-order effects of externally prestressed concrete members are evaluated by using different deviator configurations. The results indicate that, due to second-order effects, the member without any deviators exhibits quite different long-term responses other than the prestress loss, when compared to the member with two deviators at third points. On the other hand, the long-term behavior of the member with two deviators is very similar to that of the member with internal unbonded tendons, indicating that the time-dependent second-order effects are minimized by the provision of two deviators at third points.

# References

ACI Committee 318. (2019). *Building code requirements for structural concrete (ACI 318-19) and commentary (ACI 318R-19)*. Farmington Hills, MI.

Au, F. T. K., & Si, X. T. (2011). Accurate time-dependent analysis of concrete bridges considering concrete creep, concrete shrinkage and cable relaxation. *Engineering Structures*, *33*(1), 118−126. Available from https://doi.org/10.1016/j.engstruct.2010.09.024.

Bažant, Z. P. (2001). Prediction of concrete creep and shrinkage: Past, present and future. *Nuclear Engineering and Design*, *203*(1), 27−38. Available from https://doi.org/10.1016/s0029-5493(00)00299-5.

Ghali, A., Favre, R., & Elbadry, M. (2002). *Concrete structures: stresses and deformation* (3rd ed.). London and New York: Spon Press.

Huang, H., Huang, S. S., & Pilakoutas, K. (2018). Modeling for assessment of long-term behavior of prestressed concrete box-girder bridges. *Journal of Bridge Engineering, 23* (3), 04018002. Available from https://doi.org/10.1061/(ASCE)BE.1943-5592.0001210.

Lou, T., & Karavasilis, T. L. (2018). Time-dependent assessment and deflection prediction of prestressed concrete beams with unbonded CFRP tendons. *Composite Structures, 194*, 365–376. Available from https://doi.org/10.1016/j.compstruct.2018.04.013.

Lou, T., Lopes, S. M. R., & Lopes, A. V. (2015). Interaction between time-dependent and second-order effects of externally posttensioned members. *Journal of Bridge Engineering, 20*(11), 06015003. Available from https://doi.org/10.1061/(ASCE) BE.1943-5592.0000761.

Lou, T., Lopes, S. M. R., & Lopes, A. V. (2016). Time-dependent behavior of concrete beams prestressed with bonded AFRP tendons. *Composites Part B: Engineering, 97*, 1–8. Available from https://doi.org/10.1016/j.compositesb.2016.04.070.

Magura, D. D., Sozen, M. A., & Siess, C. P. (1964). A study of stress relaxation in prestressing reinforcement. *PCI Journal, 9*(2), 13–57. Available from https://doi.org/10.15554/pcij.04011964.13.57.

Pisani, M. A. (2000a). Long-term behaviour of beams prestressed with aramid fibre cables: Part 1: A general method. *Engineering Structures, 22*(12), 1641–1650. Available from https://doi.org/10.1016/S0141-0296(99)00107-8.

Pisani, M. A. (2000b). Long-term behaviour of beams prestressed with aramid fibre cables: Part 2: An approximate solution. *Engineering Structures, 22*(12), 1651–1660. Available from https://doi.org/10.1016/S0141-0296(99)00108-X.

Rodriguez-Gutierrez, J. A., & Aristizabal-Ochoa, J. D. (2007). Short- and long-term deflections in reinforced, prestressed, and composite concrete beams. *Journal of Structural Engineering, 133*(4), 495–506. Available from https://doi.org/10.1061/(ASCE)0733-9445(2007)133:4(495).

Saadatmanesh, H., & Tannous, F. E. (1999a). Long-term behavior of aramid fiber reinforced plastic (AFRP) tendons. *ACI Materials Journal, 96*(3), 297–305.

Saadatmanesh, H., & Tannous, F. E. (1999b). Relaxation, creep, and fatigue behavior of carbon fiber reinforced plastic tendons. *ACI Materials Journal, 96*(2), 143–153.

Xue, W., Liu, T., & Zeng, M. (2016). Prediction of long-term deflections for high-speed railway prestressed concrete beams. *ACI Structural Journal, 113*(4), 769–778. Available from https://doi.org/10.14359/51688621.

Youakim, S. A., & Karbhari, V. M. (2007). An approach to determine long-term behavior of concrete members prestressed with FRP tendons. *Construction and Building Materials, 21*(5), 1052–1060. Available from https://doi.org/10.1016/j.conbuildmat.2006.02.006.

# Long-term behavior of steel−concrete composite girders

**14**

## 14.1  Introduction

In steel−concrete composite girders, the composite action is achieved by connecting the steel beam and the concrete slab with shear connectors in various shapes (Davoodnabi et al., 2019; Shariati et al., 2020). By taking advantage of high steel tensile strength and concrete compressive strength, steel−concrete composite girders exhibit superior structural performance to that of steel or concrete counterparts (CEN, 2004).

The degradation of the long-term workability of steel−concrete composite structures is a prime concern in practice and has received extensive attention from researchers (Ranzi et al., 2013). Some laboratory tests were performed to investigate long-term responses of composite girders (Al-deen et al., 2011; Bradford & Gilbert, 1991; Fan et al., 2010; Gilbert & Bradford, 1995). Computer simulations are a valid supplement to the long-term tests that are expensive and time consuming. Rigorous time-dependent simulations of composite girders are challenging due to the complicated viscoelastic behavior of concrete. The age-adjusted effective modulus method (AEMM) is mostly adopted by researchers for the creep analysis of concrete (Wang & Gong, 2019) or composite structures (Huang, Wei, et al., 2018). AEMM is an improved approach of the effective modulus method through the inclusion of an aging coefficient to simulate creep at gradually varying stress. By applying AEMM, Bradford (1991) and Bradford and Gilbert (1992) developed analytical models for prediction in the time-dependent behavior of composite girders. Their models were verified with experimental results of their own tests as well as other investigators. General numerical models including finite elements were also developed by different investigators (He & Li, 2020; Henriques et al., 2019; Sakr & Sakla, 2008, 2009). Sakr and Sakla (2008, 2009) described finite element modeling along with artificial neural networks for computing the long-term deformation in composite girders. A parametric analysis was carried out and several codes of practice were investigated. In time-dependent finite element modeling, time is divided into intervals to perform a step-by-step analysis. As creep depends on the stress history, normally the concrete stresses in elements at all intervals have to be stored, leading to computational inefficiency in a creep analysis. To overcome this limitation, the creep function may be approximated by Dirichlet series terms, which were originally proposed by Zienkiewicz and Watson (1966) and extensively used in the time-dependent analysis of concrete (Huang, Huang, et al., 2018; Lou & Karavasilis, 2018) and composite structures (Henriques et al., 2018; Kwak & Seo, 2000; Tong et al., 2018; Virtuoso & Vieira, 2004).

Prestressed Members with External Fiber-Reinforced Polymer (FRP) Tendons. DOI: https://doi.org/10.1016/B978-0-443-23877-2.00014-6

The deflection of composite girders increases with time at sustained service loads. The final deflection after a long period of time may be several times the immediate deflection (Amadio et al., 2012). Preventing excessive deflection is a main task in serviceability limit-state design. Simplified equations for quantifying time-dependent deflections in reinforced concrete members are available in design codes, for example, ACI 318-19 (ACI Committee 318, 2019). However, the current codes or specifications do not provide direct design equations for calculating time-dependent deflections in composite girders. For example, in Eurocode 4 (CEN, 2004), there are no specific rules concerning long-term deflections in composite girders at the serviceability limit state. According to Eurocode 4, the effect of concrete creep might be considered by using modular ratios associated with a creep coefficient and a creep multiplier. However, the long-term deflections in composite girders appear to be dominated by concrete shrinkage rather than concrete creep (to be discussed later), while this code did not recommend any practical approach for calculating shrinkage-induced deflections. Although the AISC specification (AISC, 2016) offered a simplified model to predict shrinkage-induced deflections in composite girders, this model seems to be not convenient for design purposes (to be discussed later). Therefore it is practically important to propose a simplified equation considering key parameters to quantify long-term deflections in composite girders.

This chapter presents the results of a numerical assessment on the time-dependent behavior of steel−concrete composite girders at sustained loading (Lou et al., 2021). The effects of concrete creep and/or shrinkage, ultimate shrinkage strain, and steel rebars on the time-dependent performance of composite girders are examined. In addition, the AISC model (AISC, 2016) concerning long-term deflection is evaluated, and an equation for quantifying the time-dependent deflection in composite girders is proposed for design purposes. Finally, the second-order effects of externally prestressed steel−concrete composite girders under long-term sustained loads are analyzed.

## 14.2 Time-dependent assessment on composite girders

A single-span steel−concrete composite girder (see Fig. 14.1) is used herein as a reference girder to illustrate the results of a time-dependent assessment. The span length is 10.0 m. The concrete slab (2000 × 200 mm) contains steel rebars with an

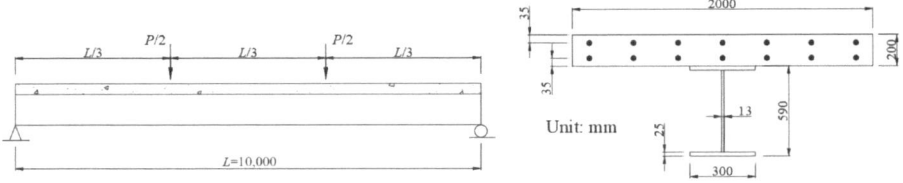

**Figure 14.1** Sketch of steel−concrete composite girders.

area of $2000 \text{ mm}^2$. The steel beam is made of HEA 600. The concrete cylinder compressive strength and modulus of elasticity at age of 28 days are 40 MPa and 35 GPa, respectively. The elastic modulus of both steel rebars and beam is 200 GPa. Third-point loads are applied at day 28, and $P = 200$ kN.

## 14.2.1 Contribution of creep and/or shrinkage

A numerical analysis considering three different time effects (creep only, shrinkage only, and creep and shrinkage) is conducted. Typical results of the analysis are presented in Fig. 14.2. All responses exhibit a quick development at an early age and the development becomes slower with time. Fig. 14.2A−D shows that the long-term deformation characteristics are dominated by concrete shrinkage. The contribution of concrete shrinkage is significantly more than that of concrete creep. In the present analysis, at day 650, the time-dependent deflection, axial shortening of concrete slab, concrete strain in the top fiber, and curvature induced by concrete shrinkage only is 72.1%, 83.8%, 72.6%, and 67.6%, respectively, of those induced by concrete creep and shrinkage. On the other hand, the percentage of those deformations due to concrete creep is 36.2%, 28.3%, 37.6%, and 40.1%, respectively. It is noted that the time-dependent deformation by combined concrete creep and shrinkage is smaller than the sum of the time-dependent deformation by concrete creep only and by concrete shrinkage only. This can be explained by the coupling between creep and shrinkage effects.

Fig. 14.2E and F shows the development of time-dependent stresses in the top and bottom flanges of the steel beam, respectively. In the top flange, the long-term (compressive) stress contributed by concrete shrinkage is much greater than that by concrete creep. On the other hand, in the bottom flange, the long-term (tensile) stress induced by concrete shrinkage is a bit smaller than that by concrete creep. The above observations can be theoretically explained as follows.

$$\sigma_{\text{top}} = E_s \kappa \left( \frac{\varepsilon_c^*}{\kappa} - d_c \right) = E_s \left( \varepsilon_c^* - \kappa d_c \right) \tag{14.1a}$$

$$\sigma_{\text{bot}} = E_s \kappa \left( d - \frac{\varepsilon_c^*}{\kappa} \right) = E_s \left( \kappa d - \varepsilon_c^* \right) \tag{14.1b}$$

where $\sigma_{\text{top}}$ and $\sigma_{\text{bot}}$ are steel stresses in top and bottom flanges, respectively; $\kappa$ is the section curvature; $\varepsilon_c^*$ is the concrete strain in the top fiber of the slab; $d$ and $d_c$ are the depths of the composite section and concrete slab, respectively. As shown in Fig. 14.2C and D, after a long period of time, concrete shrinkage induces higher values of $\varepsilon_c^*$ and $\kappa$ than concrete creep while the difference is more apparent for $\varepsilon_c^*$ than for $\kappa$. As a result, concrete shrinkage leads to a substantially higher value of $\sigma_{\text{top}}$ according to Eq. (14.1a) and may cause a lower value of $\sigma_{\text{bot}}$ according to Eq. (14.1b), when compared to concrete creep.

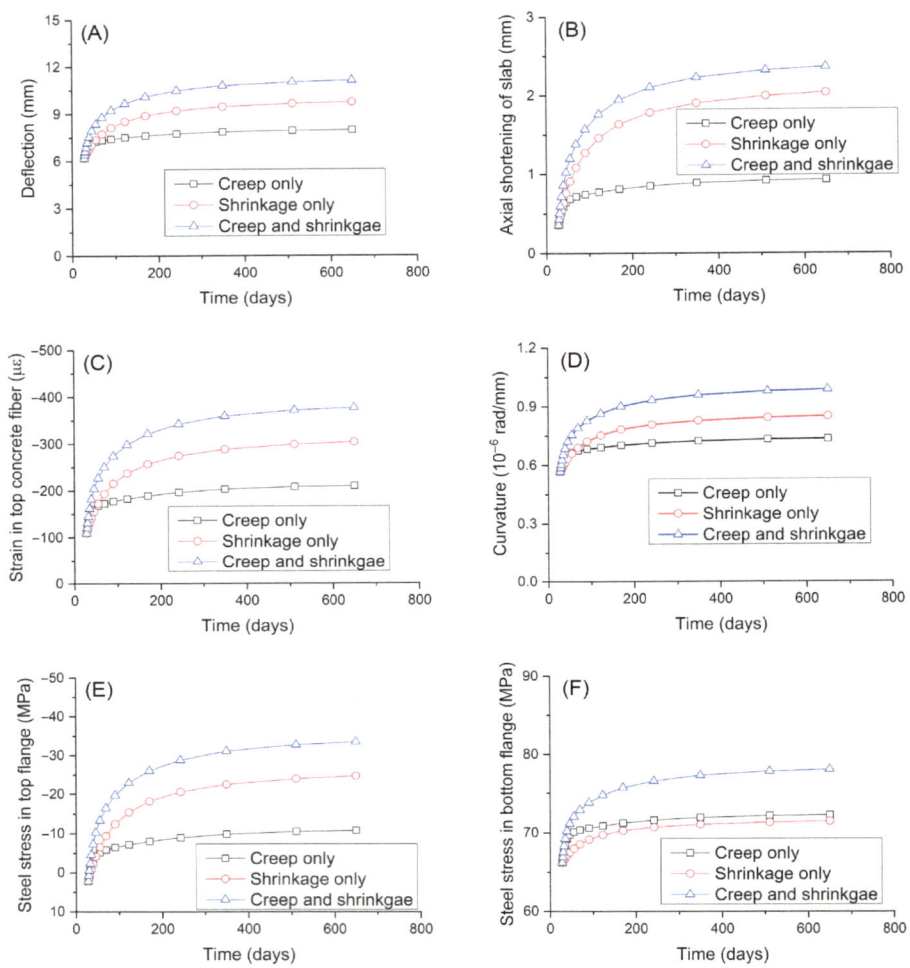

**Figure 14.2** Effect of creep and/or shrinkage on the long-term behavior of composite girders: (A) midspan deflection; (B) axial shortening of concrete slab; (C) strain in top concrete fiber at midspan; (D) midspan curvature; (E) steel stress in top flange at midspan; (F) steel stress in bottom flange at midspan.

## 14.2.2   Effect of ultimate shrinkage strain

The results of the previous section indicate that the time-dependent deformations in composite girders are dominated by the shrinkage of concrete. It is noted that shrinkage depends upon the ultimate shrinkage strain. To evaluate quantitatively the effect of ultimate shrinkage strain, three values of $\varepsilon_{sh}^{\infty}$ are used, namely, 200, 400, and 600 $\mu\varepsilon$.

Typical results of the long-term behavior of composite girders having different ultimate shrinkage strains are shown in Fig. 14.3. It is seen that the long-term deformations and steel stresses increase significantly as the ultimate shrinkage strain increases. At day 650, increasing the ultimate shrinkage strain from 200 to 600 μɛ causes increments in time-dependent deflection, axial shortening of slab, concrete strain in the top fiber, curvature, and stresses in the top and bottom flanges of steel beam by 98.2%, 116.4%, 95.0%, 89.8%, 97.7%, and 68.2%, respectively.

## 14.2.3  Effect of steel rebars

The compressive reinforcing steel is a critical parameter influencing the time-dependent behavior of concrete members. For statically determinate steel−concrete

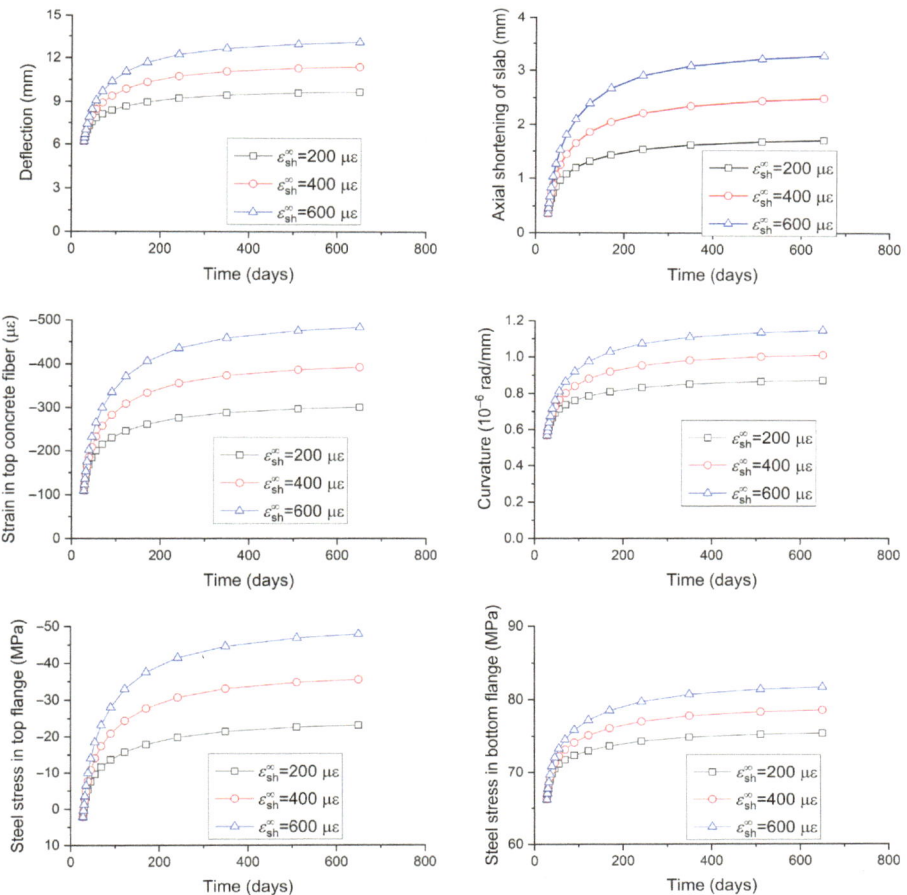

**Figure 14.3** Effect of ultimate shrinkage strain on the long-term behavior of composite girders.

composite girders, the steel rebars in the concrete slab at service conditions are generally in compression, and therefore, they may have an important effect on time-dependent performance. Three different areas of rebars are considered, namely, $A_r = 0$, 2000, and 4000 mm$^2$.

Fig. 14.4 shows the influence of the rebar area on the long-term behavior of composite girders. It is seen that the long-term deformations and stresses in steel beam decrease as the rebar area increases. At day 650, increasing the rebar area from 0 to 4000 mm$^2$ reduces time-dependent deflection, axial shortening of concrete slab, concrete strain in the top fiber, curvature, and stresses in the top and bottom flanges of steel beam by 12.0%, 11.9%, 12.3%, 12.1%, 12.4%, and 11.4%, respectively.

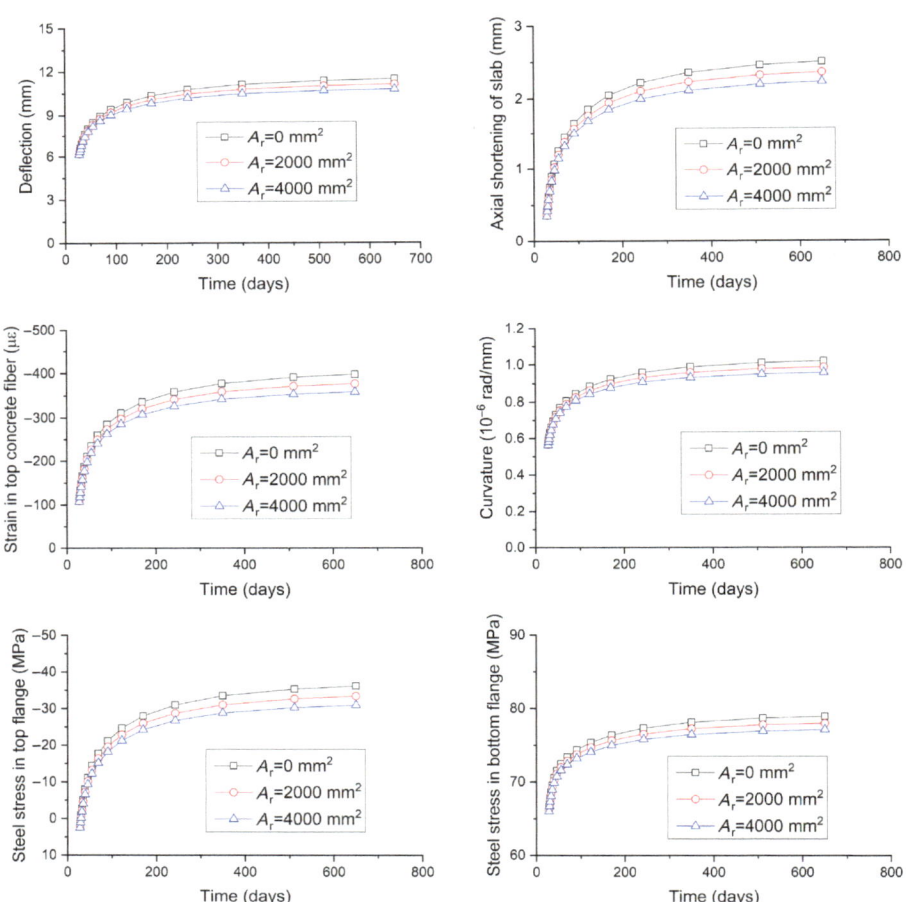

**Figure 14.4** Effect of rebar area on the long-term behavior of composite girders.

## 14.3   Evaluation of AISC model

AISC, 2016 recommended a model to quantify long-term deflections in composite girders due to concrete shrinkage. In this simplified model, the shrinkage effect is simulated using a couple of shrinkage forces acting at the end of the slab (see Fig. 14.5), which are expressed as follows:

$$P_{sh} = \varepsilon_r^{sh} E_c A_c \tag{14.2}$$

where $\varepsilon_r^{sh}$ is the restrained shrinkage strain; $A_c$ is the slab area.

Therefore the shrinkage-induced deflection at the midspan of composite girders is calculated from

$$\Delta_{sh} = \frac{P_{sh} e L^2}{8 E_s I_t} \tag{14.3}$$

where $e$ is the distance of centroidal axes of the slab and composite section; $L$ is the span length; and $I_t$ is the sectional moment of inertia (transformed).

Fig. 14.6 shows a comparison of the shrinkage-induced deflections predicted by the numerical simulation and AISC model. It is noted that the restrained shrinkage strain in Eq. (14.2) is lower than the free shrinkage strain. The value of the restrained shrinkage strain is obtained by the numerical analysis. It is seen in Fig. 14.6 that the long-term deflection by the AISC model is substantially higher than that by numerical simulations, indicating that the AISC model overestimates the shrinkage effect. It should be mentioned that this model is not quite practical because the restrained shrinkage strain is difficult to be determined in design. Moreover, the effect of concrete creep is not considered in the AISC approach. According to the results presented in Section 14.2.1, concrete creep contributes to over 30% of the time-dependent deflection. The contribution of concrete creep would be higher when a higher level of loading is applied. AISC suggested that the creep effect can be taken into consideration in a similar way as the shrinkage effect, but it is practically difficult because creep depends upon the stress history. In the following section, an equation to calculate directly time-dependent deflections in composite girders will be developed for practical purposes.

## 14.4   Prediction of long-term deflection of composite girders

The empirical equation recommended in the ACI code (ACI Committee 318, 2019) used compressive steel as a critical factor for calculating time-dependent deflections

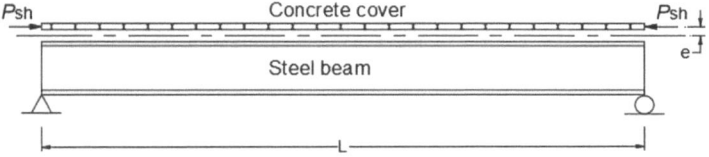

**Figure 14.5** Simplified model recommended by AISC.

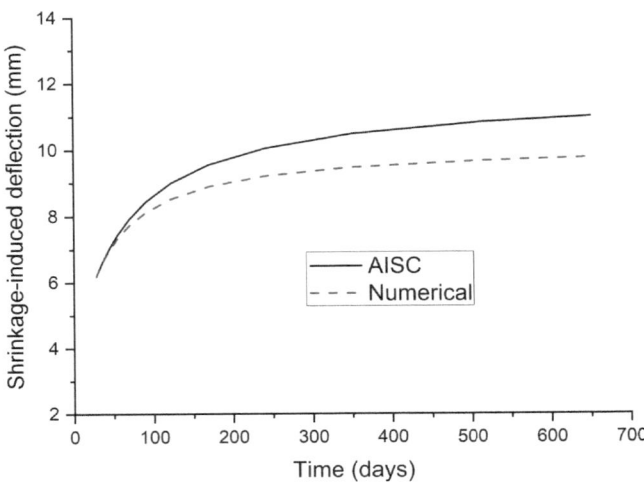

**Figure 14.6** Comparison of shrinkage-induced deflection by AISC model with numerical simulations.

of concrete members. The results presented in Section 14.2.3 show that with respect to the time-dependent deflection, the effect of rebars in steel−concrete composite girders is similar to that of compressive steel in reinforced concrete members. Therefore the ACI equation may be extended to be applicable to composite girders.

According to the ACI code, the time-dependent deflection for nonprestressed concrete members due to creep and shrinkage of concrete is computed by multiplying the immediate deflection by a factor $\lambda_\Delta$.

$$\lambda_\Delta = \frac{\xi}{1 + 50\rho'} \tag{14.4}$$

where $\xi$ is a coefficient associated with a sustained loading period, equal to 1.0, 1.2, 1.4, and 2.0 for 3-month, 6-month, 12-month, and 5-year periods of loading, respectively; $\rho'$ is the compression reinforcement ratio at midspan. Since the concrete slab of simply supported composite girders at service conditions is generally in compression, the compression reinforcement ratio in composite girders can be expressed by

$$\rho' = \frac{A_r}{A_c} \tag{14.5}$$

According to Eq. (14.4), the effect of compression reinforcement is represented by a reduction coefficient:

$$\alpha = \frac{1}{1 + 50\rho'} \tag{14.6}$$

Note that Eq. (14.6) is developed for reinforced concrete members. To check if this equation is applicable to steel−concrete composite girders, the numerical and code predictions regarding the variation of $\alpha$ with the $\rho'$ level are illustrated in Fig. 14.7 (the reference composite girder used in the previous sections is also adopted herein). The numerical results for different periods of loading are presented. It is seen that the difference between the numerical results for different periods of loading is very small, indicating the reduction coefficient $\alpha$ is almost time-independent. As $\rho'$ increases, the decrease of the $\alpha$ value predicted by the numerical analysis is considerably slower than that predicted by the ACI code. This indicates that the effect of rebars on the long-term defection of composite girders is significantly overestimated in the ACI code. Therefore the reduction coefficient of Eq. (14.6) needs to be appropriately modified in the case of composite girders. According to the fit curve to the numerically obtained data shown in Fig. 14.7, the following equation can be used to account for the effect of rebars in steel−concrete composite girders:

$$\alpha = \frac{1}{1 + 11\rho'} \tag{14.7}$$

Fig. 14.8 shows the numerically obtained curve regarding the variation of $\lambda_\Delta$ with the loading period up to 5 years for the composite girders without any rebars and with $\varepsilon_{sh}^{\infty}$ of 800 $\mu\varepsilon$. The data specified in the ACI code are also included in the graph for comparison. It is interesting to note that for 3-, 6-, and 12-month periods of loading, the ACI data are in perfect agreement with the numerical results. This

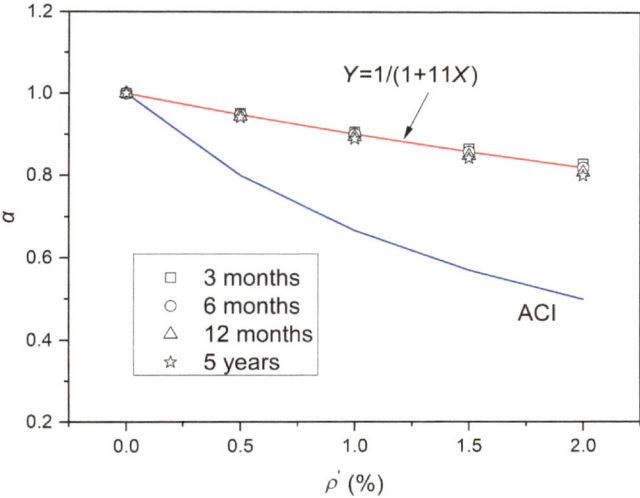

**Figure 14.7** Variation of $\alpha$ with $\rho'$.

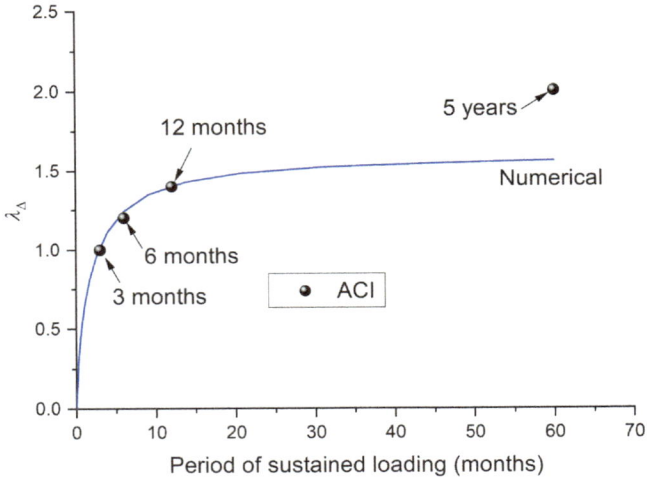

**Figure 14.8** Variation of $\lambda_\Delta$ with the period of sustained loading.

indicates that the values of $\xi$ specified in the ACI code are reasonable for loading period not greater than 12 months. However, for 5-year duration of loading, the ACI prediction is 28.4% higher than the numerical prediction. According to the numerical analysis, the value of $\lambda_\Delta$ for 5-year duration of loading is 1.56. In this case, therefore, $\xi = 1.6$ is recommended instead of 2.0 specified in the ACI code.

As illustrated previously, concrete shrinkage is critical to long-term deflections in composite girders. Therefore to extend the ACI equation to be applicable to composite girders, the following modification of Eq. (14.4) is made:

$$\lambda_\Delta = K_{sh}\frac{\xi}{1 + 11\rho'} \tag{14.8}$$

where $K_{sh}$ is a coefficient related to the ultimate shrinkage strain $\varepsilon_{sh}^\infty$. To obtain the expression of $K_{sh}$, the relationship between $\lambda_\Delta(1 + 11\rho')/\xi$ and $\varepsilon_{sh}^\infty$ for a composite girder without any rebars (i.e., $\rho' = 0$) and subjected to sustained loads of 3 months (i.e., $\xi = 1$) is plotted in Fig. 14.9. According to the fit curve, the coefficient $K_{sh}$ is expressed as follows:

$$K_{sh} = 0.226 + 985\varepsilon_{sh}^\infty \tag{14.9}$$

Fig. 14.10 shows the $\lambda_\Delta - \rho'$ relationships produced by the numerical analysis and simplified equations for different durations of loading and different values of $\varepsilon_{sh}^\infty$. It is seen that the ACI equation fails to predict satisfactorily time-dependent deflections attributed to the neglect of the importance of the shrinkage strain and also to the significant overestimate of the rebar influence. By using the modified reduction factor (and also the time-dependent factor for 5-year duration of loading)

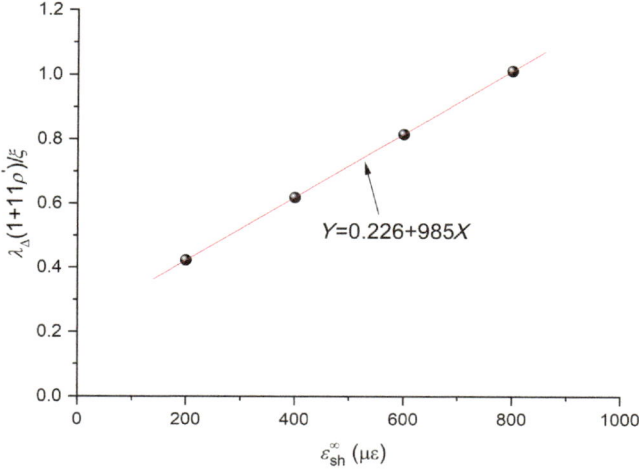

**Figure 14.9** Relationship between $\lambda_\Delta(1 + 11\rho')/\xi$ and the ultimate shrinkage strain.

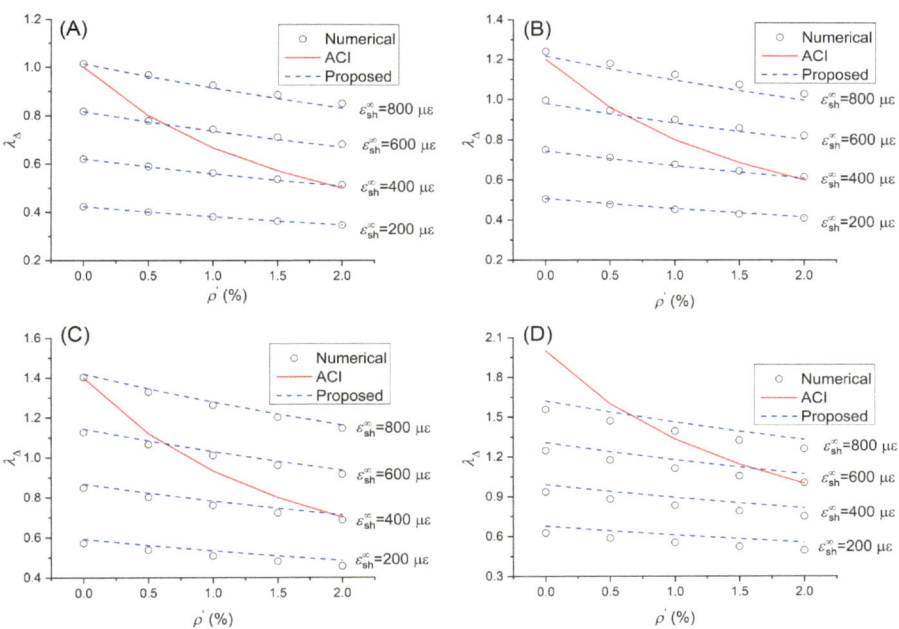

**Figure 14.10** Comparisons of $\lambda_\Delta$ versus $\rho'$ relationships by simplified equations and numerical analysis: (A) 3 months; (B) 6 months; (C) 12 months; (D) 5 years.

and by introducing the coefficient $K_{sh}$, the proposed equation predicts the $\lambda_\Delta$ values in composite girders with considerable accuracy.

The values of $\lambda_\Delta$ calculated from the ACI and proposed equations are correlated to the numerical simulations in Fig. 14.11. Eighty steel−concrete composite girders with different values of $\varepsilon_{sh}^\infty$ and $\rho'$ under different durations of loading are used for the correlation. It can be observed in the figure that the ACI equation leads to poor correlation with the numerical simulations, while the proposed equation shows an excellent fit to the $\lambda_\Delta$ values generated by the numerical analysis.

## 14.5   Time-dependent second-order effects of externally prestressed composite girders

Externally prestressed steel−concrete composite girders simply supported over a span of 20.0 m are used (see Fig. 14.12). To assess the time-dependent second-order effects of these girders, three different $S_d/L$ ratios ($S_d$ = deviator spacing; and $L$ = span length) are adopted, that is, $S_d/L = 0.0$, 1/3, and 1.0. It is noted that $S_d/L = 0.0$ corresponds to one deviator at midspan; $S_d/L = 1/3$ corresponds to two deviators at third points of the span; and $S_d/L = 1.0$ corresponds to no deviators along the span. External tendons are made of normal-relaxation prestressing steel with area, yield stress, and elastic modulus of 1500 mm$^2$, 1581 MPa, and 200 GPa, respectively. The initial prestress is 1200 MPa. The area, yield strength, and elastic modulus of steel rebars are 1500 mm$^2$, 420 MPa, and 200 GPa, respectively. The yield strength and elastic modulus of the steel beam are 275 MPa and 200 GPa, respectively. The concrete cylinder compressive strength and modulus of elasticity at age of 28 days are 40 MPa and 35 GPa, respectively. The girder is under

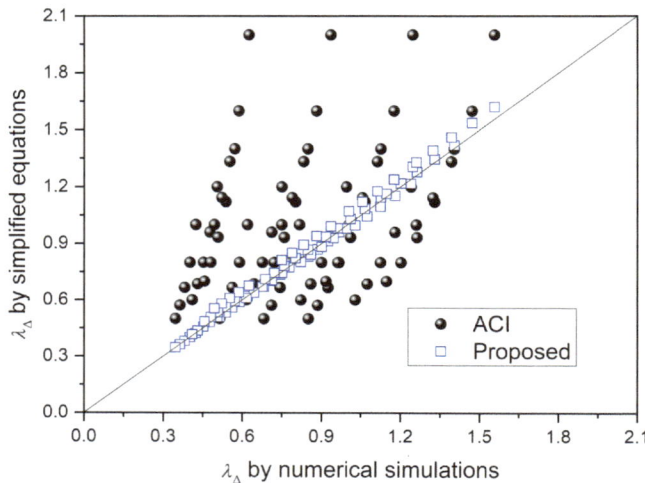

**Figure 14.11** Correlation of simplified equations with numerical simulations.

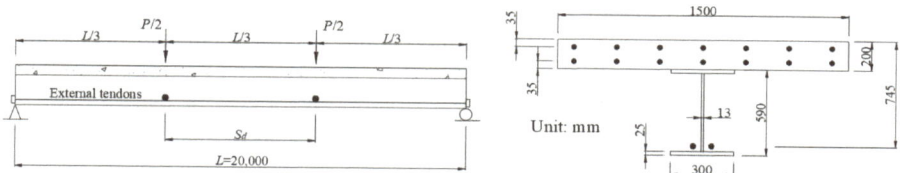

**Figure 14.12** Sketch of externally prestressed steel—concrete composite girders.

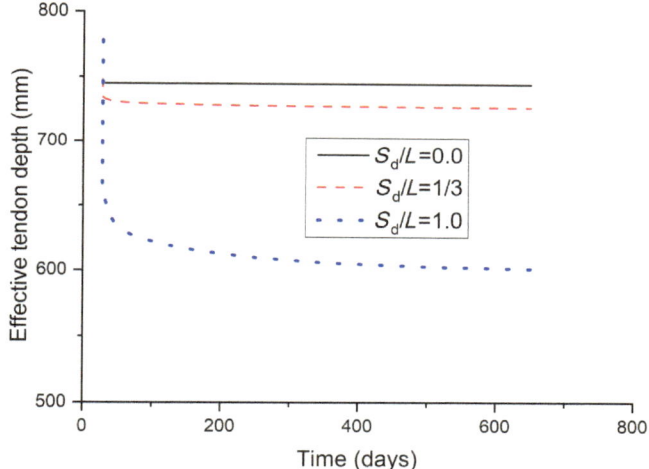

**Figure 14.13** Change in effective tendon depth at midspan over time.

sustained third-point loads ($P$) of 600 kN. Both prestress transfer and load application take place at day 28. The ultimate shrinkage strain is 400 $\mu\varepsilon$.

Fig. 14.13 shows the change in the effect depth of external tendons at midspan over time. In the case of $S_d/L = 0.0$ (i.e., one deviator at midspan), the effective tendon depth remains unchanged, keeping an initial value of 745 mm. Immediately after prestress transfer, the prestress-induced cambers of the composite girders with $S_d/L = 1/3$ and 1.0 are 30.7 and 33.2 mm, respectively, and correspondingly, the effective tendon depths for these girders are 747.9 and 778.2 mm, respectively. Immediately after the load application, the effective tendon depths of the composite girders with $S_d/L = 1/3$ and 1.0 are reduced to 734.7 and 667.9 mm, respectively. Therefore the immediate reduction in effective tendon depth is pronounced for the composite girder with $S_d/L = 1.0$ while insignificant for the composite girder with $S_d/L = 1/3$. Due to concrete creep, concrete shrinkage, and tendon relaxation, the time-dependent reduction in effective tendon depth of the composite girder with $S_d/L = 1.0$ is notable, equal to 66.2 mm at day 650. The corresponding value for the composite girder with $S_d/L = 1/3$ is marginal, equal to 7.2 mm, indicating minimal time-dependent second-order effects in the girder.

Fig. 14.14 demonstrates the typical time-dependent response of the composite girders with different $S_d/L$ ratios. It is observed that the time-dependent behavior of the composite girder with $S_d/L = 1/3$ is basically the same as that of the one with $S_d/L = 0.0$, attributed to the fact that the second-order effects in these girders are negligible. On the other hand, due to second-order effects, the composite girder with $S_d/L = 1.0$ exhibits different time-dependent behavior from that of the other two girders. It can be seen in Fig. 14.14 that $S_d/L = 1.0$ leads to substantially higher long-term deflection, compressive strain in the top concrete fiber in the slab, tensile

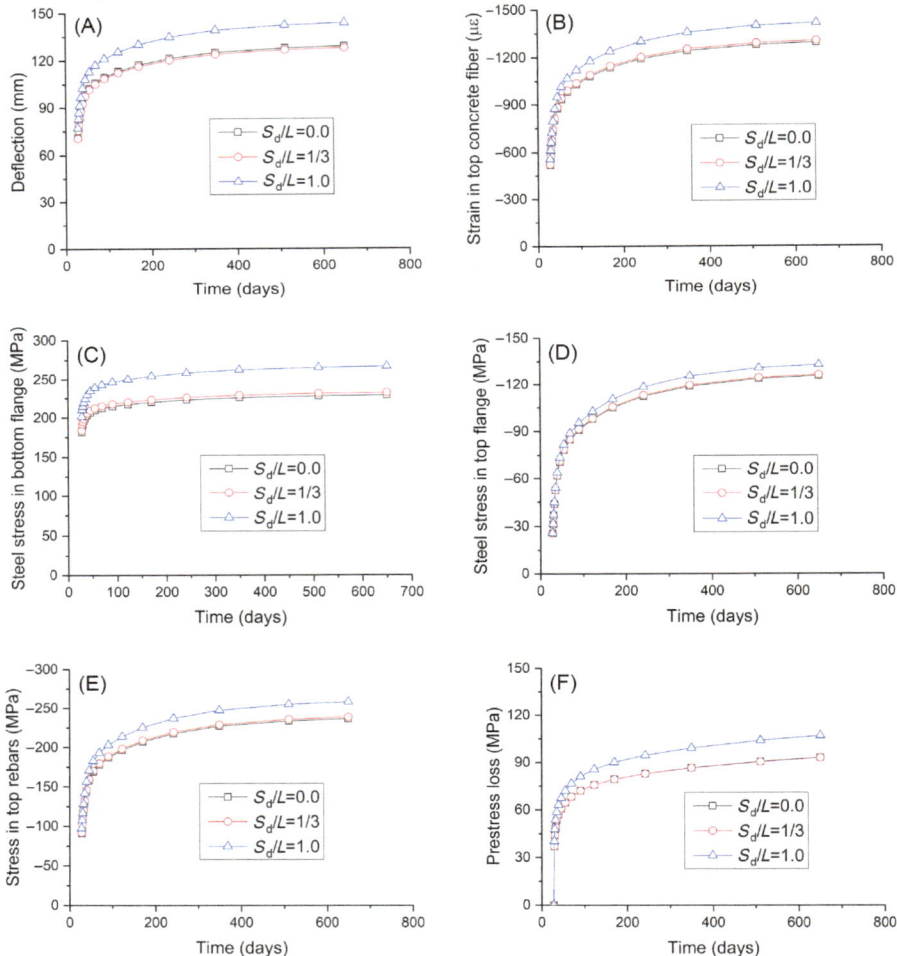

**Figure 14.14** Time-dependent behavior of externally prestressed composite girders: (A) midspan deflection; (B) strain in top concrete fiber at midspan; (C) stress in bottom flange of steel beam at midspan; (D) stress in top flange of steel beam at midspan; (E) stress in top rebars at midspan; (F) prestress loss.

stress in the bottom flange of the steel beam, compressive stress in the top flange of steel beam, compressive stress in top rebars in the slab, and prestress loss in external tendons than $S_d/L = 0.0$ or 1/3. Therefore it is concluded that the time-dependent second-order effects of externally prestressed composite girders with no deviators (i.e., $S_d/L = 1.0$) are notable. By the provision of one deviator at midspan (i.e., $S_d/L = 0.0$) or two deviators at third points (i.e., $S_d/L = 1/3$), these effects are minimized.

## 14.6 Conclusions

A time-dependent assessment is performed on steel−concrete composite girders to assess the influence of creep and/or shrinkage, the ultimate shrinkage strain, and the rebars in the slab. The results show that the time-dependent deformation contributed by concrete creep is substantially higher than that by concrete creep. The ultimate shrinkage strain affects significantly the time-dependent response of composite girders. The time-dependent deformations and stresses in steel beam decrease as the amount of rebars increases. The simplified model recommended by AISC specification is also investigated. The results demonstrate that the AISC model leads to an overestimate in predicting shrinkage-induced deflections of composite girders.

It is also indicated that the ACI code cannot predict satisfactorily long-term deflections of composite girders because on one hand it neglects the effect of shrinkage strain and on the other hand it overestimates the influence of rebars. The ACI equation is extended to be applicable to composite girders through the use of a modified reduction factor and also by introducing a new coefficient related to the ultimate shrinkage strain. The long-term deflections predicted by the proposed equation agree well with those by numerical simulations.

The time-dependent second-order effects of prestressed steel−concrete composite girders with external tendons are evaluated by using three different $S_d/L$ ratios. Notable time-dependent second-order effects are observed in the composite girder with $S_d/L = 1.0$ (i.e., no deviators). On the other hand, these effects are demonstrated negligible for the composite girder with $S_d/L = 0.0$ (i.e., one deviator at midspan) or 1/3 (i.e., two deviators at third points). Therefore providing one or two deviators can effectively minimize the time-dependent second-order effects of externally prestressed composite girders.

## References

AISC. (2016). *Specification for structural steel buildings*. ANSI/AISC 360-16. Chicago, IL.
ACI Committee 318. (2019). *Building code requirements for structural concrete (ACI 318-19) and commentary (ACI 318R-19)*. Farmington Hills, MI.

Al-deen, S., Ranzi, G., & Vrcelj, Z. (2011). Full-scale long-term experiments of simply supported composite beams with solid slabs. *Journal of Constructional Steel Research, 67* (3), 308−321. Available from https://doi.org/10.1016/j.jcsr.2010.11.001.

Amadio, C., Fragiacomo, M., & Macorini, L. (2012). Evaluation of the deflection of steel-concrete composite beams at serviceability limit state. *Journal of Constructional Steel Research, 73*, 95−104. Available from https://doi.org/10.1016/j.jcsr.2012.01.009.

Bradford, M. A. (1991). Deflections of composite steel-concrete beams subject to creep and shrinkage. *ACI Structural Journal, 88*(5), 610−614.

Bradford, M. A., & Gilbert, R. I. (1991). Time-dependent behaviour of simply-supported steel-concrete composite beams. *Magazine of Concrete Research, 43*(157), 265−274. Available from https://doi.org/10.1680/macr.1991.43.157.265.

Bradford, M. A., & Gilbert, R. I. (1992). Composite beams with partial interaction under sustained loads. *Journal of Structural Engineering, 118*(7), 1871−1883. Available from https://doi.org/10.1061/(asce)0733-9445(1992)118:7(1871).

CEN. (2004). *Eurocode 4 (EC4): Design of composite steel and concrete structures. Part 1-1: General rules and rules for buildings, part 1-1: General rules and rules for buildings.* EN 1994-1-1, Brussels, Belgium.

Davoodnabi, S. M., Mirhosseini, S. M., & Shariati, M. (2019). Behavior of steel-concrete composite beam using angle shear connectors at fire condition. *Steel and Composite Structures, 30*(2), 141−147. Available from https://doi.org/10.12989/scs.2019.30.2.141.

Fan, J., Nie, J., Li, Q., & Wang, H. (2010). Long-term behavior of composite beams under positive and negative bending. I: Experimental study. *Journal of Structural Engineering, 136*(7), 849−857. Available from https://doi.org/10.1061/(asce)st.1943-541x.0000175.

Gilbert, R. I., & Bradford, M. A. (1995). Time-dependent behavior of continuous composite beams at service loads. *Journal of Structural Engineering, 121*(2), 319−327. Available from https://doi.org/10.1061/(ASCE)0733-9445(1995)121:2(319).

He, G., & Li, X. (2020). Weak-form quadrature-element method for creep and shrinkage analysis of steel-concrete composite beams. *Journal of Engineering Mechanics, 146*(4), 04020015. Available from https://doi.org/10.1061/(asce)em.1943-7889.0001744.

Henriques, D., Gonçalves, R., & Camotim, D. (2018). *Creep analysis of steel-concrete composite beams using generalized beam theory.* Lisbon, Portugal: *ICTWS.*

Henriques, D., Gonçalves, R., & Camotim, D. (2019). A visco-elastic GBT-based finite element for steel-concrete composite beams. *Thin-Walled Structures, 145*, 106440. Available from https://doi.org/10.1016/j.tws.2019.106440.

Huang, D., Wei, J., Liu, X., Zhang, S., & Chen, T. (2018). Influence of post-pouring joint on long-term performance of steel-concrete composite beam. *Steel and Composite Structures, 28*(1), 39−49. Available from https://doi.org/10.12989/scs.2018.28.1.039.

Huang, H., Huang, S. S., & Pilakoutas, K. (2018). Modeling for assessment of long-term behavior of prestressed concrete box-girder bridges. *Journal of Bridge Engineering, 23* (3), 04018002. Available from https://doi.org/10.1061/(ASCE)BE.1943-5592.0001210.

Kwak, H. G., & Seo, Y. J. (2000). Long-term behavior of composite girder bridges. *Computers and Structures, 74*(5), 583−599. Available from https://doi.org/10.1016/S0045-7949(99)00064-4.

Lou, T., & Karavasilis, T. L. (2018). Time-dependent assessment and deflection prediction of prestressed concrete beams with unbonded CFRP tendons. *Composite Structures, 194*, 365−376. Available from https://doi.org/10.1016/j.compstruct.2018.04.013.

Lou, T., Wu, S., Karavasilis, T. L., & Chen, B. (2021). Long-term deflection prediction in steel-concrete composite beams. *Steel and Composite Structures, 39*(1), 21−33. Available from https://doi.org/10.12989/scs.2021.39.1.021.

Ranzi, G., Leoni, G., & Zandonini, R. (2013). State of the art on the time-dependent behaviour of composite steel—concrete structures. *Journal of Constructional Steel Research*, *80*, 252—263. Available from https://doi.org/10.1016/j.jcsr.2012.08.005.

Sakr, M. A., & Sakla, S. S. S. (2008). Long-term deflection of cracked composite beams with nonlinear partial shear interaction: I - Finite element modeling. *Journal of Constructional Steel Research*, *64*(12), 1446—1455. Available from https://doi.org/10.1016/j.jcsr.2008.01.003.

Sakr, M. A., & Sakla, S. S. S. (2009). Long-term deflection of cracked composite beams with nonlinear partial shear interaction - A study using neural networks. *Engineering Structures*, *31*(12), 2988—2997. Available from https://doi.org/10.1016/j.engstruct.2009.07.027.

Shariati, M., Tahmasbi, F., Mehrabi, P., Bahadori, A., & Toghroli, A. (2020). Monotonic behavior of C and L shaped angle shear connectors within steel-concrete composite beams: an experimental investigation. *Steel and Composite Structures*, *35*(2), 237—247. Available from https://doi.org/10.12989/scs.2020.35.2.237.

Tong, T., Yu, Q., & Su, Q. (2018). Coupled effects of concrete shrinkage, creep, and cracking on the performance of postconnected prestressed steel-concrete composite girders. *Journal of Bridge Engineering*, *23*(3), 04017145. Available from https://doi.org/10.1061/(asce)be.1943-5592.0001192.

Virtuoso, F., & Vieira, R. (2004). Time dependent behaviour of continuous composite beams with flexible connection. *Journal of Constructional Steel Research*, *60*(3-5), 451—463. Available from https://doi.org/10.1016/s0143-974x(03)00123-8.

Wang, W., & Gong, J. (2019). New relaxation function and age-adjusted effective modulus expressions for creep analysis of concrete structures. *Engineering Structures*, *188*, 1—10. Available from https://doi.org/10.1016/j.engstruct.2019.03.009.

Zienkiewicz, O. C., & Watson, M. (1966). Some creep effects in stress analysis with particular reference to concrete pressure vessels. *Nuclear Engineering and Design*, *4*(4), 406—412. Available from https://doi.org/10.1016/0029-5493(66)90069-0.

# Index